Vibration Monitoring and Control of the Built Environment

Vibration Monitoring and Control of the Built Environment

Guest Editor

Felix Weber

MDPI

Basel • Beijing • Wuhan • Barcelona • Belgrade • Novi Sad • Cluj • Manchester

Guest Editor
Felix Weber
Maurer Switzerland GmbH
Pfaffhausen
Switzerland

Editorial Office
MDPI AG
Grosspeteranlage 5
4052 Basel, Switzerland

This is a reprint of the Special Issue, published open access by the journal *Applied Sciences* (ISSN 2076-3417), freely accessible at: https://www.mdpi.com/journal/applsci/special_issues/511AIY434Y.

For citation purposes, cite each article independently as indicated on the article page online and as indicated below:

Lastname, A.A.; Lastname, B.B. Article Title. *Journal Name* **Year**, *Volume Number*, Page Range.

ISBN 978-3-7258-7112-4 (Hbk)
ISBN 978-3-7258-7113-1 (PDF)
https://doi.org/10.3390/books978-3-7258-7113-1

Cover image courtesy of Felix Weber
Sutong Bridge, China, 2008

Contents

About the Editor

Felix Weber

Felix Weber, Pfaffhausen, Switzerland, CEO of Maurer Switzerland GmbH, developed new vibration dampers, vibration isolators and tuned mass dampers over more than two decades. These dampers are installed on the world largest and most famous bridges such as the Franjo Tudjman Bridge (Croatia), the Alamillo Bridge (Spain), the Sutong Bridge (China), the Russky Island Bridge (Russia), the Volgograd Bridge (Russia) and in the Danube City Tower 1 (Austria). Felix Weber performed various vibration tests on these bridges to identify the supplemental damping and vibration tests on other structures as the basis for damper design. Felix Weber has published more than 140 scientific papers describing theoretical aspects of vibration control, model-based damper design and vibration test data analysis.

applied sciences

MDPI

Editorial

Monitoring and Control of Structural Vibrations

Felix Weber

Maurer Switzerland GmbH, Grossplatzstrasse 24, 8118 Pfaffhausen, Switzerland; f.weber@maurer.eu

Monitoring and control of structural vibrations are crucial to ensure structural safety and vibration comfort of civil engineering structures, such as buildings, bridges, tunnels, antennas, stadia etc. [1,2]. Structures may vibrate due to earthquakes [3], wind [4], traffic [5], and pedestrians [6]. The associated loading models are needed to accurately describe the impacts of the different excitation mechanisms on the dynamics of civil structures. The resulting structural vibrations are often measured using accelerometers, as these sensors do not require a reference point, unlike, for example, displacement transducers. However, modern monitoring systems are enriched by displacement transducers, velocity sensors (geophones), strain gauges, etc., to directly measure the states of interest [7]. The classical approach is to wire the sensors with the data recorder and to activate the acquisition of the raw or slightly lowpass filtered data based on trigger levels. Another approach is to use the sensors as so-called nodes, where some of the post-processing is conducted, and the states of interest, rather than the raw data, are wirelessly sent from node to node and eventually to the data recorder to avoid wiring [8]. These sensor networks may be preferable when wiring is difficult and the states of interest are clear. The wiring of sensors should occur when the full information of the sensor signals is of interest. In a subsequent step, the analysis of the measurement data in terms of amplitudes and frequencies or model updating techniques to identify structural parameters may be used for the model-based design of structural dampers, isolators, and vibration compensators, as well as for structural health monitoring [9,10]. This complements the further development of hydraulic and steel hysteretic dampers, curved surface sliders and lead rubber bearings, and tuned mass dampers [11,12]. These vibration dampers, isolators, and compensators can be passive, semi-active, or active, of which, the semi-active solutions combine the advantages of enhancing the resulting vibration reduction and requiring less power [13]. In the fields of measurement data analysis and structural health monitoring, the goal is to derive a digital twin of the structure to predict or estimate fatigue and damage [14].

This Special Issue presents current research in these fields. Structural vibration monitoring is addressed by Gadiraju et al., who describe a holistic modular structural health monitoring system for bridges. Amarantidou et al. investigate the strengthening of existing masonry structures against seismic excitation, considering soil–structure interaction, which is a realistic situation for many existing buildings. The research of Kodakkal et al. focuses on the question of how the modelling approach of tall buildings under wind excitation influences the prediction of the resulting storey accelerations, which is crucial for deciding whether a tuned mass damper is needed. In the contribution of Hogsberg, the tuning and efficacy of a tuned mass damper with an inerter is discussed, to better understand the impact of the inerter on the dynamics of the damper. Wang et al. analyse the loading of long-span bridges due to congested traffic using a probabilistic modelling approach, to take into consideration the random nature of this type of excitation. The semi-active tuned mass damper based on a real-time controlled semi-active damper, presented by Maslanka, is a further development of the semi-active tuned mass dampers of the Volgograd Bridge and represents the optimum solution of this semi-active control approach.

This Special Issue is enriched by two review articles. Dharmajan and Al Hamaydeh review the behaviour of steel–concrete composite structures under seismic loading conditions, and Katsimpini et al. provide a general overview of the state of art of structural control systems and identify gaps for future research.

After approximately two years, the Special Issue on "Vibration Monitoring and Control of the Built Environment" is now closed. It contributes to the documentation of recent advances in the fields of vibration monitoring and control and may motivate further research and development.

Funding: This research received no external funding.

Acknowledgments: I would like to thank the authors, peer reviewers, and staff of MDPI for their valuable contributions to this Special Issue.

Conflicts of Interest: Author Felix Weber was employed by the company Maurer Switzerland GmbH.

List of Contributions:

1. Gadiraju, D.S.; McMaster, R.; Eftekhar Azam, S.; Khazanchi, D. BIONIB: Blockchain-Based IoT Using Novelty Index in Bridge Health Monitoring. *Appl. Sci.* **2025**, *15*, 10542. https://doi.org/10.3390/app151910542.
2. Amarantidou, K.G.; Katsimpini, P.S.; Papagiannopoulos, G.; Hatzigeorgiou, G. Seismic Assessment and Strengthening of a Load-Bearing Masonry Structure Considering SSI Effects. *Appl. Sci.* **2025**, *15*, 8135. https://doi.org/10.3390/app15158135.
3. Kodakkal, A.; Péntek, M.; Bletzinger, K.-U.; Wüchner, R.; Weber, F. Systematic and Quantitative Assessment of Reduced Model Resolution on the Transient Structural Response Under Wind Load. *Appl. Sci.* **2025**, *15*, 1588. https://doi.org/10.3390/app15031588.
4. Hogsberg, J. Tuning of a Viscous Inerter Damper: How to Achieve Resonant Damping without a Damper Resonance. *Appl. Sci.* **2025**, *15*, 676. https://doi.org/10.3390/app15020676.
5. Wang, X.; Ruan, X.; Casas, J.R.; Zhang, M. Probabilistic Modeling of Congested Traffic Scenarios on Long-Span Bridges. *Appl. Sci.* **2024**, *14*, 9525. https://doi.org/10.3390/app14209525.
6. Maslanka, M. Calibration of Viscous Damping–Stiffness Control Force in Active and Semi-Active Tuned Mass Dampers for Reduction of Harmonic Vibrations. *Appl. Sci.* **2023**, *13*, 11645. https://doi.org/10.3390/app132111645.
7. Dharmajan, N.B.; Al Hamaydeh, M. State-of-the-Art Review of Structural Vibration Control: Overview and Research Gaps. *Appl. Sci.* **2025**, *15*, 7966. https://doi.org/10.3390/app15147966.
8. Katsimpini, P.; Papagiannopoulos, G.; Hatzigeorgiou, G. An In-Depth Analysis of the Seismic Performance Characteristics of Steel–Concrete Composite Structures. *Appl. Sci.* **2025**, *15*, 3715. https://doi.org/10.3390/app15073715.

References

1. Farrar, C.R.; Dervilis, N.; Worden, K. The Past, Present and Future of Structural HealthMonitoring: An Overview of Three Ages. *Strain* **2025**, *61*, e12495. [CrossRef]
2. Saaed, T.E.; Nikolakopoulos, G.; Jonasson, J.E.; Hedlund, H. A state-of-the-art review of structural control systems. *J. Vib. Control* **2013**, *21*, 919–937. [CrossRef]
3. Hu, Y.; Khatiwada, P.; Tsang, H.-H.; Menegon, S. Site-Specific Response Spectra and Accelerograms on Bedrock and Soil Surface. *CivilEng* **2023**, *4*, 311–332. [CrossRef]
4. Péntek, M.; Riedl, A.; Bletzinger, K.-U.; Weber, F. Investigating the Vibration Mitigation Efficiency of Tuned Sloshing Dampers Using a Two-Fluid CFD Approach. *Appl. Sci.* **2022**, *12*, 7033. [CrossRef]
5. Xu, M.; Zhou, X.Y.; Shen, J.; Ding, D.; Cao, S.; Cai, C.S. Performance evaluation of extreme value prediction methods for bridge traffic load effects. *Adv. Bridge Eng.* **2025**, *6*, 27. [CrossRef]
6. Yu, E.; Xu, G.; Wang, J.; Li, Y. A non-contact pedestrian load analysis and prediction model based on intelligent image recognition technology. *Measurement* **2025**, *256*, 118238. [CrossRef]

7. Mardanshahi, A.; Sreekumar, A.; Yang, X.; Barman, S.K.; Chronopoulos, D. Sensing Techniques for Structural Health Monitoring: A State-of-the-Art Review on Performance Criteria and New-Generation Technologies. *Sensors* **2025**, *25*, 1424. [CrossRef]
8. Herglotz, C.; Jabłoński, I.; Karnapke, R.; Shahin, K.; Reichenbach, M. Designing Intelligent Sensor Networks: A Comprehensive Survey of Sensing, Edge AI, and 5G+/6G Connectivity. *IEEE Access* **2025**, *13*, 172306–172325. [CrossRef]
9. Zhang, B.; Ren, Y.; He, S.; Gao, Z.; Li, B.; Song, J. A review of methods and applications in structural health monitoring (SHM) for bridges. *Measurement* **2025**, *245*, 116575. [CrossRef]
10. Weber, F.; Meier, L.; Distl, J.; Braun, C. Optimum design of curved surface sliders based on site-specific seismic input and its sensitivity. *Geosciences* **2018**, *8*, 83. [CrossRef]
11. Katsimpini, P.; Papagiannopoulos, G.; Hatzigeorgiou, G. A Thorough Examination of Innovative Supplementary Dampers Aimed at Enhancing the Seismic Behavior of Structural Systems. *Appl. Sci.* **2025**, *15*, 1226. [CrossRef]
12. Weber, F.; Distl, J.; Meier, L.; Braun, C. Curved surface sliders with friction damping, linear viscous damping, bow tie friction damping and semi-actively controlled properties. *Struct. Control Health Monit.* **2018**, *25*, e2257. [CrossRef]
13. Weber, F.; Spensberger, S.; Obholzer, F.; Distl, J.; Braun, C. Semi-Active Cable Damping to Compensate for Damping Losses Due to Reduced Cable Motion Close to Cable Anchor. *Appl. Sci.* **2022**, *12*, 1909. [CrossRef]
14. Sun, Z.; Jayasinghe, S.; Sidiq, A.; Shahrivar, F.; Mahmoodian, M.; Setunge, S. Approach Towards the Development of Digital Twin for Structural Health Monitoring of Civil Infrastructure: A Comprehensive Review. *Sensors* **2024**, *25*, 59. [CrossRef] [PubMed]

applied sciences

MDPI

Article

BIONIB: Blockchain-Based IoT Using Novelty Index in Bridge Health Monitoring

Divija Swetha Gadiraju [1,*], Ryan McMaster [1], Saeed Eftekhar Azam [2,*] and Deepak Khazanchi [1,*]

[1] College of IS&T, University of Nebraska at Omaha, Omaha, NE 68182, USA
[2] Civil and Environmental Engineering, University of New Hampshire, Durham, NH 03824, USA
* Correspondence: dgadiraju@unomaha.edu (D.S.G.); saeed.eftekharazam@unh.edu (S.E.A.); khazanchi@unomaha.edu (D.K.)

Abstract

Bridge health monitoring is critical for infrastructure safety, especially with the growing deployment of IoT sensors. This work addresses the challenge of securely storing large volumes of sensor data and extracting actionable insights for timely damage detection. We propose BIONIB, a novel framework that combines an unsupervised machine learning approach called the Novelty Index (NI) with a scalable blockchain platform (EOSIO) for secure, real-time monitoring of bridges. BIONIB leverages EOSIO's smart contracts for efficient, programmable, and secure data management across distributed sensor nodes. Experiments on real-world bridge sensor data under varying loads, climatic conditions, and health states demonstrate BIONIB's practical effectiveness. Key findings include CPU utilization below 40% across scenarios, a twofold increase in storage efficiency, and acceptable latency degradation, which is not critical in this domain. Our comparative analysis suggests that BIONIB fills a unique niche by coupling NI-based detection with a decentralized architecture, offering real-time alerts and transparent, verifiable records across sensor nodes.

Keywords: blockchain; IoT; bridge health; structural health monitoring

1. Introduction

Bridge health monitoring is an essential component of infrastructure maintenance, particularly as aging transportation networks require constant evaluation to ensure safety and performance. With the advancement of Internet of Things (IoT) technologies, it has become possible to collect continuous, real-time sensor data from bridges, enabling data-driven structural health monitoring (SHM) [1–3]. However, this abundance of data introduces critical challenges in terms of secure storage, efficient processing, and timely detection of structural anomalies. Traditional centralized data handling approaches face limitations in ensuring data integrity, scalability, and real-time decision-making [4].

Recent developments in blockchain technology offer promising solutions to these issues. Due to their decentralized, tamper-resistant architecture, blockchains provide a secure framework for recording and sharing high-frequency sensor data. When combined with smart contracts, blockchains can enable programmable, automated responses to structural health events [5–7]. Despite their advantages, blockchains have seen limited integration with lightweight, unsupervised SHM techniques that are suitable for deployment on edge devices. Existing works typically rely on heavyweight models, such as deep learning architectures, which are not optimized for real-time performance in constrained environments.

To address this gap, we propose BIONIB: Blockchain-based IoT using Novelty Index in Bridge Health Monitoring. BIONIB integrates a lightweight, unsupervised anomaly detection technique—the Novelty Index (NI)—with the high-throughput EOSIO blockchain platform. This integration enables secure storage, decentralized processing, and real-time alerting based on structural health conditions. Unlike prior SHM systems, BIONIB avoids storing raw time-series data and instead computes a compact NI representation from sensor streams, significantly improving storage and retrieval efficiency. EOSIO's parallel processing capabilities are leveraged to scale the system across a large number of sensors and bridge sites.

The main contributions of this work can be summarized as follows:

- We propose a scalable blockchain-based framework utilizing the EOSIO platform to efficiently manage and analyze IoT sensor data in bridge health monitoring. This system ensures secure and transparent storage, integrity, and traceability of structural data across distributed networks.
- We integrate a smart contract system that ingests NI values in real time and automatically triggers alerts when a threshold is crossed. This design enables prompt identification of unhealthy bridge states without relying on centralized infrastructure.
- By storing only NI values rather than full time-series sensor data, BIONIB achieves an over 99.9% data reduction, improving blockchain storage efficiency and lowering transaction overhead.
- We evaluate the scalability and efficiency of BIONIB under varying numbers of IoT sensors and blockchain nodes. Results show that throughput scales linearly while CPU utilization remains stable below 40%, validating the system's robustness and suitability for large-scale SHM deployments.

The remainder of this article is organized as follows. Section 2 reviews related work in SHM, IoT, and blockchain integration. Section 3 presents the network architecture of the proposed system. Section 4 discusses the data collection process and NI calculation. Section 5 details the BIONIB algorithm and smart contract implementation. Section 6 reports performance evaluation results. Section 7 outlines a discussion of challenges and advantages of the proposed approach. Finally, Section 8 concludes the paper and outlines directions for future work.

A preprint of this work is available on arXiv (https://arxiv.org/abs/2402.14902) (accessed on 25 September 2025.).

2. Related Work

Blockchains have recently become popular in the IoT, healthcare, smart-grid, supply-chain, and many other fields for their immutability, secure storage, and decentralization [8]. Many studies have been conducted concentrating on the applications of blockchains for specific IoT sensors and tasks [9]. In this section, we present a review of the literature related to our research.

2.1. IoT Based Blockchain

In [5], a survey is presented showing various applications of blockchains in IoT. The survey [5] also focuses on the challenges posed by the current centralized IoT models, and recent advances made both in industry and research to solve these challenges for effectively using blockchains to provide a decentralized, secure medium for the IoT. The use of blockchains in Industry 4.0 is presented in [10]. The merits and demerits of traditional security solutions in blockchains are discussed in comparison to their countermeasures and a comparison is provided [10]. Various consensus protocols like sharding [11,12] are used in IoT-based blockchains. Recent works focus on using various deep learning and scaling algorithms for blockchains for IoT applications [13–15]. These protocols demonstrate the usability of the blockchain consensus algorithms in the domain of IoT.

2.2. SHM

The evaluation of bridge health is explored through the analysis of bridge survey data as detailed in [16,17]. In [6], the recent progress of the SHM technology is reviewed for long-span bridges. In addition to the data analysis and condition assessment including techniques on modal identification, methods on signal processing, and damage identification, a case study of a SHM system of a long-span arch bridge is discussed. The authors of [1] conducted a detailed literature review of recent applications of smartphones, UAVs, cameras, and robotic sensors for structural condition monitoring and maintenance. The work in [18] employs a nondestructive evaluation test with statistical confidence and uses time-series analysis instead of frequency-domain monitoring. The work in [19] discusses partial-model-based damage identification.

The damage detection tool developed in [20,21] utilizes unsupervised machine learning, specifically the Novelty Index (NI), to detect damage in bridges using data from full-scale mock-up tests. The method was validated using field data and shows promise as a robust tool for damage detection. The NI calculation method employed in this study is based on the approaches detailed in [20,21], which compare the NI with alternative damage detection techniques and highlight its advantages as a compact, sensitive damage indicator. Readers are encouraged to consult these sources for a comprehensive understanding of NI methodology and its performance characteristics. Further discussion of the NI technique is provided in later sections.

2.3. Blockchain in SHM

The work in [22] combines IoT and blockchain-based smart contracts for SHM of underground structures to create an efficient, scalable, and secure network that enhances operational safety. The authors of [23] propose a conceptual framework to integrate vibration-based methods and blockchain for more reliable and efficient structural damage detection. The system in [4] is a blockchain network for SHM, with smart contracts for health monitoring on the Ethereum private chain for verifying authority, detecting structural damage, generating alerts, securing data, and allowing for traceability queries. In [7], sensors record monitoring information such as pressure points, temperature, and pretension force. This data is transmitted to a blockchain platform that classifies transaction criticality and securely stores the information. The authors of [24] use consortium blockchain for distributed storage, security, and traceability.

2.3.1. Why EOSIO for IoT SHM?

In [25], the authors collect and process up-to-date on-chain data from EOSIO. EOSIO achieves a significant improvement in performance when compared to Bitcoin and Ethereum [25] since it generates more blocks and hence provides more transactions per second. In [26], the authors analyzed the network traffic of three major high-scalability blockchains (EOSIO, Tezos, and XRP Ledger) for seven months and found that a small portion of transactions are used for value transfer purposes. About 96% of the transactions on EOSIO were triggered by the airdrop of a valueless token [26].

The authors of [27] propose EOSAFE, which is a static analysis framework that detects vulnerabilities in EOSIO smart contracts at the bytecode level and analyze EOSIO smart contracts against attacks. In [28], a systematic analysis is conducted on the decentralization of DPoS with data from blocks in EOSIO, the first DPoS-based blockchain system, and characterize the decentralization into two phases, block producer election, and block production. The method in [29] can record the events that occur in multiple industrial robotic arms by deploying a smart contract in the EOSIO blockchain. The authors of [30] analyzed attacks on EOSIO and found vulnerabilities in its components and outlined effective mitigations and best practices for researchers.

Our selection of EOSIO was based on the presence of a well-supported research community and a rigorous security analysis. Additionally, EOSIO offers exceptional scalability, making it suitable for IOT based SHM.

2.3.2. Comparison with Central Storage

IoT data is traditionally stored on a central server. Blockchains are particularly useful due to their better decentralization, security, and tamper resistance properties [31]. Storage efficiency refers to the effective storage and management of data over time. EOSIO [30] is one of the emerging blockchains that offers scalability and flexible operability for our implementation. In EOSIO, the storage efficiency is impacted by factors such as the number of nodes, smart contracts, and additional metadata. Conversely, centralized storage is not dependent on the number of nodes. The EOSIO blockchain's storage efficiency scales linearly with the number of nodes, making it more advantageous [30]. The time it takes to retrieve data in a blockchain network, known as data retrieval time, is influenced by the number of nodes, and the time data takes to propagate between nodes. The scalability of a blockchain network depends on its ability to handle a growing number of nodes and transactions while maintaining high performance. In contrast, the scalability of centralized storage is limited by the capacity of the central server. The EOSIO blockchain is highly scalable and can efficiently handle a larger network size, particularly when it comes to transaction processing [32]. Blockchain technology is highly secure and resistant to malicious activities due to its decentralized and immutable nature. It offers superior protection against security breaches and tampering compared to centralized systems.

2.4. Comparative Analysis of Structural Health Monitoring Approaches

To further contextualize the contribution of BIONIB, we provide a comparative analysis with existing structural health monitoring (SHM) approaches. While many recent studies have explored anomaly detection in sensor data using machine learning or signal processing techniques, few integrate these with blockchain technologies to ensure data traceability, integrity, and decentralized automation.

Table 1 summarizes key differences across various methods. Specifically, we compare BIONIB to the following:

- A deep autoencoder-based method using one-class SVM [33].
- A clustering-based technique using density-peak methods [34].
- Traditional SHM systems (e.g., threshold-based or rule-based) without blockchain [6].

The BIONIB framework offers several distinct advantages. First, it achieves real-time alerting using a lightweight unsupervised method (Novelty Index) [21], while maintaining low computational cost. Second, it integrates EOSIO smart contracts to ensure immutability and traceability of health indicators. In contrast, methods like deep autoencoders or clustering typically require significant post-processing and lack mechanisms for secure, decentralized data storage.

BIONIB also demonstrates high scalability and consistent CPU usage even under increased node or sensor load, benefiting from EOSIO's parallel processing capabilities. While other studies have introduced blockchains into SHM, such as Ethereum-based systems [4], they often suffer from low throughput and limited integration with lightweight models. BIONIB addresses this by using EOSIO, which is optimized for high-frequency IoT data streams.

Table 1. A brief comparison of BIONIB with Other SHM Approaches.

Feature	BIONIB (This Work)	Deep Autoencoder + One-Class SVM [33]	Density-Peaks Clustering [34]	Traditional SHM (No Blockchain) [6]
Damage Detection Approach	Novelty Index (NI)	Autoencoder + One-Class SVM	Unsupervised Clustering	Rule-based or threshold
Damage Localization	Sensor-level (threshold)	Yes	Yes	Limited
Blockchain Integration	EOSIO (C++ smart contracts)	Not used	Not used	Not used
Real-Time Alerting	Yes	Often post-processed	Post-processed	Rare
Data Integrity and Security	High (blockchain-based)	Low	Low	Low
Scalability and Traceability	High	Medium	Medium	Low

The integration of blockchain technology with SHM represents a significant advancement in ensuring the integrity, security, and reliability of bridge monitoring data. As demonstrated in Table 1, BIONIB leverages the EOSIO blockchain platform to provide superior data integrity and security compared to traditional approaches. Unlike conventional SHM methods that rely on centralized storage systems vulnerable to tampering and unauthorized access, BIONIB's blockchain-based approach ensures that sensor data remains immutable and transparent [4]. Furthermore, the use of the Novelty Index (NI) in BIONIB offers advantages over other machine learning approaches like deep autoencoders with one-class SVM and density-peaks clustering. While these alternative methods can effectively detect structural anomalies, they often require significant computational resources and post-processing time, limiting their real-time alerting capabilities [21]. BIONIB addresses these limitations by combining the NI algorithm with EOSIO's parallel processing capabilities, enabling efficient real-time monitoring even with an increasing number of sensors and blockchain nodes. The experimental results from the implementation demonstrate that BIONIB maintains consistent CPU usage regardless of network size, highlighting its scalability advantages for large-scale bridge monitoring deployments [14].

3. Network Architecture

In this section, we will explain the EOSIO consensus used for our implementation. Later, we will discuss the EOSIO protocol with smart contracts. Figure 1 shows the network architecture of bridge monitoring with sensors and a blockchain. Consider N number of block producer nodes and k accounts created. The blockchain layer has the EOSIO accounts and the block producer nodes, which confirm the input transactions into the blocks on the blockchain. For better understanding, EOSIO producer nodes are shown in the edge computing layer but they are also a part of the blockchain layer. Let s be the number of strain sensors collecting the data on every bridge. Each bridge sensor transmits its data to the edge computing layer which consists of the block producer nodes who calculate the NI from the received sensor data. These novelty indices are given to the EOSIO accounts who initiate a smart contract. The deployed smart contracts then output the actions, which are the transactions (containing novelty indices), which after EOSIO consensus are added to the blockchain.

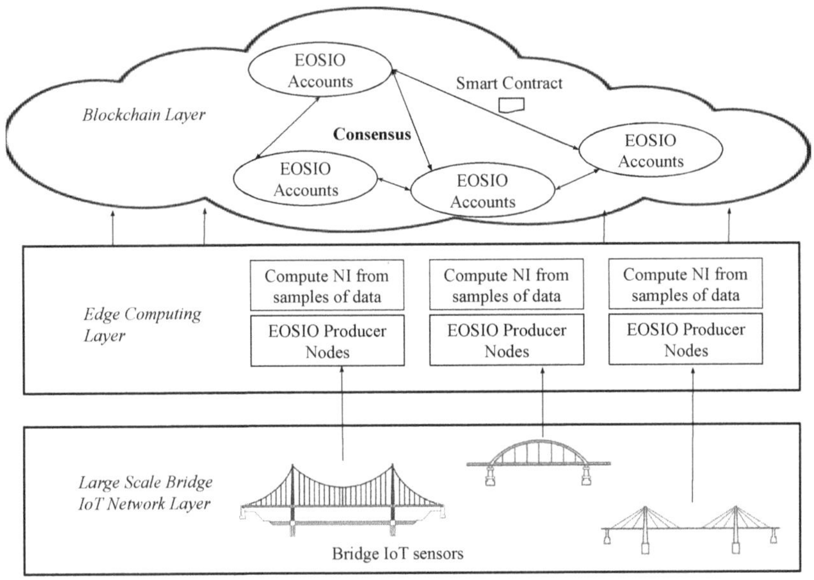

Figure 1. Network architecture illustrating the integration of blockchain technology with IoT sensors for real-time bridge health monitoring. The system ensures secure data collection, processing, and immutable storage across distributed nodes.

EOSIO Consensus

EOSIO [28,35] is a blockchain protocol designed to empower the development and deployment of decentralized applications (DApps) with a focus on scalability, flexibility, and user-friendliness. EOSIO is a blockchain platform that uses a Delegated Proof-of-Stake (DPoS) consensus mechanism. The consensus mechanism has two distinct components: producer voting and scheduling, which is executed by the DPoS layer 2, and block production and validation, which is carried out by the native consensus layer 1. Additionally, EOSIO uses asynchronous Byzantine Fault Tolerance (aBFT) in Layer 1 to ensure safety and consistency in the validation process. In Delegated Proof-of-Stake (DPoS), a small group of elected nodes, called block producers, are responsible for verifying transactions and adding blocks to the blockchain. Compared to the traditional Proof-of-Work (PoW) consensus mechanism, this design improves scalability and transaction throughput. EOSIO uses parallel processing, allowing multiple transactions to be processed at the same time, resulting in increased transaction speed and efficiency. It is ideal for applications that require high performance and low latency. Additionally, EOSIO introduces a resource model that includes bandwidth, computation, and storage.

Layer 1 uses Asynchronous Byzantine Fault Tolerance (aBFT), an advanced consensus algorithm, to validate blocks produced by elected producers and record them on the blockchain permanently. The layer ensures that each block is signed by the corresponding producer by utilizing the proposed schedule of producers from Layer 2. To ensure byzantine fault tolerance, this layer follows a two-stage block confirmation process, where each block must be confirmed twice by a two-thirds supermajority of producers from the currently scheduled set. The layer proposes a Last Irreversible Block (LIB) in the first stage, and the proposed LIB is confirmed as final in the second stage, making the block irreversible. Additionally, any changes made to the producer schedule are notified at the beginning of every schedule round. The EOSIO consensus mechanism attains algorithmic finality, a departure from the probabilistic finality characteristic of Proof-of-Work models. Active producers on the EOSIO blockchain use signatures to validate each block. The schedule determines authorized signers for each block at specific time slots. Smart contracts can alter the schedule, but changes are implemented only after the block has undergone two stages of confirmation. EOSIO smart contracts allow us to define specific rules and conditions for transactions. In the proposed approach for bridge health monitoring, EOSIO blockchain integration involves deploying smart contracts to handle the secure storage of real-time IoT data and execute predefined logic for data analysis. This incorporation enhances the overall security, transparency, and efficiency of the bridge health monitoring system.

4. Data Collection and Analysis

4.1. Collection of Data

Field tests were conducted at the Midwest Roadside Safety Facility (MwRSF) to evaluate barrier performance during vehicle impacts on a bridge deck mock-up. The authors of [20] measured the bridge's response to vehicle loads at different speeds and under various levels of damage before and after the crash test. Figures 2 and 3 show a cantilevered bridge deck with a concrete barrier at the end. It comprises a grade beam, deck, barrier, and overlay. The mock-up mimics a real bridge. The grade beam replicates a girder, and the design matches realistic forces. The overhang and barrier meet AASHTO's and NCHRP's design specifications. The bridge deck had 40 BDI strain transducers placed across its width at eight locations. Each section was equipped with two longitudinal and three transverse strain sensors, out of which, 24 sensors were used for the study. Figure 3 illustrates the locations of the sensors on one half of the bridge. The other half of the bridge follows the same sensor placement. The proposed damage detection method for a bridge was tested using a pickup truck and a dump truck. The bridge was tested in different conditions, with the pickup truck traveling at different speeds, while the dump truck moved slowly, both empty and loaded. The initial tests were carried out on the bridge in its original state, followed by tests with different levels of damage. Strain data were collected for each test, and the damage scenarios were designed to mimic real-world situations.

Figure 2. Sensors installed on the underside of the bridge deck for structural health monitoring.

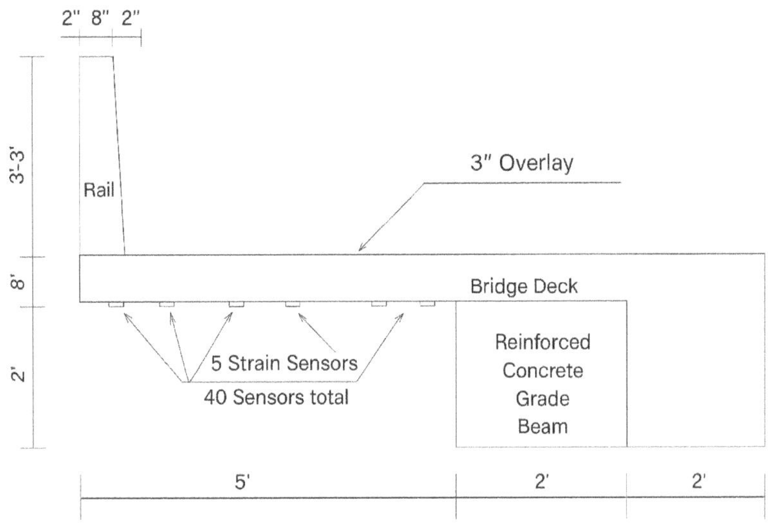

Figure 3. Illustrationof half of the bridge showing deployed sensors, including measurements, rail, and reinforced concrete beam. The other half follows the same sensor pattern.

4.2. Novelty Index Calculation

In this work we use NI calculations based on [20,21]. To detect damage, features are extracted from response data. Linear modal properties like resonant frequencies and mode shape curvatures are used to fit a linear, dynamic model to both healthy and damaged structure data. However, these features have limitations, such as being unable to detect nonlinear changes in system response and being influenced by environmental conditions. Recent research has focused on developing damage-sensitive features such as novelty analyses using Singular Value Decomposition (SVD), and Independent Component Analysis (ICA). These features transform measured, nonlinear response data into low-dimensional features, serving as damage indicators and input parameters for novelty indices. SVD and ICA offer computational simplicity compared to traditional vibration modes. It is important to note that strain transducers were used for live load response measurement and Operational Modal Analysis (OMA) was not utilized. SVD can be applied to both linear and nonlinear systems and is robust to measurement noise, while ICA is a potential alternative known for data reduction and handling measurement noise effectively. Proper Orthogonal Decomposition (POD) generates orthonormal vectors ordered by importance. It has an intuitive link with PCA, which transforms correlated variables into principal components. In PCA used for POD, the first POM captures the most variability in the data. SVD and Karhunen–Lo'eve Decomposition (KLD) are other POD techniques. Numerically, PCA and SVD provide similar results. SVD was chosen to develop POMs in this study, but PCA could also be used. Then we use datasets from bridge mock-ups, where $U = [u_1, u_2, \cdots, u_{ns}]$ represents snapshot matrices of vehicle passage data. The SVD of these matrices allows for the extraction of damage features, represented as $U = L\Sigma R$, where L, Σ, and R are orthonormal matrices. The first Proper Orthogonal Mode (POM), ϕ_1, contains information relevant to the presence of structural damage and is used as one of the damage features.

This work in [20] proposes a detection framework to differentiate between healthy and damaged states in bridge response data from vehicle passages. Figure 4 shows how NI summarizes the state of the bridge based on bridge sensor data samples. The experiment is conducted initially on a healthy bridge and then on an unhealthy bridge. The scatter plot shows the clustering of NIs calculated from the strain transducers. In Figure 4, H represents the healthy bridge state without any damages. The damage levels D_1, D_2, and D_3 are the bridge condition after introducing damage to the structure. The damage levels are in the increasing order of their severity. The concrete bridge deck is saw-cut to introduce the three damage levels as described in [21]. The damage level D_3 is a complete saw cut deck damage and hence the NI is closer to 1. However, even at D_3 the bridge has not completely failed and we could conduct the experiments with load on the bridge. NI is calculated first, and the feature extraction is performed using SVD. Then healthy feature vectors, denoted by h are stored. Next, we calculate the healthy feature vector mean for POM

$$\bar{\chi} = \frac{\Sigma_{i=1}^{h} \chi^i}{h}. \tag{1}$$

After this, we calculate $\mathcal{N}^i$ for each data set with a Euclidean norm

$$\mathcal{N}^i = ||\chi^i - \bar{\chi}||. \tag{2}$$

The detailed procedure of the usage of NI in BIONIB is explained in the subsequent sections.

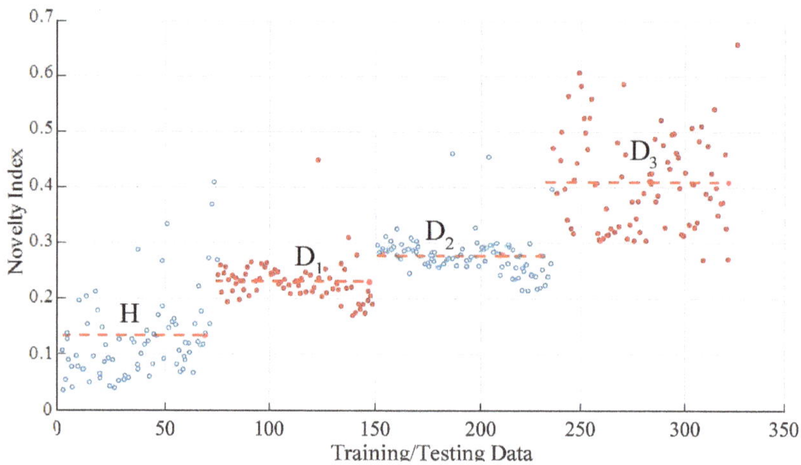

Figure 4. Novelty Index (NI) values plotted for collected data points across four conditions: Healthy (H) and three damage levels (D_1, D_2, D_3).

5. Proposed Approach: BIONIB

5.1. EOSIO Integration with IoT Data

BIONIB utilizes EOSIO, a secure, transparent, and decentralized platform to store and access NI data from IoT sensors. This ensures data integrity and immutability while facilitating efficient data management and analysis. Bridge IoT sensors collect measurement data, such as strain from bridges. This data is transmitted to a blockchain system for processing and analysis. The collected data undergoes processing to extract relevant features and calculate an NI. Feature extraction technique, SVD is used to extract information from the sensor data. The NI is calculated based on the extracted features to assess the deviation or abnormality of the bridge's condition from a baseline. A smart contract is developed using the EOSIO Contract Development Toolkit (EOSIO.CDT) [36]. The smart contract includes actions and data structures necessary for adding and retrieving NI data to and from the blockchain. The compiled smart contract is deployed to the EOSIO blockchain using a deployment transaction. This transaction specifies the account deploying the contract.

Once deployed, users can interact with the smart contract by invoking its actions. An action within the smart contract is created to add the NI data to the blockchain. This action includes parameters such as the bridge identifier and the calculated NI. The smart contracts are invoked and the action adds the NI data to the blockchain by sending a transaction. The transaction is processed by EOSIO nodes in the network, validated, and included in a block by block producers through the consensus mechanism. Once included in a block and confirmed by subsequent blocks, the transaction becomes immutable and part of the blockchain's permanent record. Users, applications, or external systems can query the blockchain to retrieve NI data stored by the smart contract. EOSIO APIs enable easy access and transparency to stored data. Nodeos handles the blockchain data persistence layer, peer-to-peer networking, and contract code scheduling. The command 'cleos' is a command line tool that interfaces with the REST APIs exposed by nodeos.

Using a smart contract has several advantages such as structured data management, automation, decentralized trust, and efficiency. However, it also comes with some drawbacks like development complexity, execution costs, and maintenance requirements. Conversely, without a smart contract, users may have to rely on centralized systems and experience higher costs and processing times [30].

5.2. Smart Contract EOSIO Using NI

The first step in creating an EOSIO smart contract is writing the contract code in C++, which defines the actions, data structures, and logic that govern the behavior of the smart contract [25]. Once the smart contract code has been developed, it needs to be deployed to the EOSIO blockchain. Deployment involves uploading the compiled contract code to the blockchain network. This step requires an account with the necessary permissions to deploy contracts. Once deployed, the smart contract can be invoked by users or other smart contracts on the blockchain. Invoking a smart contract typically involves calling one of its actions, which triggers the execution of the corresponding code within the contract. When an action is invoked, the EOSIO blockchain executes the corresponding code in the smart contract. This code can read from and write to the blockchain's state, perform computations, and interact with other contracts and external data sources. During execution, the EOSIO blockchain validates the transaction and ensures that it meets the requirements specified by the smart contract code. Once the transaction is validated and executed, it is included in a block by the block producer nodes on the EOSIO blockchain. The block is then added to the blockchain, and the transaction becomes part of the immutable ledger. During execution, smart contracts can emit events to provide information about their state or the outcome of certain operations. These events can be logged and monitored by external systems or other smart contracts for tracking and auditing purposes. Figure 5 depicts the workflow of the EOSIO smart contract. The EOSIO smart contract is called and deployed using the cleos commands v2.0.13. At the end of the execution of each smart contract, the novelty indices are added to the table, which is confirmed on the blockchain using nodeos v2.0.13. Nodeos is responsible for block producers, blockchain state, transaction validation, and consensus. Algorithm 1 shows the process of collection of IoT data, calculation of NI, and confirming it onto the EOSIO blockchain using a smart contract.

Algorithm 1 BIONIB

Input: IoT bridge data
Measure IoT sensor response data
Feature extraction using SVD
Store all healthy feature vectors h
Calculate healthy feature vector mean for Proper Orthogonal Modes (POM)

$$\bar{\chi} = \frac{\Sigma_{i=1}^{h} \chi^{i}}{h}$$

Calculate $\mathcal{N}^i$ for each data set with Euclidean norm

$$\mathcal{N}^i = ||\chi^i - \bar{\chi}||$$

Calculate all $\mathcal{N}^i$s at EOSIO block producer nodes (*'nodeos'*)
Block producer nodes initiate a smart contract with blockchain accounts (*'cleos create account'*)
while $i \neq 0$
Contract Name: SensorID_timestamp (*'cleos set contract'*)
Actions:
  addnovelty
    Input: account_name, $\mathcal{N}^i$
    Require: Authentication of user
      Add $\mathcal{N}^i$ to the output table (*'cleos push action'*)
      Check if the NI is close to 1:

$$\textbf{if } |\mathcal{N}^i - 1| < \epsilon \text{ then}$$

        Display message: "Unhealthy bridge detected!"
      else
        Continue
      end if
Consensus among EOSIO nodes
Add $\mathcal{N}^i$ on the blockchain as input transactions

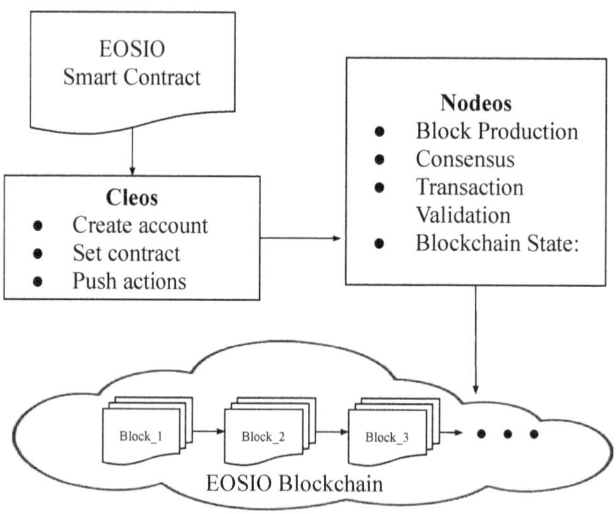

Figure 5. Step-by-step workflow of EOSIO smart contract integration in BIONIB. This illustration shows how bridge sensor data is processed on the EOSIO blockchain. It highlights key EOSIO components: account creation, contract deployment, action triggering, and block confirmation.

To improve clarity for practitioners, we emphasize EOSIO's modular structure: nodeos manages consensus and block production, cleos facilitates contract interaction, and EOSIO.CDT enables smart contract development in C++. Each NI is handled via specific addnovelty actions linked to identifiable accounts and timestamps.

The algorithm, BIONIB starts by collecting sensor response data and extracting relevant features. Initially, IoT sensors placed on bridges continuously record structural responses, such as strain. This raw data is then processed using SVD, a dimensionality reduction technique that extracts key damage-sensitive features known as Proper Orthogonal Modes (POMs). From these features, a baseline is established by computing the mean feature vector of healthy-state data, denoted as $\bar{\chi}$. To detect abnormal conditions, a Novelty Index (NI) is calculated for each new dataset by measuring the Euclidean distance between the current feature vector and the baseline. This computation effectively quantifies the deviation of the bridge's structural response from its healthy state. Mathematically, the mean feature vector $\bar{\chi}$ is calculated as Equation (1) and and the NI is given by Equation (2).

The calculated NI values are transmitted to EOSIO block producer nodes, which are responsible for blockchain operations. Using the EOSIO command-line interface tool `cleos`, blockchain accounts are created for each bridge or sensor node using the `create account` command. Then, smart contracts are deployed via `set contract`, and transactions containing the NI values are submitted using `push action`.

Each smart contract stores the NI along with metadata such as sensor ID and timestamp. Upon receiving the NI, the smart contract verifies the input, performs authentication, and logs the data into a blockchain table. If the NI value is sufficiently close to 1 (i.e., $|NI - 1| < \epsilon$), the system flags the bridge as potentially unhealthy and generates a real-time alert.

Once the data is verified, EOSIO's Delegated Proof-of-Stake (DPoS) consensus mechanism validates the transaction. The data is then permanently recorded on the blockchain by the `nodeos` module, ensuring both immutability and traceability. This step completes the secure storage process, allowing users and systems to retrieve the NI data on-demand through blockchain queries.

The BIONIB algorithm offers an approach to bridge health monitoring by leveraging the capabilities of the EOSIO blockchain to facilitate secure and efficient data processing and unhealthy bridge detection.

6. Performance Analysis

In this section, we discuss the performance of BIONIB in various scenarios. We begin by discussing the implementation details, followed by a thorough examination of the performance metrics.

6.1. Implementation Details

Our experimental setup leverages real-time sensor data collected from a bridge in Nebraska State, covering multiple days with varying traffic loads and bridge conditions, including both healthy and unhealthy states. Executed on a computer running Ubuntu 18.04, equipped with 6169 MB of base memory, an AMD Ryzen 7 5800H processor, and 16.0 GB RAM, alongside a 64-bit processor architecture, our implementation utilizes Python 3.7, Pandas 1.3.5, Numpy 1.21.6, and Matplotlib 3.5 for NI calculations. the experimental setup is given in Table 2. The integration of the NI within EOSIO smart contracts is facilitated by CPP within the EOSIO mainchain framework.

Table 2. Experimental Setup.

Parameters	Value
Total Sensors	51 strain transducers
Maximum EOSIO nodes	50
Maximum smart contracts per node	50
Memory buffer size	150 MB
Sampling rate	256
Data rate	100 Mbps
Propagation delay of data	5 ms

Comparison of With and Without NI

In BIONIB we show the implementation with NI on IoT data. The following Table 3 summarizes the advantages of using NI instead of storing all the IoT data on the blockchain. Consider a blockchain with N nodes, k accounts created, and s number of IoT sensors. For every epoch of the blockchain, m files from the memory buffer are processed by the smart contract. Each of the m files has data of $i = 1, 2, \ldots, s$ sensors in its columns. A NI run on m files gives s number of NI values. Let B be the size of the block. Assume that we know all the block details to retrieve data using cleos commands. For example, if we have 51 sensors and 500 files, we get 51 NI values. However, each of the 500 files has 50 data columns and 4396 rows entries making the total data 224,196 values per file. Each of these values with or without an NI is stored on the blockchain ledger in a distributed fashion. Data retrieval is performed using a multi-index table. This uses the find method on the table, which takes the primary key of the row that you want to retrieve and will return an iterator (reference) to the row. The size of multi-index table created depends on the number of actions performed. To remove data from a table, the erase method is used, which takes a reference to the iterator. Table 3 provides a comparison of executing BIONIB with and without NI.

Table 3. Comparision of BIONIB with and without NI.

Parameter	BIONIB with NI	BIONIB Without NI
Storage Efficiency	O (ns)	O (nms)
Data Retrieval Time	O (ks)	O (kms)
Scalability with bridges	O (n)	O (n)
Latency per epoch	O (Bs)	O (Bms)

Comparing our method with and without the NI can provide valuable insights into the efficiency and effectiveness of our approach. In Table 3, we can see the benefits of incorporating the NI, such as improvements in storage efficiency, data retrieval time, scalability with bridges, and latency per epoch. The proposed scheme significantly enhances storage efficiency by reducing the data stored on the blockchain. This optimization helps to improve the overall system performance by utilizing resources more effectively.

6.2. *Performance for Increasing Blockchain Nodes*

In this section, we analyze the performance of BIONIB as we increase the number of blockchain nodes. This investigation is crucial to understand how our approach scales as the network infrastructure expands. The EOSIO-based method used in BIONIB confirms two blocks per second. This is consistent throughout the experimentation and is depicted in

Figure 6. The blocks confirmed by the blockchain linearly increase. Each block can consist of a varying number of transactions. Figure 7 displays a trend of transaction throughput increasing proportionally with the number of nodes. This observation highlights the scalability of our approach, indicating that it can handle transactions efficiently even in larger network configurations.

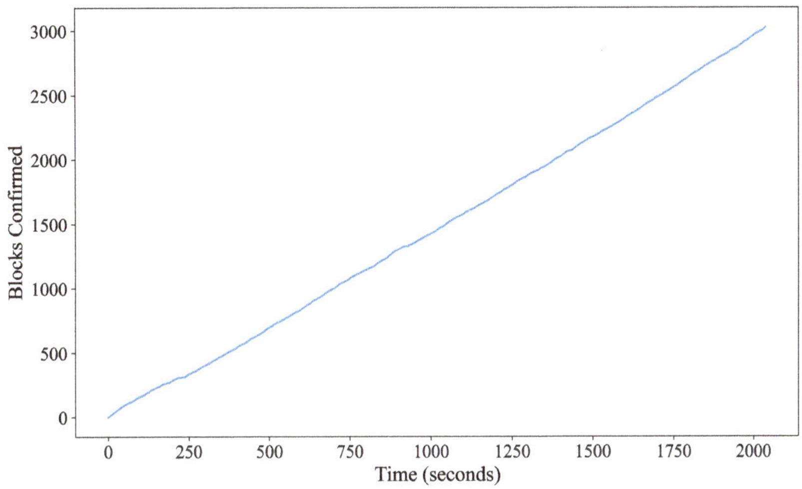

Figure 6. Performance of BIONIB showing blocks confirmed per second, with a new block confirmed approximately every 0.5 s.

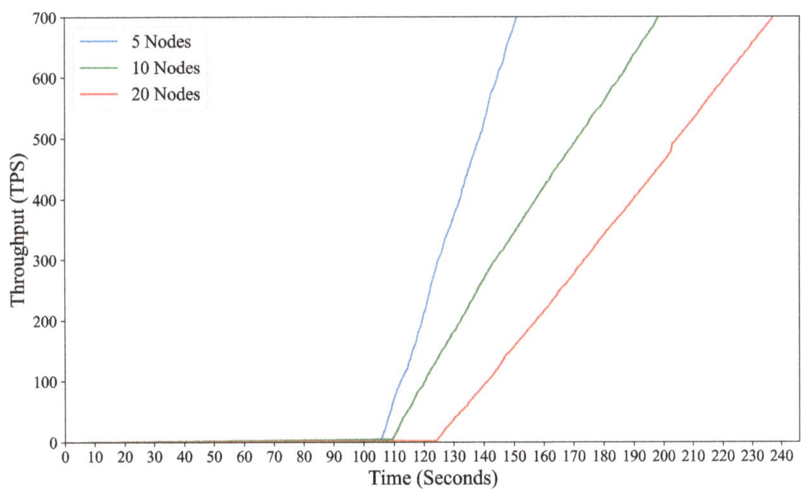

Figure 7. Performance of BIONIB showing transaction throughput scaling proportionally with an increasing number of blockchain nodes.

Our analysis has revealed some interesting insights into the usage of resources. Figure 8 shows the block CPU usage as a percentage of 5000 µs during the first 30 s of transaction execution. It is remarkable that the CPU usage remains stable despite an increasing number of blockchain nodes. This is demonstrated in Figure 8d. This consistency in CPU usage can be attributed to the parallel processing capability of the EOSIO blockchain framework. By distributing computational tasks across multiple nodes, EOSIO ensures efficient resource utilization, enabling the CPU resources to handle transaction processing demands effectively.

In Figure 9, we observe the impact of increasing the number of blockchain nodes on the average latency of the system. As the number of nodes in the blockchain network grows, the process of reaching consensus becomes more complex and time-consuming. Consensus mechanism, DPoS requires all participating nodes to agree on the validity of transactions and the state of the blockchain. With more nodes involved, the consensus algorithm must coordinate a larger number of participants, leading to increased communication overhead and computational requirements. Consequently, the average latency, which measures the time taken for transactions to be confirmed and added to the blockchain, tends to rise as the network expands. This is depicted in Figure 9, where the average latency increases with the number of blockchain nodes. The higher latency observed with more nodes underscores the trade-off between decentralization and transaction speed inherent in blockchain networks. While a larger number of nodes enhances the network's resilience and security by distributing control among multiple parties, it also introduces delays in transaction processing due to the consensus process's increased complexity. To enhance blockchain performance, it is important to know the link between node number and latency. Our study of BIONIB with increasing nodes shows it is scalable and efficient. It can handle high transaction throughput and manage resources, making it suitable for large-scale deployments in bridge health monitoring applications. By managing resources, adjusting consensus parameters, and implementing scalability solutions, developers can mitigate latency effects caused by network growth, ensuring efficient transaction processing.

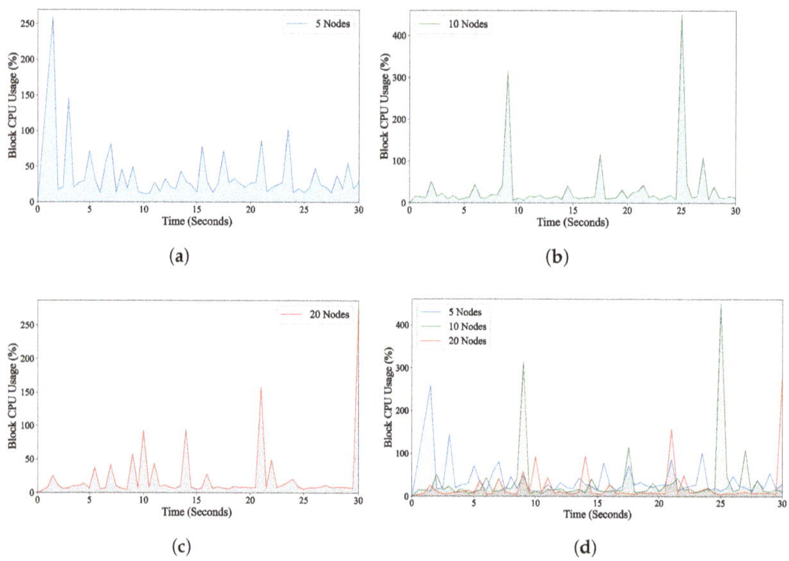

Figure 8. The percentage of CPU usage for block confirmation is fairly consistent for 5, 10, and 20 blockchain nodes. (**a**) 5 blockchain nodes, (**b**) 10 blockchain nodes, (**c**) 20 blockchain nodes, (**d**) Comparison of block CPU usage for 5, 10, and 20 blockchain nodes.

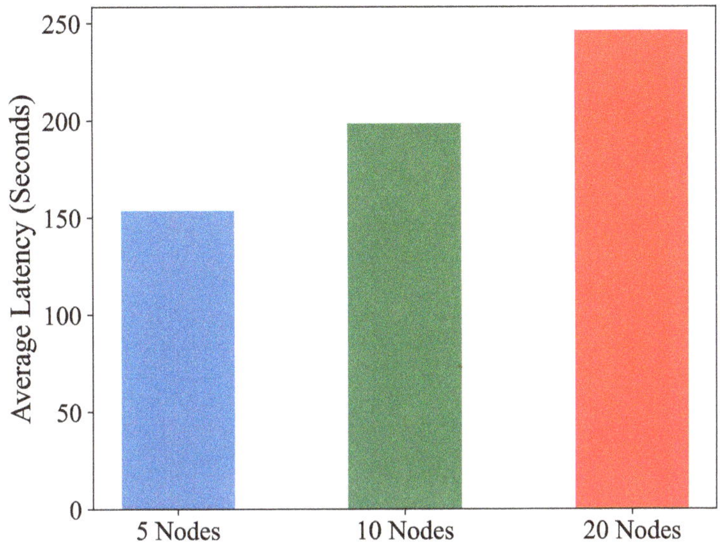

Figure 9. Average latency of BIONIB for 5, 10, and 20 blockchain nodes, showing increased latency as the number of nodes grows.

6.3. Performance with Increasing of IoT Sensors

In this section, we explore the impact of increasing the number of sensors on the performance of our proposed methodology, BIONIB. This analysis is crucial for understanding how our approach manages changes in data volumes that arise with a growing number of sensors. This is a common scenario in bridge health monitoring applications.

We have observed a clear correlation between the number of sensors and the volume of transactions confirmed by our system, as shown in Figure 10. With each increase in the number of sensors, there is a corresponding rise in the number of transactions processed and confirmed by our method. This observation highlights the system's capability to handle and process sensor data efficiently, ensuring timely and reliable transaction confirmation even as the data volume escalates. The increasing trend in confirmed transactions demonstrates the scalability and robustness of our approach, proving its ability to adapt to changing data volumes without compromising performance. By systematically scaling the number of sensors, we gain valuable insights into the system's capacity to accommodate growing data loads, assuring its suitability for real-world deployment scenarios where sensor networks may expand over time. Moreover, the positive relationship between the number of sensors and confirmed transactions underscores the effectiveness of our approach in harnessing sensor data to derive meaningful insights and facilitate informed decision-making in bridge health monitoring. As the number of sensors increases, the volume of data available for analysis also increases, enabling more comprehensive and accurate assessments of bridge health and structural integrity.

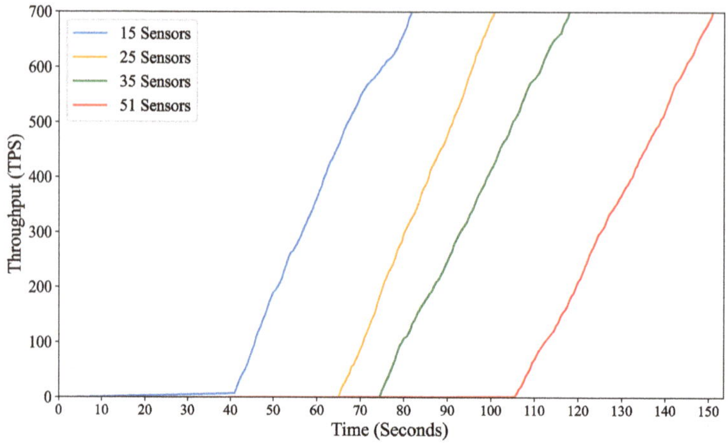

Figure 10. The throughput versus time of BIONIB is plotted and is observed that the throughput scales with an increasing number of sensors.

In Figure 11, we observe the CPU resource utilization in our system as the number of sensors is scaled. The plot shows a consistent pattern of CPU usage among different sensor configurations, highlighting the robust parallel processing capability of the EOSIO blockchain framework. The stable CPU usage demonstrates the efficiency of the EOSIO blockchain in distributing computational tasks across multiple nodes, ensuring optimal resource utilization even under fluctuating data volumes and network conditions.

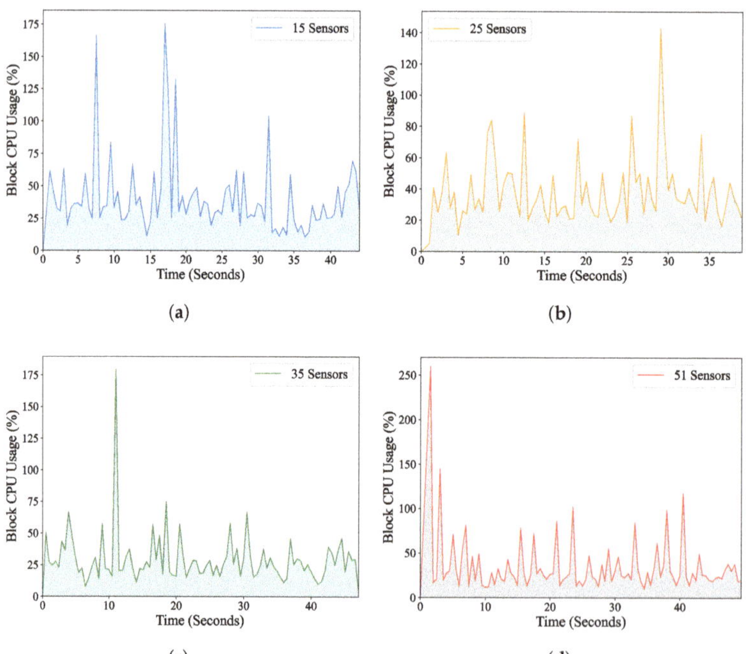

Figure 11. The percentage of CPU usage for block confirmation is fairly consistent for 15, 25, 35, and 51 bridge IoT sensors. (**a**) 15 bridge IoT sensors, (**b**) 25 bridge IoT sensors, (**c**) 35 bridge IoT sensors, (**d**) 51 bridge IoT sensors.

This analysis examines the relationship between the number of sensors and the average latency of our system, as shown in Figure 12. The plot shows that as the number of sensors increases, the average latency also rises due to the larger volume of data generated by the growing sensor network. This increase in data volume adds computational overhead, resulting in longer transaction confirmation times. This is partly because of more time taken to calculate the NI with more sensor data. To accommodate increasing data volumes, it is crucial to optimize system resources and network architecture. The observed increase in latency highlights the scalability challenges associated with handling large-scale sensor deployments within decentralized systems. While the EOSIO blockchain framework provides robust parallel processing capabilities, increasing data volumes can increase latency.

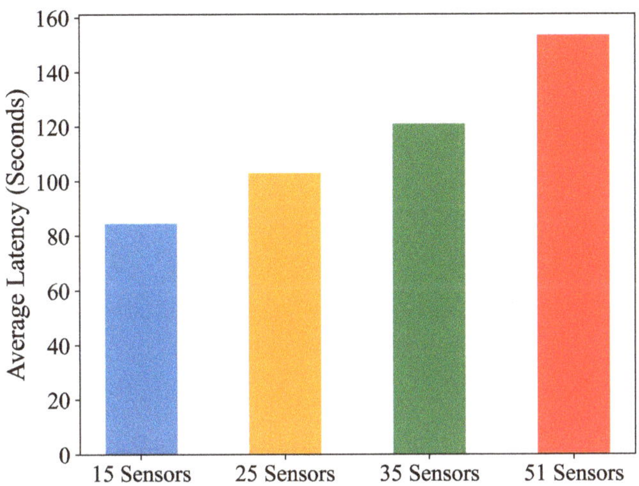

Figure 12. The average latency is plotted for 15, 25, 35, and 51 bridge IoT sensors. The latency increases with an increase in the IoT sensors.

7. Discussion

We observe a scalability and latency tradeoff. As the number of sensors increases, the system's scalability is tested by its ability to handle growing data volumes and transaction throughput. However, this scalability may lead to increased latency as processing and validating larger amounts of data can result in longer transaction confirmation times. Another tradeoff exists between resource utilization and performance. The efficient resource utilization of the EOSIO blockchain framework is demonstrated by the consistent CPU usage observed across different sensor configurations. However, stable resource utilization levels may require sacrificing some performance metrics, such as transaction throughput or latency. A balance between maximizing system throughput and minimizing resource consumption needs to be achieved. A tradeoff also exists between data processing efficiency and data volume. An increase in the number of sensors results in a larger volume of data generated, which must be processed and validated by the system. The EOSIO blockchain framework offers robust parallel processing capabilities, but handling large data volumes may introduce latency issues.

In this work, we conducted experiments using EOSIO smart contracts integrated with an NI approach for analyzing bridge sensor data. Our experimentation was conducted within a localized setup. However, for scenarios involving larger-scale data and a higher number of bridges, we propose the utilization of the TELOS implementation of the EOSIO blockchain specifically tailored for IoT data management.

In this study, we focus on storing scalar NI values on the blockchain rather than raw sensor data, which considerably reduces blockchain storage requirements and improves system scalability. However, this design inherently limits the ability to reprocess original sensor data for event validation, forensic investigations, or application of updated damage models—capabilities often required in real-world monitoring scenarios. The NI calculation used here builds upon the established methods in [20,21], which validate the NI as a sensitive and compact damage metric. To balance storage efficiency with forensic needs, potential extensions include integrating off-chain archival storage for raw data, linked via blockchain records, enabling selective retrieval and re-analysis while preserving the benefits of decentralized data integrity on-chain. This trade-off and its impact on practical deployment are discussed in detail to inform future work on hybrid storage architectures combining blockchain and traditional databases.

8. Conclusions

This research demonstrates how EOSIO blockchain technology can be used to monitor the health of bridges. By utilizing data from IoT sensors, a technique called the NI is used to obtain valuable insights. This technique is based on clustering and helps to identify unusual patterns in the sensor data. The integration of smart contracts enhances security and enables predictions based on the NI. The proposed scheme, BIONIB has the potential to securely store and analyze sensor data to enhance bridge health monitoring systems. The effectiveness of the scheme is evaluated by analyzing real-world IoT sensor data from healthy and unhealthy bridges. The evaluation also includes scenarios with varying numbers of sensors. This work contributes to the growing field of blockchain applications, particularly in the infrastructure monitoring context.

In the future, this approach can be extended to multiple bridges throughout the state by leveraging the virtual private blockchain feature of the EOSIO blockchain. The researchers also plan to conduct further analysis beyond NI calculations for bridge health monitoring. In future work, we will incorporate state-of-the-art unsupervised deep learning techniques (e.g., autoencoder-based damage localization) into the BIONIB framework while maintaining blockchain-based security and traceability. Our comparative analysis suggests that BIONIB fills a unique niche by coupling NI-based detection with a decentralized architecture, offering real-time alerts and transparent, verifiable records across sensor nodes.

Author Contributions: Conceptualization, All Authors; methodology, D.S.G. and S.E.A.; software, D.S.G. and R.M.; validation, All authors; formal analysis, D.S.G. and R.M.; investigation, D.S.G., D.K., and S.E.A.; resources, All authors; data curation, S.E.A. and R.M.; writing—original draft preparation, D.S.G.; writing—review and editing, S.E.A. and D.K.; visualization, D.S.G. and R.M.; supervision, D.K. and S.E.A.; project administration, D.K. and S.E.A.; funding acquisition, D.K. and S.E.A. All authors have read and agreed to the published version of the manuscript.

Funding: This work was partially supported by contracts W912HZ21C0060 and W912HZ23C0005 from the U.S. Army Corps of Engineers, Engineering Research and Development Center (ERDC).

Data Availability Statement: The data supporting the findings of this study are available from the authors upon reasonable request.

Conflicts of Interest: The authors declare that they have no conflict of interest.

References

1. Sony, S.; Laventure, S.; Sadhu, A. A literature review of next-generation smart sensing technology in structural health monitoring. *Struct. Control Health Monit.* **2019**, *26*, e2321. [CrossRef]
2. Skordilis, E.; Moghaddass, R. A deep reinforcement learning approach for real-time sensor-driven decision making and predictive analytics. *Comput. Ind. Eng.* **2020**, *147*, 106600. [CrossRef]
3. Yi, L.; Deng, X.; Yang, L.T.; Wu, H.; Wang, M.; Situ, Y. Reinforcement-learning-enabled partial confident information coverage for IoT-based bridge structural health monitoring. *IEEE Internet Things J.* **2020**, *8*, 3108–3119. [CrossRef]
4. Xu, J. ; Liu, H.; Han, Q. Blockchain technology and smart contract for civil structural health monitoring system. *Comput.-Aided Civ. Infrastruct. Eng.* **2021**, *36*, 1288–1305. [CrossRef]
5. Ali, M.S.; Vecchio, M.; Pincheira, M.; Dolui, K.; Antonelli, F.; Rehmani, M.H. Applications of blockchains in the Internet of Things: A comprehensive survey. *IEEE Commun. Surv. Tutorials* **2018**, *21*, 1676–1717. [CrossRef]
6. Chen, Z.; Zhou, X.; Wang, X.; Dong, L.; Qian, Y. Deployment of a smart structural health monitoring system for long-span arch bridges: A review and a case study. *Sensors* **2017**, *17*, 2151. [CrossRef]
7. Alenazy, W.M. Blockchain-Enabled internet of things for unsupervised structural health monitoring in potential building structures. In *Handbook of Research on Innovations and Applications of AI, IoT, and Cognitive Technologies*; IGI Global: Hershey, PA, USA , 2021; pp. 158–170.
8. Abou Jaoude, J.; Saade, R.G. Blockchain applications–usage in different domains. *IEEE Access* **2019**, *7*, 45360–45381. [CrossRef]
9. Wu, M.; Wang, K.; Cai, X.; Guo, S.; Guo, M.; Rong, C. A comprehensive survey of blockchain: From theory to IoT applications and beyond. *IEEE Internet Things J.* **2019**, *6*, 8114–8154. [CrossRef]
10. Bodkhe, U.; Tanwar, S.; Parekh, K.; Khanpara, P.; Tyagi, S.; Kumar, N.; Alazab, M. Blockchain for industry 4.0: A comprehensive review. *IEEE Access* **2020**, *8*, 79764–79800. [CrossRef]
11. Gadiraju, D.S.; Lalitha, V.; Aggarwal, V. Secure regenerating codes for reducing storage and bootstrap costs in sharded blockchains. In Proceedings of the 2020 IEEE International Conference on Blockchain (Blockchain), Rhodes, Greece, 2–6 November 2020; pp. 229–236.
12. Cai, X.; Geng, S.; Zhang, J.; Wu, D.; Cui, Z.; Zhang, W.; Chen, J. A sharding scheme-based many-objective optimization algorithm for enhancing security in blockchain-enabled industrial internet of things. *IEEE Trans. Ind. Inform.* **2021**, *17*, 7650–7658. [CrossRef]
13. Khan, P.W.; Byun, Y.C.; Park, N. IoT-blockchain enabled optimized provenance system for food industry 4.0 using advanced deep learning. *Sensors* **2020**, *20*, 2990. [CrossRef]
14. Gadiraju, D.S.; Lalitha, V.; Aggarwal, V. An optimization framework based on deep reinforcement learning approaches for prism blockchain. *IEEE Trans. Serv. Comput.* **2023**, *16*, 2451–2461. [CrossRef]
15. Aljuhani, A.; Kumar, P.; Alanazi, R.; Albalawi, T.; Taouali, O.; Islam, A.N.; Kumar, N.; Alazab, M. A Deep Learning Integrated Blockchain Framework for Securing Industrial IoT. *IEEE Internet Things J.* **2023**, *11*, 7817–7827. [CrossRef]
16. Du, A.; Ghavidel, A. Parameterized deep reinforcement learning-enabled maintenance decision-support and life-cycle risk assessment for highway bridge portfolios. *Struct. Saf.* **2022**, *97*, 102221. [CrossRef]
17. Gadiraju, D.S.; Muthiah, S.R.; Khazanchi, D. A Deep Reinforcement Learning based Approach for Bridge Health Maintenance. In Proceedings of the 2023 International Conference on Computational Science and Computational Intelligence (CSCI), Las Vegas, NV, USA, 13–15 December 2023; pp. 43–48. [CrossRef]
18. Da Silva, S.; Dias Junior, M.; Lopes Junior, V. Structural health monitoring in smart structures through time series analysis. *Struct. Health Monit.* **2008**, *7*, 231–244. [CrossRef]
19. Xiao, F.; Mao, Y.; Tian, G.; Chen, G.S. Partial-Model-Based Damage Identification of Long-Span Steel Truss Bridge Based on Stiffness Separation Method. *Struct. Control Health Monit.* **2024**, *2024*, 5530300. [CrossRef]
20. Ardani, S.; Akintunde, E.; Linzell, D.; Azam, S.E.; Alomari, Q. Evaluating pod-based unsupervised damage identification using controlled damage propagation of out-of-service bridges. *Eng. Struct.* **2023**, *286*, 116096. [CrossRef]
21. Akintunde, E.; Azam, S.E.; Rageh, A.; Linzell, D.G. Unsupervised machine learning for robust bridge damage detection: Full-scale experimental validation. *Eng. Struct.* **2021**, *249*, 113250. [CrossRef]
22. Jo, B.W.; Khan, R.M.A.; Lee, Y.S. Hybrid blockchain and internet-of-things network for underground structure health monitoring. *Sensors* **2018**, *18*, 4268. [CrossRef]
23. Gordan, M.; Ismail, Z.; Mohd Rahim, F.A.; Chao, O.Z.; Ibrahim, Z.; Hashim, H.; Kiong, L.C. Defining a conceptual framework for vibration-based damage detection platforms using blockchain. *J. Civ. Eng. Mater. Appl.* **2021**, *5*, 25–33.
24. Zhang, D.M.; Nie, C.; Zhang, J.Z.; Huang, H.W.; Huang, X. Consortium blockchain-based tunnel data bank for traceable sharing and treatment of structural health monitoring data. *Autom. Constr.* **2024**, *167*, 105720. [CrossRef]
25. Zheng, W.; Zheng, Z.; Dai, H.N.; Chen, X.; Zheng, P. XBlock-EOS: Extracting and exploring blockchain data from EOSIO. *Inf. Process. Manag.* **2021**, *58*, 102477. [CrossRef]
26. Perez, D.; Xu, J.; Livshits, B. Revisiting transactional statistics of high-scalability blockchains. In Proceedings of the ACM Internet Measurement Conference, Virtual, 27–29 October 2020; pp. 535–550.

27. He, N.; Zhang, R.; Wang, H.; Wu, L.; Luo, X.; Guo, Y.; Yu, T.; Jiang, X. EOSAFE: Security analysis of EOSIO smart contracts. In Proceedings of the 30th USENIX Security Symposium (USENIX Security 21), Vancouver, BC, USA, 11–13 August 2021; pp. 1271–1288.
28. Liu, J.; Zheng, W.; Lu, D.; Wu, J.; Zheng, Z. From decentralization to oligopoly: A data-driven analysis of decentralization evolution and voting behaviors on EOSIO. *IEEE Trans. Comput. Soc. Syst.* **2022**, *10*, 2752–2763. [CrossRef]
29. Fernández-Carrasco, J.Á.; Echeberria-Barrio, X.; Paredes-García, D.; Zola, F.; Orduna-Urrutia, R. ChronoEOS 2.0: Device Fingerprinting and EOSIO Blockchain Technology for On-Running Forensic Analysis in an IoT Environment. *Smart Cities* **2023**, *6*, 897–912. [CrossRef]
30. He, N.; Wang, H.; Wu, L.; Luo, X.; Guo, Y.; Chen, X. A Survey on EOSIO Systems Security: Vulnerability, Attack, and Mitigation. *Front. Comput. Sci.* **2025**, *19*, 196806. [CrossRef]
31. Gadiraju, D.S.; Aggarwal, V. Prism blockchain enabled Internet of Things with deep reinforcement learning. *Blockchain Res. Appl.* **2024**, *5*, 100205. [CrossRef]
32. Gadiraju, D.S.; Khazanchi, D. Blockchain-Driven Framework for DRL-Controlled UAV Bridge Health Monitoring. *IEEE Intell. Syst.* **2025**, *40*, 19–28. [CrossRef]
33. Pathirage, C.S.N.; Li, J.; Li, L.; Hao, H.; Liu, W.; Ni, P. Structural damage identification based on autoencoder neural networks and deep learning. *Eng. Struct.* **2018**, *172*, 13–28. [CrossRef]
34. Diez, A.; Khoa, N.L.D.; Makki Alamdari, M.; Wang, Y.; Chen, F.; Runcie, P. A clustering approach for structural health monitoring on bridges. *J. Civ. Struct. Health Monit.* **2016**, *6*, 429–445. [CrossRef]
35. Grigg, I. Eos—An Introduction. White Paper. 2017. Available online: https://www.semanticscholar.org/paper/EOS-An-Introduction-Grigg/c7a922dffb06ee1a75e1a5527c557da21b3c1d90 (accessed on 25 September 2025).
36. EOSIO. EOSIO GitHub Organization. 2024. Available online: https://github.com/EOSIO (accessed on 25 September 2025).

applied sciences

MDPI

Article

Seismic Assessment and Strengthening of a Load-Bearing Masonry Structure Considering SSI Effects

Kyriaki G. Amarantidou *, Panagiota S. Katsimpini *, George Papagiannopoulos and George Hatzigeorgiou

School of Science and Technology, Hellenic Open University, 26 335 Patras, Greece; papagiannopoulos@eap.gr (G.P.); hatzigeorgiou@eap.gr (G.H.)
* Correspondence: std147855@ac.eap.gr (K.G.A.); katsimpini.panagiota@ac.eap.gr (P.S.K.)

Abstract

This article examines the seismic assessment and strengthening of a traditional load-bearing masonry structure subjected to strong motion data, with particular emphasis on the effects of soil–structure interaction (SSI). The case study is the Archaeological Museum of Lemnos (AML)—a three-storey building with a composite load-bearing system of timber-framed stone masonry. Over time, the structure has undergone irreversible modifications, primarily involving reinforced concrete (RC) interventions. The building's seismic performance was evaluated using two finite element models developed in the SAP2000 software (v. 25.3.00). The first model simulates the original structure, strengthened by grout injections, while the second represents the current condition of the structural system following RC additions. Soil–structure interaction was also investigated, given that the local soil is classified as Category D according to Eurocode 8 (EC8). Each model was analyzed under two different support conditions: fixed-base and SSI-inclusive. A suite of appropriate accelerograms was applied to both models, in compliance with Eurocode 8 using the SeismoMatch software, and linear time-history analyses were conducted. The results underscore the significant impact of SSI on the increase of peak tensile stress and interstorey drift ratios (IDRs), and highlight the influence of different strengthening techniques on the seismic response of historic load-bearing masonry structures.

Keywords: timber-reinforced masonry; soil–structure interaction (SSI); linear time history analysis; strengthening interventions

1. Introduction

Although relatively rare, destructive earthquakes can cause extensive damage or even lead to the partial or total collapse of historical load-bearing masonry buildings that have not been retrofitted in accordance with modern seismic codes.

These structures typically exhibit inherent structural weaknesses, such as inadequate connections between individual load-bearing components and the low mechanical strength of materials, which render them particularly vulnerable to seismic forces. Failure mechanisms—such as in-plane and out-of-plane bending of masonry walls, leading to wall overturning, disaggregation and/or collapse of external wall corners [1]—have been repeatedly observed in recent major earthquakes, including those in Lesbos (2017) [2] and Turkey (2023) [3,4], thus confirming the need for further investigation into the seismic performance of such structures.

To this end, this paper examines the Archaeological Museum of Lemnos—a three-storey load-bearing stone masonry building with a timber-tiled roof, located in the island's capital,

Myrina. The structure features a regular floor plan and a distinctive composite system of timber-framed load-bearing masonry. Analogous construction typologies are found in the traditional architecture of various regions, including Lesvos [5], Lefkada [6], Turkey [4], the Himalayas [7], Chile [8], and Portugal (Pombalino System) [9].

Generally, in timber-framed load-bearing masonry buildings, the timber framing functions as a secondary seismic response mechanism (substructure). Whether implemented as floor-level ties, reinforcement at critical masonry points (e.g., around openings), or as a timber space truss embedded within the masonry—often in combination with internal timber-framed partition walls (bagdadi)—this system enhances the compressive strength of the masonry (by up to 20%), prevents vertical cracking, and reduces overall cracking (by up to 50%) [10]. It also increases the shear capacity and deformability of the masonry prior to failure [10]. Notably, despite the large deformations they may undergo, such structures have demonstrated the ability to withstand major seismic events [7]. In addition, timber-framed interior partition walls contribute—among other things—to increasing the stiffness and overall strength of the structure.

It should be noted that significant interventions have been carried out on the building under study, including the addition of reinforced concrete (RC) elements (an RC slab at the first-floor level and a bond beam at the roof level), which have altered its original structural system at the diaphragm levels [11].

Regarding the seismicity of the broader study area—the northeastern Aegean Sea, where the western end of the North Anatolian Fault extends—significant earthquakes have been recorded, including the Agios Efstratios event on 20 February 1968 (Mw 7.0–7.4) and the Lemnos earthquake on 24 May 2014 (Mw 6.9). However, seismic acceleration data are only available for the latter, with a recorded peak ground acceleration (PGA) of 0.11 g [12].

According to a seismic hazard assessment study for the North Aegean region [13], PGAs of up to 0.3 g are estimated in areas near Lemnos under earthquake scenarios with magnitudes of Mw 7.0 or higher. Notably, during the 12 June 2017 earthquake on the neighboring island of Lesvos, accelerations reached values as high as 0.2 g—approaching the design PGA value specified by the Eurocode 8 response spectrum for the broader region [2].

To assess the seismic response of the AML to strong ground motions, three major seismic events were selected: the Northridge earthquake (Pacific Palisades—Sunset station, Los Angeles, CA, USA, 1994, Mw = 6.7) [14]; the Kahramanmaraş earthquake (Turkey, 2023, Mw = 7.7) [15]; and the Kocaeli earthquake (Düzce station, Turkey, 1999, Mw = 7.51) [14]. To adapt these seismic records to the target spectrum calculated for the AML, the procedure outlined in EC8 was followed. The response spectra of the selected ground motions were scaled to match the EC8 design spectrum within the period range of 0.2 T to 2 T, where T is the fundamental period of the system in each seismic direction.

Additionally, in order to compare the original and the existing structural systems, two finite element models were developed, one excluding and one incorporating RC elements. It should be noted that interventions involving reinforced concrete (e.g., slabs and bond beams) can have a significant impact on load-bearing masonry buildings, depending on factors such as the adequacy of slab-to-masonry connections, the proper use of reinforcement [4], and whether the intervention is accompanied by masonry strengthening [2].

Finally, given that the building is founded on class D soft soil, as defined by EC8, soil–structure interaction (SSI) effects were also investigated for each model. The results of the analyses are compared with those of the study by Genç et al., 2023 [16] which analyzed the seismic behavior of a historic building under both fixed-base conditions and soil–structure interaction (massed and massless) across three different soil categories: hard, medium, and soft. The results are also compared with the study of [17]. Also, Requena-Garcia-

Cruz [18] examined the impact of soil-structure interaction (SSI) effects on the seismic analysis of cultural heritage buildings, such as the Mosque-Cathedral of Córdoba in Spain. An investigation of Özmen & Sayin [19] was conducted into the variations in the seismic response of a historical masonry church, utilizing four different SSI models in addition to a fixed base model that did not consider SSI. Furthermore, the research of Altiok and Demir [20] examines the seismic response of the historical Lala Mehmet Pasha minaret by taking into account Soil Structure Interaction (SSI). The influence of SSI on the out-of-plane performance of an ancient structure in Iran, known as Arge-Tabriz has been investigated by Fathi et al., [21]. Tzanakis et al., [22] explored the seismic performance of St. Titus Church located in Heraklion, Crete, Greece, along with the necessity for its seismic retrofitting. Additionally, the impacts of soil-structure interaction have been considered.

This research tackles significant deficiencies in the seismic evaluation of historic masonry structures by offering a thorough analysis that concurrently takes into account original construction methods, contemporary strengthening measures, and the effects of soil-structure interaction, elements that are seldom analyzed together in the current literature. The engineering significance of these structures is underscored by their dual function as cultural heritage assets and operational public buildings, rendering their seismic safety crucial for both the protection of lives and the preservation of heritage. Historic load-bearing masonry edifices, such as the Archaeological Museum of Lemnos, are especially susceptible due to their age, material deterioration, and construction techniques that predate modern seismic regulations. Nevertheless, they must continue to fulfill modern roles while preserving their historical authenticity. The originality of this study arises from its direct comparison of two strengthening approaches, traditional grout injection against modern reinforced concrete solutions, utilizing the same analytical methods and realistic seismic inputs that adhere to Eurocode 8 standards [23]. By integrating soil-structure interaction effects for Category D soils and concentrating on quantitative metrics like interstorey drift ratios and peak tensile stress, this research offers valuable insights for engineers and conservators striving to reconcile structural safety with heritage preservation needs, while also establishing a replicable framework for evaluating similar composite timber-framed stone masonry systems globally.

2. Case Study

2.1. Historical Background

The Archaeological Museum of Lemnos is located in Myrina, on the coastal front of the "Romeikos Gialos" area, between the cape of the Meteorological Station and the Castle peninsula (Figure 1). Settlement remains from the Late Neolithic period have been discovered in this area, while the earliest settlement of the "city" of Myrina was established on the Castle peninsula during the Early Iron Age. The museum is situated near the prehistoric settlement of Myrina, and together the two sites form an important archaeological complex. Positioned directly adjacent to the Ecclesiastical Museum of the Holy Metropolis of Lemnos, the Myrina Gymnasium (Figure 2), and the Pantelideion Building (formerly the Girls' School), the AML constitutes an integral part of the architectural ensemble of Romeikos Gialos, developed between the 19th and early 20th centuries.

The building was constructed in the late 19th century to serve as the seat of the Turkish Administration, while the idea of housing the Archaeological Museum was first proposed in the 1930s. The plan, however, was not realized until several years later, following the end of World War II. In 1956, the building was repaired, and the first exhibition of the AML was held in 1961, following the repatriation of the island's archaeological finds, which had been transferred for safekeeping during the war to the Museum of Mytilene (Lesvos) and the National Archaeological Museum of Athens. Thirty years later, in 1991, a new exhibition was held following additional repair work on the building (Figure 3a).

Figure 1. Map of the town of Myrina, showing the location of the Archaeological Museum of Lemnos (AML) (Adapted from map data provided by the Hellenic Ministry of Culture and Sports—Archaeological Cadastre (www.arxaiologikoktimatologio.gov.gr, accessed on 5 June 2025).

Figure 2. Photograph from 1950. From left to right: the Archaeological Museum of Lemnos, the Gymnasium of Myrina, and the Ecclesiastical Museum of the Holy Metropolis of Lemnos [24].

(a) (b)

Figure 3. Main façade of the Archaeological Museum of Lemnos: (**a**) Repair works carried out in 1991; (**b**) The main façade after the completion of repair works in 2014 (Source: Digital archive of the Ephorate of Antiquities of Lesvos (Hellenic Ministry of Culture and Sports), used with permission (Prot. No. 252765/16-06-2025)).

In May 2014, the island was struck by a strong earthquake that caused damage to both the museum and its exhibits (Figures 3b and 4). Although the building's repairs were completed in October 2014, it was decided not to reopen the exhibition on the second floor, as the museum required renovation to meet modern museological standards.

(a) (b)

Figure 4. Cracks on the spandrel (**a**) and at the junction of the external masonry and transverse internal timber wall on the second floor (**a**,**b**), caused by the 2014 earthquake (Source: Digital archive of the Ephorate of Antiquities of Lesvos (Hellenic Ministry of Culture and Sports), used with permission (Prot. No. 252765/16-06-2025)).

In 2019, the Greek Ministry of Culture approved a project entitled "Modernization of the Archaeological Museum of Lemnos". The scope of the project included the strengthening of the building's foundations and stone masonry, floor repairs, replacement of wooden window frames and door frames, application of new coatings, and the installation of a lift for people with disabilities. The re-exhibition of the museum's antiquities is scheduled for summer 2025.

2.2. Intervention Phases

Over the course of its history, the building has undergone five distinct phases of structural and functional transformation. The first construction phase dates to the 19th century and corresponds to its original function as the seat of the Turkish Administration. The original architectural form of the building remains unchanged to this day [25].

The second construction phase refers to the period associated with the building's first museum exhibition in 1961. In preparation for this exhibition, extensive and largely irreversible interventions were carried out in 1956. These included the replacement of the original timber floors at the ground and first levels with new reinforced concrete (RC) slabs, and the strengthening of the second-level floor with the addition of metal beams.

The third construction phase took place thirty years later, in 1991, during the re-exhibition of the Archaeological Museum's collection. At that time, several significant and mostly reversible interventions were made to the building, including roof repairs, the construction of new sanitary facilities, the installation of electrical and mechanical systems, and the addition of new wooden flooring.

The fourth construction phase followed the earthquake of 24 May 2014, which had a magnitude of Mw 6.9 [12]. During this phase, a reinforced concrete bond beam (senaz) was constructed to strengthen the masonry walls, and both roof waterproofing and roof tile repairs were carried out.

The fifth and final construction phase is currently underway and is set to be completed in time for the opening of the museum's new exhibition, scheduled for the summer of 2025. It involves significant and irreversible interventions, including the demolition of exterior walls on the southwest elevation to accommodate a lift for people with disabilities; construction of a new reinforced concrete floor at ground level; foundation strengthening through grout injection; repointing of the stone masonry walls and application of new traditional-style renderings; reconstruction of the external staircases; replacement of the wooden staircase; installation of new timber flooring on the second floor; and general renovation works in the interior spaces.

2.3. Architectural Documentation

The Archaeological Museum is an impressive three-storey structure comprising a low ground floor and two upper levels, with a total area of approximately 606 square meters and a height of 13.09 m. The plot includes outdoor storage areas totaling approximately 210 square meters, along with a courtyard that surrounds the building. Three of the building's elevations are exposed, while the fourth—on the northeastern side—borders the courtyard of the Myrina Gymnasium. The main façade faces the seafront to the northwest and is fronted by an open space enclosed by a low fence.

Although the museum building is not officially designated as a protected monument, it is a representative example of its period, featuring elements of restrained neoclassical architecture typical of urban buildings of the late 19th and early 20th centuries. All three levels share a similar spatial configuration, with a symmetrical layout that consists of a large central hall flanked by rooms on either side. The primary exhibition spaces are located on the upper floors, while auxiliary rooms occupy the ground level. The floors are internally connected by a staircase located to the southwest of the hall, and externally by staircases positioned at ground level on both main façades [25].

Morphologically, the building adheres to the eclectic architectural style prevalent in Myrina—and particularly in the Romeikos Gialos area—at the end of the 19th century. Its design blends restrained neoclassical elements with features of local vernacular architecture. Numerous openings characterize the building's façades, which follow symmetrical design principles, most rigorously applied to the two principal elevations—the northwestern and southeastern. The fundamental neoclassical tripartite division into base (lower ground floor), main storeys (two floors) and crown (roof) is retained. Distinctive features of the building's form include the corner quoin stones and two monumental, semi-circular external stone staircases, symmetrically positioned at the centre of the main façades, with the second-floor balconies located directly above them [25].

2.4. Structural Description

2.4.1. Ground Floor

The exterior and interior walls of the ground floor are made of load-bearing stone masonry, except for one wall constructed with clay bricks (optoplinthi). The external load-bearing masonry is 65 cm thick, the internal is 50 cm, and the clay-brick wall 30 cm. The external masonry is built as three-leaf rubble construction with lime mortar applied on all wythes. Large dressed stones are positioned at the corners of the walls, while semi-dressed stones and clay bricks are used as infill between them, contributing to the stability and integrity of the masonry. The core is filled with rubble stone and mortar, forming a solid mass between the outer wythes [5]. The internal stone masonry is constructed as two-leaf rubble masonry (Figure 5a).

(a) (b)

Figure 5. Structural features of AML: (**a**) The internal masonry construction method of the ground floor; (**b**) Metal anchor tie bar on the exterior face of the masonry (Source: Digital archive of the Ephorate of Antiquities of Lesvos (Hellenic Ministry of Culture and Sports), used with permission (Prot. No. 252765/16-06-2025)).

The stone used in the masonry, of volcanic origin and gray in color, was sourced from quarries near the settlements of Thanos and Romanou in Lemnos. According to the geotechnical study of the Castle of Myrina [26] and after the identification of the building stones, it was found that they belong to the categories of lavas found in the local quarries (dacites or ignimbrites) with an average strength of fbc = 50 MPa.

The mortar used in the building is cohesive and contains lime and large dark-colored coarse aggregates (brown and gray), collected from streambeds where volcanic rock erosion products accumulate. A small amount of ceramic material grains was also identified. The plaster applied to the ground-floor masonry is a lime-rich mortar of medium consistency, containing medium-grained and coarse-grained aggregates. The tensile strength of the interior masonry plaster was estimated at 78 kPa [27].

No timber lacing was found in the ground floor masonry; however, iron ties and anchor bars—visible on the façades (Figure 5b)—were used at floor level to strengthen the walls. These ties—metal bars running along the exterior of the masonry—are not connected to the wooden elements of the floor or the timber framing present on the second floor. Timber ties were found embedded within the internal stone masonry, oriented transversely across the wall at door openings (Figure 5a).

2.4.2. First Floor

Regarding the horizontal load-bearing elements, the original timber floor at the first level was replaced in 1956 (second construction phase) with a 15 cm thick reinforced concrete slab, supported by 35 cm wide horizontal RC beams. The ground floor connects to the first floor via an internal staircase, made of reinforced concrete and ceramic tiles (second construction phase), or externally through the central entrance and via the external circular staircase on the main façades.

The exterior masonry of the first floor consists of 50 cm thick load-bearing stonework and appears to be more carefully constructed than that of the ground floor, incorporating a specific technique involving clay bricks [5]. On the interior side of the masonry, above the openings, there are arched spandrels made of clay bricks (Figure 6a). The interior partition walls are made of 30 cm thick clay brick masonry, added during the second construction phase following the installation of the reinforced concrete slab. The plaster on the interior face of the first-floor masonry is 1 cm thick and contains dark-colored aggregate in a 1:1 weight ratio, along with a significant amount of plant fibers [27]. The first level is connected to the second by a wooden staircase (Figure 6b).

(a) (b)

Figure 6. Structural features of the AML building: (**a**) Interior view of the external wall at the first level; (**b**) The two internal staircases of the building—a reinforced concrete staircase at the ground floor and a wooden staircase at the first floor. (Source: Digital Archive of the Ephorate of Antiquities of Lesvos, Hellenic Ministry of Culture and Sports; used with permission, Prot. No. 252765/16-06-2025).

2.4.3. Second Floor

At the second level, the timber floor was reinforced with metal IPE beams during the second construction phase, while an additional timber layer (floorboards) was installed during the third phase. The floor system comprises simply supported timber joists with a cross-section of 10 × 10 cm, spaced at intervals of 40–50 cm. Floorboards and ceiling boards, each approximately 2.4 cm thick, are nailed above and below the joists, respectively.

At the this level, the building features a distinctive hybrid system that combines stone masonry with integrated timber framing. This construction technique has been examined and thoroughly documented by N. Karydis (2001) in his study of the traditional settlement of Eressos in Lesvos [5]. The walls consist of 50 cm thick two-leaf masonry, reinforced on the inner face with a dense timber framework, locally known as "friggia". This system consists of vertical timber posts, typically with a cross-section of 10 × 10 cm, spaced 40–100 cm apart depending on the wall configuration (Figure 7a). In areas requiring increased stiffness—such as between openings—horizontal members and diagonal braces of similar cross-section are installed. N. Karydis (2001) [5] (pp. 6–7) notes that vertical posts are generally aligned with the corners of openings, while horizontal ties correspond to lintels and sills. The timber framework is self-supporting and structurally capable of carrying roof loads independently of the masonry, while also connecting both wall faces transversely at openings [5] (p. 8).

(a)

(b)

Figure 7. Structural features of the AML building at 2nd level: (**a**) Interior view of the external wall; (**b**) Baghdati internal wall. (Source: Digital Archive of the Ephorate of Antiquities of Lesvos, Hellenic Ministry of Culture and Sports; used with permission, Prot. No. 252765/16-06-2025).

The interior walls of the second floor are of the baghdati type—timber framed partitions with a total thickness of 17 cm (Figure 7b). These lightweight walls complete the aforementioned timber frame system and contribute to the overall stiffness of the structure. The plaster used is of the same composition and strength as that of the first floor [27].

The building has a four-pitched timber roof of a distinctive structural configuration, covered with Roman-style clay tiles. Several additional timber members were added during the second, third, and fourth construction phases. However, the primary structural elements of the roof today include vertical posts, principle rafters and tie beams (14 × 14 cm cross-section), along with diagonal braces (7 × 14 cm cross-section).

The structure is founded on stone foundation strips aligned with the ground-floor load-bearing masonry walls. These foundations reach a depth of 82 cm from the axis of the ground level floor and feature a cross-sectional widening of 20 cm on either side of the masonry in the final 60 cm of depth.

As described in detail in this section, the building's structural system is of particular interest. It has a rigid base formed by load-bearing masonry with relatively small openings. The first floor is reinforced with arched spandrels to improve load distribution onto the piers surrounding the openings, while the second floor has two mechanisms for absorbing seismic forces: the stone masonry itself and the system of timber elements (timber wall frames, baghdati-type walls and timber flooring). These systems collectively facilitate the transfer of roof loads to the lower levels. In the event of masonry failure during an earthquake, the loads are transferred to the timber frame system. Table 1 depicts the repair methods of the structure under study.

Table 1. Repair Methods.

	Model 1	Model 2
Three-leaf rubble masonry (ground floor)	Grout injections	Repointing works
Two-leaf rubble masonry (1st and 2nd floors)	Grout injections	Repointing works
1st level's floor	Double layered timber flooring —timber joists	15 cm thick reinforced concrete slab supported by 35 cm wide horizontal RC beams
2nd level's floor	Double layered timber flooring —timber joists	Triple layered timber flooring—timber joists
Openings	-	Demolition of exterior walls to accommodate a lift for people with disabilities
Roof beams	Wooden tie beams	Reinforced concrete bond beam (senaz)

The steel MRF study of [17] indicates that the incorporation of infill masonry walls (IMWs) can lead to a reduction in $IDR_{\_Med}$ by as much as 32.26% and enhance collapse performance by 12.48%, thereby illustrating the efficacy of wall-based retrofitting methods. The present research's comparison of grout injection strengthening with RC interventions offers a complementary analysis of various retrofitting philosophies applicable to masonry structures. While the steel study emphasizes the addition of infill walls to frame systems, the investigation of the authors explores how different strengthening methods (traditional versus modern) influence the seismic response of load-bearing masonry. Both studies enhance the understanding that the effectiveness of retrofitting is significantly influenced by the interplay between the chosen strengthening technique and the original structural framework.

3. Materials and Numerical Modeling

3.1. FE Models

The structural composition of the Archaeological Museum of Lemnos comprises a complex system that incorporates both traditional materials and later reinforced concrete additions. Each floor level exhibits different masonry types and construction techniques, which were carefully taken into account in the development of the numerical models.

The ground floor is constructed with three-leaf rubble masonry, while two-leaf masonry is used for the interior walls. On the first floor, the exterior walls are also constructed with two-leaf masonry, whereas the interior partitions are made of clay brick masonry. The second floor features a dual construction system: two-leaf masonry integrated with a traditional embedded timber frame system (friggia). Although structurally significant, this system was excluded from direct modelling due to its complexity and the variability in the performance of timber elements, which depends on factors such as age and the quality of joinery. Timber components were, however, included in the model to represent floor joists, ties, and wall plates. Subsequent intervention phases introduced reinforced concrete (RC) elements into the structure, including a floor slab with beams, RC bond beams, and IPE steel beams supporting the second-storey floor.

The numerical modeling and structural analysis were performed using SAP2000 [28]. For modeling purposes, the following material and section typologies were identified and assigned geometrical and mechanical properties:

- Three-leaf rubble masonry (ground floor exterior walls).
- Two-leaf rubble masonry (ground floor interior walls and first-floor exterior walls).

- Two-leaf rubble masonry with timber frames (second-floor exterior walls, modeled with adjusted unit weight to reflect embedded timber content).
- Clay brick masonry (ground floor and first-floor interior walls).
- Steel beams (IPE).
- Timber elements (tie beams, floor joists, and roof wall plates).
- Reinforced concrete elements (slab with beams, bond beam).

Two structural models were simulated to evaluate the seismic behavior of the building (Figure 8):

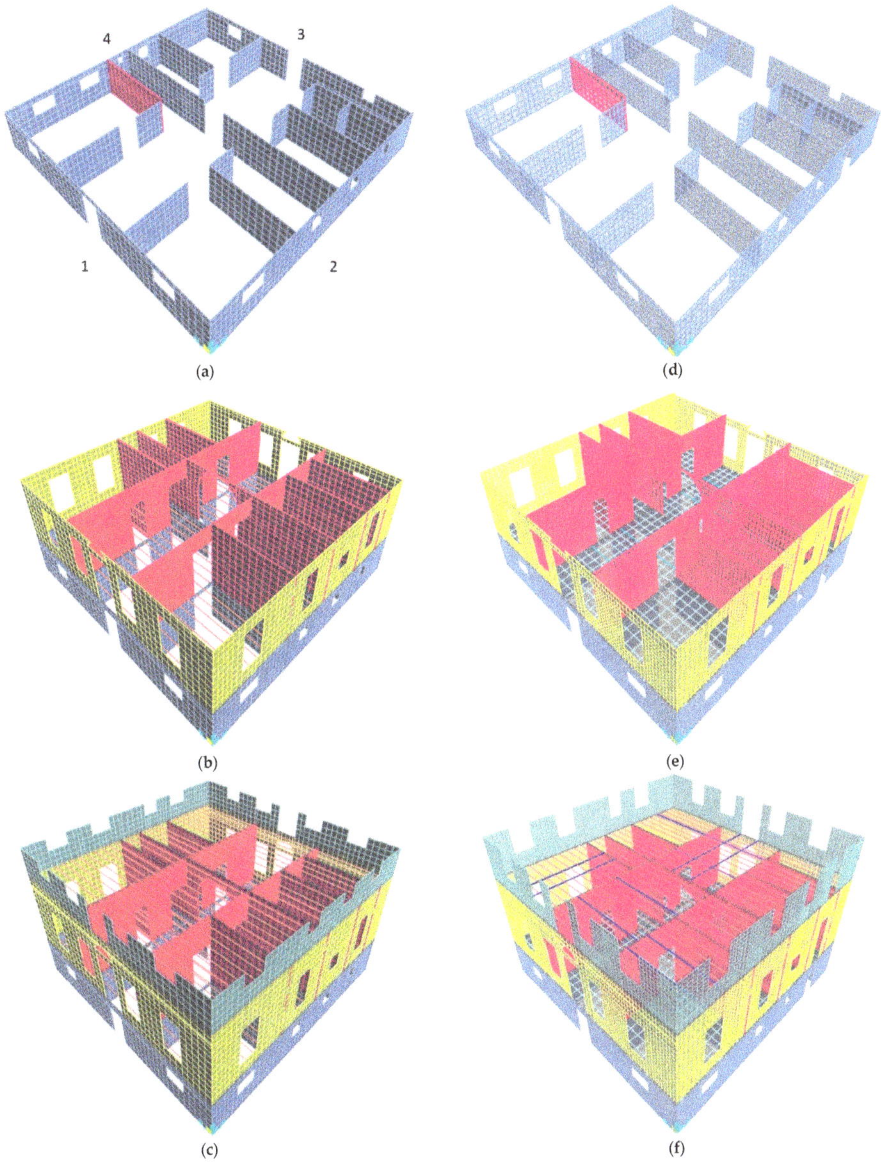

Figure 8. FE Models 1 and 2: M1: Ground floor with numbered building elevations (**a**), first floor (**b**) and second floor (**c**); M2: Ground floor (**d**), first floor (**e**) and second floor (**f**).

The first model (Model 1 or M1) represents the building with its original structure (wooden floors), assuming that the load-bearing masonry has been strengthened through grout injection. Although constructed at a later phase, the first-floor interior walls are modeled as clay brick masonry, to support the overlying timber joists.

The second model (Model 2 or M2) represents the current state of the building (fifth construction phase). In this case, a 20% increase in masonry strength is applied [29] to account for improvements due to repointing works.

In both models, masonry is treated as a homogeneous, isotropic material that combines the properties of its constituent materials (stone and mortar). The rationale for adopting an isotropic approach probably arises from practical modeling factors and the constraints of existing material property data for historical masonry. Assessing the complete anisotropic material properties of century-old masonry would necessitate comprehensive and possibly destructive testing, which may not be practical for a functioning museum structure. Furthermore, the intricate geometry and diverse construction methods employed throughout the building could complicate the establishment of uniform directional properties. However, the mechanical contribution of the embedded timber frames is not explicitly modeled; only their self-weight is taken into account in the structural analysis. The geometric data for both models were derived from the architectural plans provided in [25].

To facilitate the modeling process and reduce computational complexity, a series of simplifications and assumptions were adopted:

- Architectural details were simplified and elements (e.g., openings) were moved by 5–10 cm in order to reduce the number of structural nodes. Decorative elements on the façades were not simulated.
- The roof was not modeled, but its loads were applied at the points of contact between the roof trusses and the masonry.
- Interior and exterior staircases, as well as permanent operational loads associated with the building's function as a museum, were not included in the simulation.
- The marble balcony on the main façade was not modeled; however, its loads were applied as pairs of forces and moments at its four support points.
- The masonry walls were modeled with finite shell elements. The eccentricity of the masonry at each floor was taken into account—since the ground floor walls are thicker than those of the upper floors—and was modeled using the 'Area Thickness Overwrites' command.
- All timber elements (roof and floor joists and ties), steel IPE beams and the RC bond beam at roof level were modeled as linear frame elements, with frame releases applied at their ends to reflect realistic connection behavior.
- The 'Insertion Point' command was used on all frame elements to accurately model their eccentric positioning relative to the global geometry.
- The reinforced concrete floor slab was modeled using shell finite elements.
- Timber floorboards were not modeled; however, the weight of the timber floors was applied as distributed loads on the joists.
- The discretization of shell elements was performed using the 'Divide Areas' command with a maximum element size of 0.5×0.5 m. Frame elements were meshed using Assign Automatic Frame Mesh.

3.2. Mechanichal Properties

The mechanical properties of all structural elements were calculated according to [1,23,26,27,29–37] and they are presented in Tables 2–5.

Table 2. Mechanical properties of masonry.

Section	Model 1				Model 2			
	f_c (MPa)	$E_{50\%}$ (MPa)	f_t (MPa)	γ (KN/m³)	f_c (MPa)	$E_{50\%}$ (MPa)	f_t (MPa)	γ (KN/m³)
Three—leaf Masonry	2.06 [32]	800 [22]	0.185 [31]	23	2.48 [22]	900 [22]	0.22 [31]	22
Two—leaf Masonry	4.93 [32]	2500 [28]	0.44 [31]	23	6.42 [22]	3210 [28]	0.58 [31]	22
Clay Brick Masonry	1.76 [28]	879 [28]	0.12 [31]	18	1.76 [28]	879 [28]	0.12 [31]	18

Table 3. Mechanical properties of concrete materials.

Concrete	f_{ck} (MPa)	γ (KN/m³)	E (Gpa)
C25/25 [30]	20	25	30
B160 [26]	9	24	27

Table 4. Mechanical properties of steel materials.

Steel	f_{yk} (MPa)	f_u (MPa)	E (GPa)
Sthl I [30]	240	340	200
B500C [25]	500	650	200
S235 [29]	215	360	210

Table 5. Mechanical properties of wood.

Wood	γ (KN/m³)	E_1//(MPa)	E_2 ⊥ (MPa)	E_3 ⊥ (GPa)
D30 [30]	6.4	8000	640	640

In the above tables, f_c stands for compressive strength, f_t for tensile strength, f_{yk} and f_u (steel) for characteristic yield strength and ultimate tensile strength respectively, E denotes the modulus of elasticity, and γ the unit weight. The Poisson's ratio is assumed to be 0.2 for all materials, except for wood and steel S235, for which a value of 0.3 is adopted.

3.3. Loads

Dead loads are assumed to be 0.5 kN/m² for wooden floorboards and 1.2 kN/m² for ceramic tiles [38]. In addition, the self-weight of structural elements has also been taken into account. Live loads are assumed to be 5.0 kN/m² for the floors and 0.5 kN/m² for the roof [38].

3.4. Soil Properties and Soil–Structure Interaction (SSI)

According to the geotechnical report carried out for the AML, the wider coastal zone of Romeikos Gialos is covered by Holocene Quaternary (al) deposits, coastal deposits and dunes. The deposits consist of alluvial clay, argillic materials, sands, and weathered products of older sedimentary and volcanic rocks [39]. The subsoil of the building's foundations is classified as category D: Deposits of loose-to-medium cohesionless soil (with or without some soft cohesive layers), or of predominantly soft-to-firm cohesive soil [23].

In general, soil–structure interaction (SSI) can significantly affect structures built on soft or saturated soils, as it can lead to an increase in their fundamental period and consequently alter their dynamic behavior. According to [40], taking SSI into account is particularly important for structures founded on soils with high groundwater levels and high plasticity (PI > 40), typically corresponding to soil categories C or D, for which reduction factors to the shear wave velocity (Vs) and the shear modulus (G) of the soil are recommended.

For the purposes of the present study, the soil–structure interaction effect was investigated by using formulas [41], which model the foundation-soil system as a discrete system consisting of frequency independent springs, dashpots and masses. More specifically, for each model examined in this study, an additional case was analyzed. In these cases, the fixed supports (as defined in the initial models) (M1fix, M2fix) were replaced by link elements with specific mechanical properties (M1ssi, M2ssi).

To apply the aforementioned formulas, the length of the foundation strips was calculated and then divided by the number of masonry joints at the foundation level. The width of the foundation strips is 1.05 m, the calculated length is 0.27 m, so $\alpha = 0.27$ m. The shear wave velocity was taken as 150 m/s, the soil density as $\rho = 1900$ kg/m^3 [39], and the Poisson's ratio as $\nu = 0.3$. Based on these values, the shear modulus G was calculated as 42,750.00 KPa. Applying a reduction coefficient of 0.37 [40], the reduced shear modulus becomes 15,817.50 KPa, corresponding to a reduced shear wave velocity of 91.24 m/s. The values of static stiffness (K) and damping (C) of each link element are presented in Table 6. The soil was assumed to be massless.

Table 6. Stiffness and damping coefficients for FSI models.

	Static Stiffness K	Damping C
Vertical	28,273.80 KN/m	66 KNs/m
Horizontal	22,788.90 KN/m	10.84 KNs/m
Rocking	1705.45 KNm/rad	2.99 KNms/rad
Torsion	2480.15 KNm/rad	0.92 KNms/rad

To ascertain the fundamental frequency of the site and evaluate possible resonance effects with the structure of the Archaeological Museum of Lemnos, calculations were conducted utilizing the established soil parameters. With a shear wave velocity of Vs = 150 m/s and presuming a typical soil layer depth of H = 30 m for Category D soils, the fundamental frequency of the soil deposit was computed using the quarter-wavelength approximation: $f_0 = Vs/(4H) = 150/(4 \times 30) = 1.25$ Hz. The shear modulus G = 42,750.00 KPa was obtained from the equation $G = \rho Vs^2$, where $\rho = 1900$ kg/m^3 and Vs = 150 m/s, confirming the relatively soft characteristics of the soil typical of Category D classification. The computed site frequency of 1.25 Hz is within the typical range for historic masonry buildings (0.8–2.0 Hz), indicating potential resonance conditions that could significantly enhance the seismic response. This frequency analysis lays the physical groundwork for understanding why the SSI effects led to increased peak tensile stress and interstorey drift ratios, as the structure-soil system likely underwent dynamic amplification when the building's fundamental frequency neared the site frequency, especially following various strengthening interventions that may have modified the original structural frequency.

4. Type of Analysis Performed

4.1. Static Analysis

Based on the structural analysis using the load combination 1.35G + 1.5Q, the total weight of M1 was calculated as 14,778.3 kN, and that of of M2 as 16,056.92 kN. Under this load combination, no exceedance of the maximum compressive strength was observed. However, with regard to tensile strength, exceedance points were identified in Model 2, specifically at the connection points between the second-floor steel beams and the masonry wall.

4.2. Modal Analysis

In the modal analysis, the periods and participating mass ratios were calculated for each model. According to seismic codes [23,29], the cumulative effective modal mass of the modes considered in the spectral analysis must amount to at least 75% of the total mass of the structure, as specified in [29], and 90% as required by [23]. For the purposes of the present study, 200 modes were considered for each model (Tables 7–10):

Table 7. M1fix—modal participating mass ratios.

StepNum	Period	SumUX	SumUY	SumUZ
Unitless	Sec	Unitless	Unitless	Unitless
1	0.18602	0.00%	5.20%	0.00%
2	0.13667	12.09%	5.20%	0.00%
41	0.05178	87.83%	87.10%	21.49%
133	0.03032	90.03%	88.84%	84.80%
190	0.02547	90.49%	90.00%	86.03%
200	0.02468	90.56%	90.13%	86.17%

Table 8. M1ssi—modal participating mass ratios.

StepNum	Period	SumUX	SumUY	SumUZ
Unitless	Sec	Unitless	Unitless	Unitless
1	0.189562	0.02%	16.77%	0.00%
2	0.168865	88.18%	16.77%	0.00%
3	0.143698	88.18%	94.25%	0.01%
9	0.117733	91.98%	94.31%	0.03%
15	0.091661	93.43%	94.38%	98.23%

Table 9. M2fix—modal participating mass ratios.

StepNum	Period	SumUX	SumUY	SumUZ
Unitless	Sec	Unitless	Unitless	Unitless
1	0.287152	0.00%	0.00%	0.42%
3	0.156688	5.01%	3.73%	0.43%
5	0.150461	73.52%	3.74%	0.99%
8	0.122583	75.61%	57.18%	1.00%
59	0.05362	90.05%	89.25%	77.62%
81	0.045667	90.48%	90.03%	83.13%
200	0.029105	92.11%	91.54%	87.77%

Table 10. M2ssi—modal participating mass ratios.

StepNum	Period	SumUX	SumUY	SumUZ
Unitless	Sec	Unitless	Unitless	Unitless
1	0.287175	0.00%	0.00%	0.44%
2	0.272051	0.02%	5.90%	0.45%
3	0.194507	91.14%	5.90%	0.45%
4	0.166549	91.16%	93.56%	0.46%
17	0.102015	93.04%	94.23%	84.39%
200	0.031548	97.52%	97.82%	87.40%

The following results are extracted from Tables 7–10:

- Compared to M1fix and M1ssi, the fundamental period increases by 54.37% for M2fix and 51.49% for M2ssi, respectively.
- The fundamental period of M1ssi is 1.90% higher than that of M1fix, while M2ssi exhibits only a 0.01% increase relative to M2fix, indicating no significant change in the fundamental natural period in either case.
- The ratios $T_1M1ssi/T_1M1fix = 1.02$ and $T_1M2ssi/T_1M2fix = 1.00$ are both ≤ 1.08; therefore, the structures can be analyzed as if founded on rigid soil [42]. Nevertheless, the effect of SSI was still examined in order to evaluate its influence on the maximum stress values.

4.3. Time History Analysis

For the linear time history analysis of the structure, three strong motion time histories were selected based on the guidelines in [23], and for each time history, three accelerograms were used: two for the horizontal directions and one for the vertical direction.

The seismic records selected were those from the Northridge earthquake (Pacific Palisades—Sunset station, Los Angeles, CA, USA, 1994, Mw = 6.7) [14], the Kahramanmaraş earthquake (Turkey, 2023, Mw = 7.7) [15], and the Kocaeli earthquake (Düzce station, Turkey, 1999, Mw = 7.51) [14]. The process of selecting earthquake records for the seismic evaluation of the Archaeological Museum of Lemnos illustrates a methodical strategy that integrates historical seismic information with ground motions pertinent to the region. However, a more robust site-specific justification would enhance the analysis. The incorporation of the Northridge earthquake (1994, Mw = 6.7) offers a significant historical context as one of the most thoroughly documented seismic occurrences, providing high-quality accelerometric data that acts as a standard for comparison and guarantees the inclusion of well-analyzed ground motion characteristics. Moreover, the choice of two Turkish earthquakes—Kahramanmaraş (2023, Mw = 7.7) and Kocaeli (1999, Mw = 7.51)—demonstrates the geographic closeness and seismotectonic resemblance to the Lemnos site. Both the Turkish seismic zones and the North Aegean region are affected by the intricate tectonic interactions between the Eurasian and African plates, including the westward extension of the North Anatolian Fault system into the Aegean Sea. The Kahramanmaraş earthquake supplies contemporary ground motion data utilizing modern recording techniques, while the Kocaeli earthquake provides insights from a significant strike-slip event that shares similar fault mechanisms with those anticipated in the North Aegean region. Nevertheless, the authors ought to offer clearer justification concerning tectonic similarity, the appropriateness of source-to-site distance, the compatibility of original recording site conditions with the Category D soils at Lemnos, and whether the spectral characteristics remain representative after scaling to align with Eurocode 8 requirements, which would bolster the credibility of employing these records for evaluating this vital cultural heritage structure. The ground motion records were modified to match the target response spectrum specified in [23], within the period range of 0.2T to 2T for each direction, following the methodology outlined therein. Spectral matching as shown in Figure 9 was performed using the Seismomatch software (v. 2025) [43].

According to the spectrum defined in [23], the building is classified as Importance Class III, with an importance factor of $\gamma I = 1.2$. A behavior factor of $q = 1.5$ was adopted, appropriate for masonry structures. The seismic zone is classified as Zone Z2 with a design ground acceleration of $Ag = 0.24$ g, and a vertical design acceleration of $Av = 0.9 \times Ag$. The site soil is categorized as Type D ($S = 1.35$), and a constant damping ratio of $\zeta = 5\%$ was assumed in the analyses.

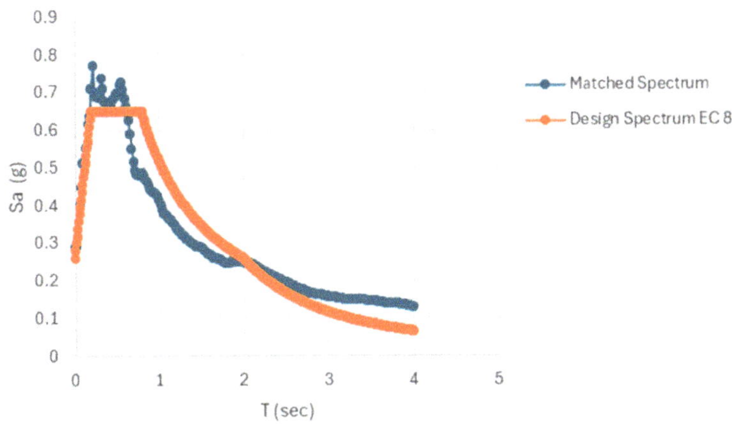

Figure 9. Spectral matching of the mean spectrum of selected earthquakes (x—axis) to the target spectrum (EC8).

For each earthquake record, two separate load cases were created by reversing U1 and U2 (Load Case Data—Linear Direct Integration History) for each horizontal accelerogram. The following load cases were created:

- Kahramanmaraş1 (Kah1) and Kahramanmaraş2 (Kah2),
- Northridge1 (Nor1) and Northridge2 (Nor2) and
- Kocaeli1 (Koc1) and Kocaeli2 (Koc2)

4.4. Analysis Challenges

According to [44], the combination of actions for seismic design situations applied to the building under study is G + 0.3Q + E. However, it was not possible to extract this load combination from the analysis software, nor was it possible to obtain envelope stresses, i.e., the maximum (SMAX) and minimum (SMIN) stress values for each elevation of the building.

For this reason, time series were extracted for selected shell elements on each elevation, in order to determine the approximate time points at which maximum stress values occur on each elevation. Additionally, time series were obtained for two shell elements in each model—one on the ground floor and one on the second floor—on the elevation that exhibits the highest SMAX exceedance. This enables direct comparison of the stresses developed across the models. To assess the interstorey drift ratio (IDR), four joints were selected at each floor level along the same vertical axis on the same elevation. It should be noted that the load combination G + 0.3Q does not result in any stress exceedances.

5. Results

5.1. Maximum SMAX Stress Values per Elevation

The distribution of maximum tensile stresses (SMAX) was evaluated across all four elevations for each model. The results offer an overview of the seismic stress demands imposed on the building and facilitate a comparative assessment between fixed-base and SSI conditions, as well as between the two different strengthening methods (Tables 11 and 12). Figures 10 and 11 illustrate the SMAX stress distribution for the Kahramanmaraş1 event at selected elevations.

Table 11. SMAX for shells 7342 (M1) and 6724 (M2) on ground floor.

M1	SMAX			M2	SMAX			% Change
Shell 7342	At (s)	Value (kPa)	% Change	Shell 6724	At (s)	Value (kPa)	% Change	M1 vs. M2
1fkah1 1skah1	11.22 11.23	323.40 371.80	14.97%	2fkah1 2skah1	11.52 12.14	394.20 318.80	−19.13%	21.89% −14.25%
1fkah2 1skah2	12.42 11.23	272.00 400.70	47.32%	2fkah2 2skah2	12.74 11.55	474.00 413.30	−12.81%	74.26% 3.14%
1fnor1 1snor1	11.33 9.38	231.70 415.00	79.11%	2fnor1 2snor1	11.67 10.22	467.40 724.60	55.03%	101.73% 74.60%
1fnor2 1snor2	5.86 7.86	283.70 445.80	57.14%	1fnor2 2snor2	13.62 7.89	447.90 564.00	25.92%	57.88% 26.51%
1fkoc1 1skoc1	8.28 9.96	262.80 506.80	92.85%	2fkoc1 2skoc1	11.11 11.13	465.70 659.00	41.51%	77.21% 30.03%
1fkoc2 1skoc2	8.98 9.38	264.50 404.20	52.82%	2fkoc2 2skoc2	9.95 10	480.30 639.10	33.06%	81.59% 58.11%

Table 12. SMAX for shells 7587 (M1) and 6724 (M2) on ground floor.

M1	SMAX			M2	SMAX			% Change
Shell 7587	At (s)	Value (kPa)	% Change	Shell6939	At (s)	Value (kPa)	% Change	M1 vs. M2
1fkah1 1skah1	11.93 11.94	1024.00 2286.00	123.24%	2fkah1 2skah1	19.22 19.21	1014.00 1191.00	17.46%	−0.98% −47.90%
1fkah2 1skah2	11.04 11.02	777.40 1397.00	79.70%	2fkah2 2skah2	11.09 11.09	1019.00 1030.00	1.08%	31.08% −26.27%
1fnor1 1snor1	8.96 9.34	1078.00 1919.00	78.01%	2fnor1 2snor1	11.22 10.67	1088.00 1425.00	30.97%	0.93% −25.74%
1fnor2 1snor2	8.96 9.34	1091.00 1971.00	80.66%	1fnor2 2snor2	11.22 10.67	1137.00 1499.00	31.84%	4.22% −23.95%
1fkoc1 1skoc1	9.31 9.31	976.00 1731.00	77.36%	2fkoc1 2skoc1	9.13 10.18	1112.00 1470.00	32.19%	13.93% −15.08%
1fkoc2 1skoc2	9.21 11.26	1133.00 2131.00	88.08%	2fkoc2 2skoc2	9.34 11.3	941.50 1409.00	49.65%	−16.90% −33.88%

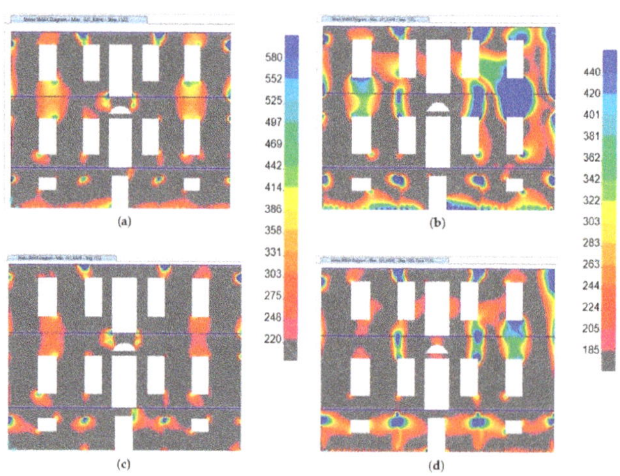

Figure 10. SMAX stress results on the main (1) elevation. The minimum values on the chromatic scales correspond to the tensile strength at the ground floor, while the maximum values represent the tensile strength at the 1st and 2nd floors. Stress exceedance at 1st floor and 2nd floor occurs only in areas indicated by blue coloration: M1fix (**a**); M2fix (**b**); M1ssi (**c**); M2ssi (**d**).

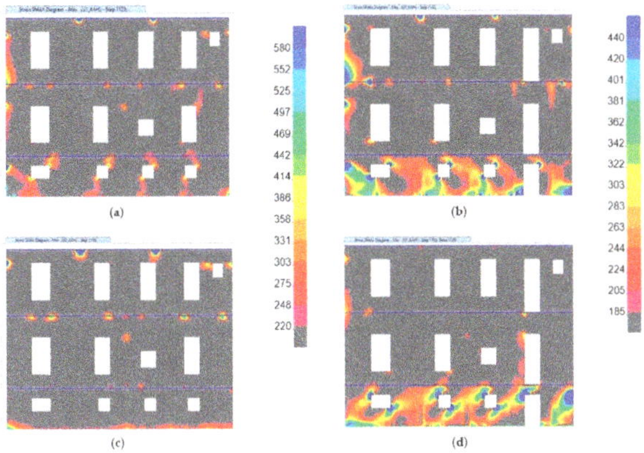

Figure 11. SMAX stress results on elevation 3. The minimum values on the chromatic scales correspond to the tensile strength at the ground floor, while the maximum values represent the tensile strength at the 1st and 2nd floors: M1fix (**a**); M2fix (**b**); M1ssi (**c**); M2ssi (**d**).

5.2. Interstorey Drift Tables and Residual Interstorey Drift Tables

On the same elevation, four joints were monitored for displacements along the axis where maximum deformations occur (Figure 12). Table 13 presents the height distributions for peak interstorey drift ratios (IDR) calculated with reference to the top joint. For this assessment, the deformation limits adopted in [29] were used: 4‰ for in-plane and 8‰ for out-of-plane deformations.

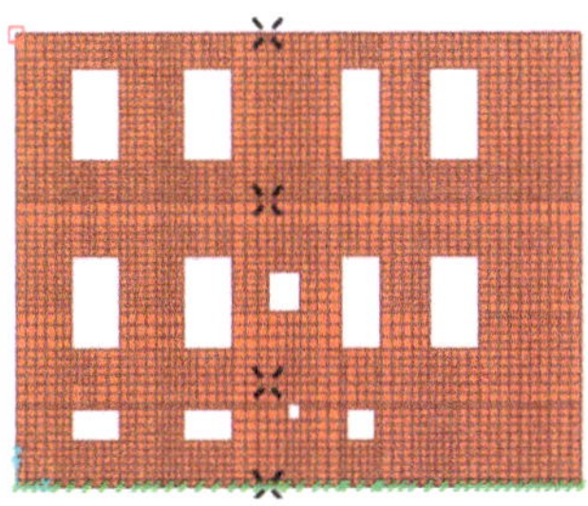

Figure 12. Shells 7342/6724 of Models M1/M2 on ground floor and 7587/6939 of Models M1/M2 on second floor.

Table 13. Peak IDR_U2 (out-of-plane deformation) for the top selected joint on elevation 4.

Model - Event	Max IDR ‰	Time of Max IDR (s)	Model - Event	Max IDR ‰	Time of Max IDR (s)
M1fix - Kah1	−3.137	11.82	**M2fix - Kah1**	−7.639	19.08
M1fix - Kah2	2.355	11.04	**M2fix - Kah2**	−7.456	11.27
M1ssi - Kah1	6.806	11.94	**M2ssi - Kah1**	−9.530	19.08
M1ssi - Kah2	4.191	11.02	**M2ssi - Kah2**	7.385	11.09
M1fix - Nor1	3.198	8.96	**M2fix - Nor1**	−7.778	10.54
M1fix - Nor2	−3.274	10.20	**M2fix - Nor2**	−8.229	10.54
M1ssi - Nor1	−6.313	10.21	**M2ssi - Nor1**	−10.124	12.47
M1ssi - Nor2	−6.512	10.21	**M2ssi - Nor2**	−10.409	12.47
M1fix - Koc1	−3.306	9.40	**M2fix - Koc1**	−7.477	10.05
M1fix - Koc2	3.383	9.21	**M2fix - Koc2**	6.248	11.29
M1ssi - Koc1	−6.232	10.00	**M2ssi - Koc1**	9.668	10.18
M1ssi - Koc2	6.354	11.26	**M2ssi - Koc2**	−9.633	11.41

It should be noted that U1 and U2 refer to the X and Y axes, respectively, while elevation 4 is oriented towards the XZ plane. Therefore, U1 refers to in-plane deformations and U2 refers to out-of-plane deformations.

6. Discussion

The analyses conducted under both static and dynamic loading conditions reveal critical differences in the structural behavior of Models 1 and 2, and highlight the influence of soil–structure interaction (SSI) on their seismic performance.

With regard to the static analysis, soil–structure interaction results in varying tensile stress distributions across the building elevations, thereby preventing the derivation of reliable conclusions. However, a comparison of compressive stresses within the building's external masonry indicates that models incorporating SSI effects exhibit lower stress levels.

The two models, M1 and M2, were subjected to particularly high seismic loads. Elastic dynamic time history analysis of the selected events revealed significant damage across all four elevations of both models. This damage includes the ramming of the external masonry by floor joists and transverse masonry, and pertains to the influence of roof loads on the external walls, as well as to both in-plane and out-of-plane failure mechanisms. These mechanisms indicate potential wall collapse, as evidenced in the SMAX tables for Elevation 4.

However, the observed failure pattern of the second floor does not accurately reflect its actual structural behavior, as the wooden elements were not simulated. Consequently, the influence of the embedded timber frames (friggia) and the timber-framed internal walls was not taken into account, rendering the second-floor masonry more vulnerable to seismic actions.

Regarding SSI influence, the conclusions according to Tables 11 and 12 are summarized below:

Ground floor:

- In Model M1, the maximum SMAX value is observed in case M1ssiKoc1, exhibiting a 93% increase compared to M1fKoc1. Stress increases are observed across all M1ssi models, ranging from 15% to 93%.
- In Model M2, the maximum SMAX value is observed in case M2ssiNor1, exhibiting a 55% increase compared to M2fNor1. The Kah1 and Kah2 events exhibit stress reductions under SSI conditions. Overall, the observed increases range from 26% to 55%, while the reductions range from 13% to 19%.
- A comparison between Models M1 and M2 reveals an increase in stress across all M2 cases, with the exception of M2ssiKah1, which presents a 14.25% decrease relative to M1ssiKah1. Additionally, all M2 cases exhibit a delay in the timing of peak stress occurrence compared to M1.

2nd floor:

- In Model M1, the maximum SMAX value is observed in case M1ssiKah1, exhibiting an increase of 123.24% compared to M1fKah1. All cases show increases ranging from 78% to 125%, indicating that stress increases on the first floor are significantly higher than those on the ground floor.
- In Model M2, the maximum SMAX value is observed in case M2ssiNor2, exhibiting an increase of 32% compared to M2fNor2. However, the greatest increase occurs in the Koc2 event, reaching 50%. Unlike the ground floor, no event exhibits stress reduction under SSI conditions. Overall, stress increases in this model range from 1% to 50%.
- A comparison between Models M1 and M2 shows that all M2fix models exhibit increased stress, except for cases M2fkah1 and M2fKoc2. In contrast, M2ssi models

show a reduction in stress compared to M1ssi. Finally, similar to the ground floor, all M2 cases—except for M2fKoc1—exhibit a delay in the timing of peak stress occurrence compared to M1.

The findings indicate that the influence of soil–structure interaction (SSI) on stress amplification is more significant in Model M1, which exhibits a shorter fundamental period due to increased stiffness resulting from masonry consolidation via grout injection. Model M2, by comparison, which features greater mass, reduced masonry stiffness, and a longer fundamental period than M1, also experiences stress increases under SSI (with the exception of the Kahramanmaraş earthquake), albeit to a lesser extent.

It is also important to note that, on the second floor, soil–structure interaction results in stress reduction across all combinations for Model M2. Interestingly, the second-floor masonry in Model M1 is reinforced with timber laces at both the floor and roof levels, yet exhibits greater stiffness and strength compared to M2. In contrast, in Model M2 there is a reinforced concrete beam at roof level. Nevertheless, stress exceedances are observed in both models, while the structural system of M2 appears to be more vulnerable under the seismic scenarios examined in this study.

Concerning interstorey drift ratios (IDRs), SSI increases the relative floor displacements—typically by nearly double—across most seismic combinations and structural models. The sole exception is case M2Kah2, in which the maximum IDR value remains unchanged.

Regarding in-plane deformation, no IDR value exceeded the 4‰ threshold, with drift ratios remaining low (around 0.2‰). However, out-of-plane deformation exhibited significantly higher values on the second floor, particularly in cases M2ssiKah1, M2ssiKoc1, M2ssiKoc2, M2fixNor2, and M2ssiNor1 and Nor2, where the 8‰ threshold was surpassed. The IDR differences for each seismic scenario are listed below in detail:

Model 1

- In the Kahramanmaraş earthquake, the maximum IDR on the second floor is recorded in case M1ssiKah1, reaching 6.8‰—the highest IDR value among all M1 seismic cases, in agreement with the values reported in Table 11. The increases observed in the SSI models are 117% for the Kah1 event and 78% for Kah2.
- In the Northridge earthquake, the maximum IDR on the second floor is recorded in case M1ssiNor2, reaching 6.5‰. The increases observed in the SSI models are 97% for the Nor1 event and 99% for Nor2.
- In the Kocaeli earthquake, the maximum IDR on the second floor is recorded in case M1ssiKoc2, reaching 6.35‰. The increases observed in the SSI models are 89% for the Koc1 event and 88% for Koc2.

Model 2

- In the Kahramanmaraş earthquake, the maximum IDR on the second floor is recorded in case M2ssiKah1, reaching 9.53‰. With regard to the SSI models, the Kah1 event exhibits a 25% increase, while SSI has virtually no effect in the Kah2 event, as the IDR increases by only 1%.
- In the Northridge earthquake, the maximum IDR on the second floor is recorded in case M2ssiNor2, reaching 10.41‰—the highest IDR value among all M2 seismic cases, in agreement with the values reported in Table 11. The increases observed in the SSI models are 30% for the Nor1 event and 27% for Nor2.
- In the Kocaeli earthquake, the maximum IDR on the second floor is recorded in case M2ssiKoc1, reaching 9.69‰ (Koc2 exhibits a nearly identical value of 9.63‰). The increases observed in the SSI models are 29% for the Koc1 event and 54% for Koc2.

A comparison between models M1 and M2 regarding the effects of SSI on IDRs leads to the same conclusion previously drawn concerning maximum stresses—namely, that SSI has a greater influence in model M1. Overall, the increase in IDRs observed in the SSI models is in line with [16], where the incorporation of soil–structure interaction (SSI) into the analysis of historical load-bearing masonry structures revealed significant increases in drift ratios.

Finally, comparing models M1 and M2 in terms of IDR values, it is observed that, for the same seismic combination, maximum relative floor displacements exhibit significant increases on the second floor in M2. It is also observed, that the largest percentage increases are recorded in fix–based models. Specifically, the greatest percentage change is recorded in the M1fixKah2 and M2fixKah2 models (217%), with a range of values for all combinations spanning from 85% to 217%. The SSI models show an average percentage change in IDR of approximately 60%, with a range between 40% and 76%. It should be noted that this analysis refers to percentage changes; however, as mentioned above, the highest values are recorded in the Mssi models. The detected concentration of maximum displacements on the second floor in M2 could indicate that the RC interventions have led to an irregular distribution of stiffness, fostering soft-story behavior. This alarming finding suggests that the modern strengthening techniques may have inadvertently created new vulnerabilities. This analysis emphasizes the critical need to consider both the strengthening approaches and the soil-structure interaction effects when reviewing retrofit strategies for historic masonry structures, as the combination of these factors can profoundly impact the evaluation of structural performance and safety.

The study of [17] and the present study underscore the vital significance of accurate soil classification in seismic analysis. The steel MRF study reveals that "assuming different soil types can influence the results of $IDR_{\_Med}$ and $RDR_{\_Med}$, which can subsequently affect the performance of the building," with soil type C increasing maximum $IDR_{\_Med}$ by 66.76% in comparison to soil type B for a 5-story MRF under DBE loading. In a similar vein, the present investigation of the Archaeological Museum indicates substantial SSI effects for Category D soils, where the soft soil conditions greatly enhance peak tensile stress and interstorey drift ratios. This parallel observation bolsters the reliability of the approach in explicitly modeling SSI effects, as both studies validate that neglecting proper soil characterization can lead to significant underestimation of seismic demands.

The comprehensive approach of the steel MRF study [17], which utilizes 3456 performance curves derived from nonlinear dynamic analyses (NDAs) and incremental dynamic analyses (IDAs), establishes a benchmark for the validation level of analyses that enhances confidence in the research outcomes. In contrast, the present study's dependence on linear time-history analyses employing equivalent linear soil modeling (with a 37% reduction in stiffness) could gain from a similar validation process through nonlinear analyses, especially considering the recognized limitations in addressing "gapping, sliding, and uplift" effects that are crucial for Category D soils. The methodology of the steel study indicates that a more thorough validation of the linear spring approach might reinforce the conclusions regarding the relative effectiveness of various strengthening strategies for the Archaeological Museum and this constitutes a limitation.

7. Conclusions

The findings of the present study indicate the critical role of SSI in the seismic behavior of load-bearing masonry buildings. Significant increases in both stresses and displacements were recorded in the models with elastic supports (link elements), clearly demonstrating the influence of SSI on the seismic response. Additionally, most cases exhibited a delay in the timing of peak structural response, indicating a dynamic coupling between the soil and the superstructure. On the contrary, excluding SSI effects from the seismic assessment of historical masonry buildings could result in a significant underestimation of seismic demand and, consequently, of the necessary strengthening interventions.

Finally, with regard to retrofitting interventions, the model that includes reinforced concrete (RC) additions (Model 2) showed higher stress concentrations on the story beneath the RC diaphragm (i.e., the ground floor), as well as significantly higher interstorey drifts on the second floor. Despite the potential benefits of RC-based interventions—such as improved box–type and diaphragm behavior etc.—their effectiveness depends on construction quality, proper detailing, and, most importantly, on adequate mechanical connection with the masonry. These measures should always be accompanied by direct strengthening of the masonry itself.

This study highlights the importance of this distinction by comparing two strengthening strategies: one focused on masonry strengthening (Model 1), and another emphasizing diaphragm strengthening through reinforced concrete (RC) intervention (Model 2). Based on the analysis results, the masonry-centered approach (Model 1) demonstrated better seismic performance, emphasizing its suitability for the seismic protection of historical masonry structures.

Author Contributions: Conceptualization, K.G.A.; methodology, K.G.A.; software, K.G.A.; writing—original draft preparation, K.G.A.; writing—review and editing, K.G.A.; visualization, P.S.K.; supervision, P.S.K., G.P. and G.H. All authors have read and agreed to the published version of the manuscript.

Funding: This research received no funding.

Data Availability Statement: The original contributions presented in this study are included in the article. Further inquiries can be directed to the corresponding authors.

Conflicts of Interest: The authors declare no conflicts of interest.

Abbreviations

The following abbreviations are used in this manuscript:

AML	Archaeological Museum of Lemnos
SSI	Soil–Structure Interaction
Mi	Model i, i = 1, 2
Kahi	Kahramanmaraş, i = 1, 2
Nori	Northridge, i = 1, 2
Koci	Kocaeli1, i = 1, 2
Mif	Model i, fix -based supports
Missi	Model i, SSI (link elements) supports

References

1. Verras, D. *Seismic Damage, Repair and Strengthening—Assessment of Seismic Damage, Repair and Strengthening of Traditional and Monumental Buildings, SMA 61/2*; HOU (Hellenic Open University): Patras, Greece, 2004; Volume II, pp. 136–154, 193–223. (In Greek)
2. Vlachakis, G.; Vlachaki, E.; Lourenço, P.B. Learning from failure: Damage and failure of masonry structures, after the 2017 Lesvos earthquake (Greece). *Eng. Fail. Anal.* **2020**, *117*, 104803. [CrossRef]
3. Garini, E.; Gazetas, G.; Giarlelis, C.; Moretti, M.; Palieraki, V.; Pitilakis, D.; Stefanidou, S.; Thanopoulos, P.; Thanop-oulos, P.; Tsiatas, G.; et al. The February 6th Kahranmanmaraş Earthquake Sequence in Turkiye, HAEE, NTUA, TCG. Technical Report, 2023. Available online: https://www.researchgate.net/publication/379550289_The_February_6th_2023_Kahranmanmaras_Earthquake_sequence_in_Turkiye_HAEE_NTUA_TCG?channel=doi&linkId=660eadfeb839e05a20bd4ac4&showFulltext=true (accessed on 2 December 2024). [CrossRef]

4. Vintzileou, E.; Palieraki, V. Damage to stone masonry buildings in historical centers due to the 2023 earthquake sequence in Turkey. *Earthq. Spectra* **2024**, *40*, 2306–2338. [CrossRef]
5. Karydis, N. Analysis of the Traditional Settlement of Eressos in Lesvos. Athens, Greece, September 2001. Available online: https://oasp.gr/node/618 (accessed on 23 May 2025). (In Greek)
6. Makarios, T.; Demosthenous, M. Traditional Buildings of Lefkada with the Dual Structural System and Their Seismic Behavior. In Proceedings of the 3rd Panhellenic Conference on Earthquake Engineering and Engineering Seismology, Athens, Greece, 5–7 November 2008. Available online: http://library.tee.gr/vufind/Record/1%2F29964 (accessed on 23 May 2025). (In Greek)
7. Dhandapany, D.; Menon, A. Seismic Behavior of Timber-Laced Masonry Structures in the Himalayan Belt. In *Structural Analysis of Historical Constructions*, 1st ed.; RILEM Bookseries; Springer: Dordrecht, The Netherlands, 2019; Volume 18, pp. 605–613. [CrossRef]
8. Jiménez, B.; Saloustros, S.; Pelà, L. Seismic vulnerability index method for hybrid timber–masonry structures. Numerical calibration and application to the city of Valparaíso, Chile. *J. Build. Eng.* **2021**, *44*, 103185. [CrossRef]
9. Cardoso, R.; Lopes, M.; Bento, R. Seismic assessment of 'Pombalino' buildings. In Proceedings of the 13th World Conference on Earthquake Engineering, Vancouver, BC, Canada, 1–6 August 2004. Available online: https://www.iitk.ac.in/nicee/wcee/thirteenth_conf_Canada/ (accessed on 23 May 2025).
10. Vintzileou, E. Effect of timber ties on the behavior of historic masonry. *J. Struct. Eng.* **2008**, *134*, 961–972. [CrossRef]
11. Doudoumis, I.N. Analytical modelling of traditional composite timber-masonry walls. *Adv. Mater. Res.* **2010**, *133–134*, 441–446. [CrossRef]
12. ITSAK–OASP. The North Aegean Earthquake of 24/5/2014 12:25 Mw=6.9; May 2014. Available online: https://oasp.gr/userfiles/file/2014_24%CE%9C%CE%91%CE%A5_%CE%9B%CE%97%CE%9C%CE%9D%CE%9F%CE%A3/2014_05_24_LHMNOS_EQ_FINAL(3).pdf (accessed on 20 May 2025). (In Greek)
13. Stylianou, E.; Maravas, G.; Kouskouna, V.; Papoulia, J. Seismic hazard assessment in the North Aegean Trough based on a new seismogenic zonation. *Bull. Geol. Soc. Greece* **2016**, *50*, 1443–1452. [CrossRef]
14. Pacific Earthquake Engineering Research Center. PEER Ground Motion Database. Available online: https://ngawest2.berkeley.edu/ (accessed on 26 January 2024).
15. TADAS. Turkish Accelerometric Database and Analysis System. Available online: https://tadas.afad.gov.tr/login (accessed on 26 January 2024).
16. Genç, A.F.; Ertürk Atmaca, E.; Günaydin, M.; Altunişik, A.C.; Sevim, B. Evaluation of soil structure interaction effects on structural performance of historical masonry buildings considering earthquake input models. *Structures* **2023**, *54*, 869–889. [CrossRef]
17. Kazemi, F.; Asgarkhani, N.; Jankowski, R. Enhancing seismic performance of steel buildings having semi-rigid connection with infill masonry walls considering soil type effects. *Soil Dyn. Earthq. Eng.* **2024**, *177*, 108396. [CrossRef]
18. Requena-Garcia-Cruz, M.V.; Romero-Sánchez, E.; Morales-Esteban, A. Contribution of the soil-structure interaction to the seismic behaviour of the Mosque-Cathedral of Córdoba. *J. Build. Eng.* **2025**, *110*, 113061. [CrossRef]
19. Özmen, A.; Sayin, E. Different Soil-Structure Interaction Modelling Strategies for Seismic Analysis of a Masonry Church. *Int. J. Archit. Herit.* **2024**, 1–22. [CrossRef]
20. Altiok, T.Y.; Demir, A. Collapse mechanism estimation of a historical masonry minaret considered soil-structure interaction. *Earthq. Struct.* **2021**, *21*, 161–172. [CrossRef]
21. Fathi, A.; Sadeghi, A.; Azadi, M.R.E.; Hoveidae, N. Assessing the soil-structure interaction effects by direct method on the out-of-plane behavior of masonry structures (case study: Arge-Tabriz). *Bull. Earthq. Eng.* **2020**, *18*, 6429–6443. [CrossRef]
22. Tzanakis, M.J.; Papagiannopoulos, G.A.; Hatzigeorgiou, G.D. Seismic response and retrofitting proposals of the St. Titus Chruch, Heraklion, Crete, Greece. *Earthq. Struct.* **2016**, *10*, 1347–1367. [CrossRef]
23. *EN 1998-1:2004*; Eurocode 8: Design of Structures for Earthquake Resistance—Part 1: General Rules, Seismic Actions and Rules for Buildings. CEN: Brussels, Belgium, 2004.
24. Lemnos, K.G. *Historical and Cultural Heritage*; Self-published: Athens, Greece, 2010; Volume 2, p. 189. (In Greek)
25. Taxiarchis, K. *New Building Design Study of the Archaeological Museum of Lemnos*; Technical report Submitted to the Hellenic Ministry of Culture and Sports; Ephorate of Antiquities of Lesvos: Mytilene, Greece, 2018. (In Greek)
26. EDAFOS; Consultants S.A. *Geotechnical Study of the Castle of Myrina*; Technical Report Submitted to Ephorate of Antiquities of Lesvos; Ephorate of Antiquities of Lesvos: Mytilene, Greece, 2022. (In Greek)
27. Dogani, I.; Kalamboka, N.; Galanou, A. *Analysis of Original Mortars and Proposal of Repair Mortars for the Archaeological Museum of Myrina*; Technical Report Submitted to Ephorate of Antiquities of Lesvos; Ephorate of Antiquities of Lesvos: Mytilene, Greece, 2023. (In Greek)
28. *SAP2000: Static and Dynamic Finite Element Analysis of Structures*; Version 25.3.00; Computers and Structures, Inc.: Berkeley, CA, USA, 2023.
29. Ministry for Climate Crisis and Civil Protection. *KADET—Regulation for the Structural Assessment and Design of Masonry Buildings; Government Gazette 2493/B/2023*; Ministry for Climate Crisis and Civil Protection: Athens, Greece, 2023. (In Greek)

30. *EN 338:2016*; Structural Timber—Strength Classes. CEN: Brussels, Belgium, 2016.
31. Ministry for the Environment, Physical Planning and Public Works. *KTX 2008—Regulation for Reinforcement Steel Technology in Concrete Structures; Government Gazette 1416B/2008*; Ministry for the Environment, Physical Planning and Public Works: Athens, Greece, 2008. (In Greek)
32. Ministry for Climate Crisis and Civil Protection. *KAN.EPE.—Code of Interventions for Buildings of Reinforced Concrete (3rd Revision)*; Ministerial Decision ΥΠ 449; Government Gazette 3197/B/2022; Ministry for Climate Crisis and Civil Protection: Athens, Greece, 2022. (In Greek)
33. *EN 1996-1-1:2005*; Eurocode 6: Design of Masonry Structures—Part 1-1: General Rules for Reinforced and Unreinforced Masonry Structures. CEN: Brussels, Belgium, 2005.
34. *EN 1993-1-1:2005*; Eurocode 3: Design of Steel Structures—Part 1-1: General Rules and Rules for Buildings. CEN: Brussels, Belgium, 2005.
35. *EN 1992-1-1:2004*; Eurocode 2: Design of Concrete Structures—Part 1-1: General Rules and Rules for Buildings. CEN: Brussels, Belgium, 2004.
36. Karantoni, F. *Masonry Structures*; A. Papasotiriou & Co.: Athens, Greece, 2004; p. 90. (In Greek)
37. Spyrakos, K. *Masonry Structures: Assessment and Strengthening for Seismic Loads*; Ergonomos IKE: Athens, Greece, 2019; p. 257. (In Greek)
38. *EN 1991-1-1:2002*; Eurocode 1: Actions on Structures—Part 1-1: General Actions—Densities, Self-Weight, Imposed Loads for Buildings. CEN: Brussels, Belgium, 2002.
39. Geotechnical Research TEE. *Geotechnical Foundation Report for the New Storage Building of the Archaeological Museum of Lemnos*; Technical Report Submitted to the Ephorate of Antiquities of Lesvos; Ephorate of Antiquities of Lesvos: Mytilene, Greece, 2017. (In Greek)
40. *EN 1998-5:2004*; Eurocode 8: Design of Structures for Earthquake Resistance—Part 5: Foundations, Retaining Structures and Geotechnical Aspects. CEN: Brussels, Belgium, 2004.
41. Mulliken, J.S.; Karabalis, D.L. Discrete model for dynamic through-the-soil coupling of 3-D foundations and structures. *Earthq. Eng. Struct. Dyn.* **1998**, *27*, 687–710. [CrossRef]
42. Beskos, D. *Dynamic Analysis of Structures—Special Topics in Structural Dynamics and Earthquake Engineering, SMA 50/2*; Hellenic Open University (HOU): Patras, Greece, 2003; Volume II, p. 110. (In Greek)
43. Seismosoft. Seismomatch. Earthquake Software for Response Spectrum Matching. Version 2025. Available online: www.seismosoft.com (accessed on 25 March 2025).
44. *EN 1990:2002*; Eurocode 0: Basis of Structural Design. CEN: Brussels, Belgium, 2002.

applied sciences

MDPI

Article

Systematic and Quantitative Assessment of Reduced Model Resolution on the Transient Structural Response Under Wind Load

Anoop Kodakkal [1,*], Máté Péntek [1], Kai-Uwe Bletzinger [1], Roland Wüchner [1] and Felix Weber [2]

1 Lehrstuhl für Statik und Dynamik, Technische Universität München, Arcistr. 21, 80333 Munich, Germany; mate.pentek@tum.de (M.P.); kub@tum.de (K.-U.B.); r.wuechner@tu-braunschweig.de (R.W.)
2 Maurer Switzerland GmbH, Grossplatzstraße 24, 8118 Pfaffhausen, Switzerland; f.weber@maurer.eu
* Correspondence: anoop.kodakkal@tum.de

Abstract: The wind-induced response of structures is typically studied in wind tunnels either on scaled models or using numerical approaches under similar transient load conditions. In early design phases—where the potential for impactful change is most significant—information is often limited. As a result, studies are frequently conducted on simplified or reduced-resolution structural models. Typical applications for dimensionally reduced engineering models include early design phases, deciding on the need for high-fidelity analyses, and verifying wind tunnel models, which are often constructed using beams with lumped masses. In this contribution, the validity of these approaches is tested. Various limitations intrinsically arising from such modeling assumptions, showcased on a generic high-rise under dynamic wind load conditions, are highlighted. The systematic parametric analysis focuses on the variations in transient structural responses, particularly displacement and accelerations at the top of a building. Various wind loading cases are studied, with the reduction of the resolution taking place either in the original or in modal space. Results indicate that a considerable reduction is possible, but characteristic design values tend to deteriorate in cases of a high reduction, particularly when higher mode contributions are truncated. It is observed that the top-floor acceleration and displacement can be captured with considerable accuracy with three lumped masses for tall buildings. It is critical to study the impact of simplifying models starting at the highest level of detail possible. Here, a three-DoF model was able to capture the displacement up to a deviation of 11% and accelerations up to 20%. These approximate models are useful for initial design stages, optimization, uncertainty quantification, etc., where fast, cheap, and moderately accurate model evaluations are necessary.

Keywords: wind loading; structural modeling; structural response; numerical methods; reduced-resolution models; computational wind engineering

1. Introduction

Structural wind engineering is the art and science of modeling and analyzing structures under natural wind flows. It aims at estimating the accurate responses of these structures under natural wind loads. Wind tunnel measurements traditionally represent the established method for quantifying wind loads [1]. These typically result from high-resolution pressure time histories (in the case of high-frequency pressure integration—HFPI) or global structural forces (from high-frequency force balance—HFFB). Additionally, codes and standards provide further insights into the quantification of the effects of wind flow on structures. These, however, have severe restrictions and limitations for special constructions (such as high-rise, large span, lightweight, etc.) and inherently cannot provide sufficiently detailed information on local/global loads and respective responses in high resolution (in time and space). In recent decades, computational wind engineering (CWE) has gained a lot of attention from the structural wind engineering practice, where wind loads are estimated by modeling the wind flows around the structure using computational methods like computational fluid dynamics (CFD).

To obtain structural responses, like displacement and accelerations, an aeroelastic model that incorporates the structural properties of the system is used in wind tunnels. The structures are equivalently represented in these scaled-down aeroelastic models. The sizing effect needs to be taken into account during such modeling. For tall buildings, a commonly used simplification for the structural system is to use lumped mass models. The number of degrees of freedom in such lumped mass models is chosen by the researcher/modeler. Wind tunnel testing with aeroelastic models commonly uses lumped mass models for tall buildings. Newer techniques like hybrid aeroelastic pressure balance (HYPB) use a single degree of freedom (SDoF) [2] for the structural system in a wind tunnel, capturing the two-way coupling of structure and wind. In order to capture higher vibration modes in these aeroelastic tests, MDoF is used in previous studies. Vortex-induced vibrations (VIVs) and resonance were investigated with aeroelastic models in a wind tunnel to model super high-rise buildings using MDoF in [3,4]. A series of wind tunnel tests with the MDoF model of Shanghai World Financial Centre is carried out in [5] for vibration measurements.

In computational wind engineering, this is captured by coupling the CFD model developed for wind with a structural model. This is referred to as a typical fluid–structure interaction problem. Recent advancements in algorithms and computational resources have made it possible to perform fully coupled simulations in complex wind flows and complex structures. Large Eddy simulation (LESs) of wind effects on tall buildings with and without nearby buildings are presented in [6], and wind effects on tall steel buildings are presented in [7]. Multiple turbulence models were numerically investigated for predicting wind effects on tall buildings in [8,9]. The wind effects on structures may be captured by a two-way coupled or one-way coupled simulation. The accuracy of these modeling choices depends on the flexibility of the structure. In tall building simulations, it is observed that a one-way coupled simulation is capable of capturing important characteristics sufficiently. In CWE, the wind loads evaluated from the CFD may be imposed on a finite element (FE) structural model. It is common practice to model the structure in FE software tools for evaluating the building response and, hence, for design. The FE model of a tall building comprises multiple structural elements and is an accurate representation of the structure. These complex FEM models are expensive to evaluate in many cases.

In a design scenario, it is not possible for the designer to know exact structural properties like the mass, stiffness, and damping of the structural system accurately in the early stages. It is a common norm to start with simplified models in such scenarios. Also, in wind tunnel testing, it is not possible to capture all the structural behaviors, and a common approach is to use MDoF systems. An MDoF system is the best representation of a tall building and is widely used in computational studies for evaluating the wind effects on structures. A more complex and accurate structural model may be built at a later time as more information and decisions on the structural system are made in a design scenario. However, the accuracy of these simplified models has not been not well studied in the literature. Here, in this contribution, a systematic study is carried out to quantify the effects of simplified modeling in structures under transient loads with a focus on the wind.

The current study aims to assess (dynamic) wind load in various scenarios (angle of attack, degree of turbulence) obtained from CFD simulations, which represent the input load case in the form of force time history on the structural model. A generic highrise structure (CAARC B—a well-established model in the wind engineering community proposed by the International Association of Wind Engineering—IAWE [10,11]) is chosen due to the vast literature available related to the airflow around it as well as necessary information regarding the choice and set up of the structural model. Originally, the structural model is only prescribed a certain geometry (uniform rectangular along height), mass distribution (uniformly along height), eigenfrequencies of the first three modes (weak and strong bending, torsion), and type of mode shape (linear, as the initial study way intended for HFFB in experimental wind tunnels). Additional remarks on structural damping are given for the assessment of responses. This study takes these existing prescriptions and uses them as a base model: continuous and uniform mass and stiffness distribution along the height chosen so that target parameters (eigenfrequencies) are met. The study of slight variations of this model, aiming to assess other scenarios typical for the engineering practices for such constructions, such as three intervals along the height with varying mass and stiffness distribution, is included. A model of the building with a full FEM model is presented, comparing the responses with approximate MDoF systems.

Wind loading is considered as a given input, with the variations described previously. Structural models and respective modifications will be analyzed as briefly outlined. The systematic parametric study is further enhanced with the investigation of the spatial resolution of the structural model itself. This last step aims to quantify the effect of structural detailing through the usage of various (decreasing) numbers of structural elements. It is aimed to highlight what minimum requirements related to proper analysis are needed, such that certain target response parameters (e.g., time and magnitude of maximum displacements and accelerations) are captured with sufficient accuracy. The insights are linked to modal analysis and a subsequent reduction of the number of modes in order to identify minimally relevant ones. The numerical toolchain that is developed and used permits a flexible and thorough parametric assessment of these aspects. The accuracy assessment of these simplified models can be useful for both simplified numerical modeling and the use of simplified models in wind tunnel tests.

The rest of this paper is organized as follows: Section 2 describes the methodology used for modeling and analyzing fluids and structures. Section 3 describes the details of the numerical study and results, and discusses the obtained results. Section 4 presents the conclusions and outlooks.

2. Methodology

This section focuses on the detailed methodological description and underlying assumptions for each of those modules.

2.1. Wind Load Modeling and Simulation

Wind load is modeled using CFD analysis with the open-source Kratos Multiphysics tool [12–14]. This involves a finite element method (FEM) formulation for flow problems based upon a variational multi scale (VMS) formulation [15]. The computational domain setup is the one presented in ref. [16], which was set up for the CAARC B building [10,11]. Tetrahedral elements are employed to discretize the fluid domain. The computational domain setup for the current simulation is illustrated in Figure 1. The figure also outlines the boundaries of the fluid domain and illustrates the applied boundary conditions.

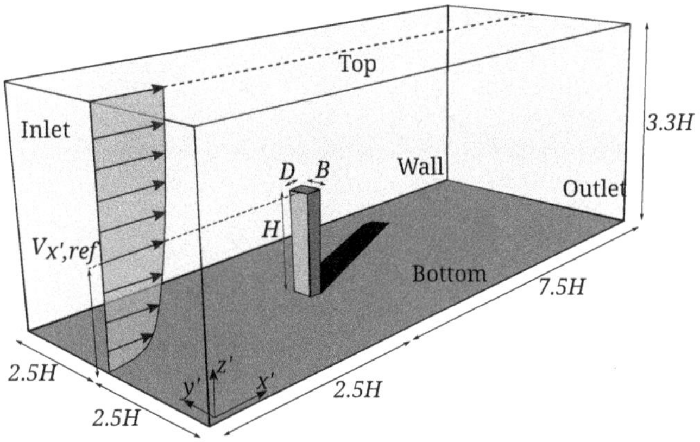

Geometry & flow conditions

D	30 m
B	45 m
H	180 m
α	0.25
$V_{x',ref}$	40 m/s (±10%)
$TI_{Vx',ref}$	0.143 (±10%)
$L_{Vx',ref}$	175 m (±20%)

CFD domain conditions

Inlet	imposed velocity
Outlet	imposed pressure
Bottom	wall condition
Structure	no slip condition
Top	slip condition
Wall	slip condition

Figure 1. Setup for wind load modeling.

Inlet Boundary Condition

The wind characteristics are taken care of by the inlet boundary condition. The wind effects in the atmospheric boundary layer (ABL) are modeled by decomposing the wind velocity field as a Reynolds decomposition, $V = \overline{V} + V'$, where $\overline{V}$ is the steady mean profile component and V' is the unsteady turbulent fluctuations component. The mean wind profile, denoted as $\overline{V}$, represents the averaged contribution to the overall wind field V. This profile undergoes changes over the time scale of several hours to days, as indicated in [17]. Therefore, for the numerical simulation, the mean profile is assumed to remain constant. The turbulent fluctuations of wind V' represent the wind gusts with a time span of seconds. The gust wind speed triggers transient conditions of peak wind loading. If the structural eigenfrequencies of large structures align with frequencies in the gust-induced wind load pattern, resonant effects may occur. Separate models are employed for each term, $\overline{V}$ and V'.

An exponential profile is adopted for the study for the mean wind profile. The wind velocity at any height z is given by

$$\overline{V}(z) = V_{\text{ref}}\left(\frac{z}{z_{\text{ref}}}\right)^{\alpha} \tag{1}$$

where z_{ref} is the reference height and α is the exponent which depends on the terrain. To capture wind statistics at the location of interest, wind fluctuations $V'(t, y, z)$ are modeled. Various numerical models are available to address this issue [18–20]. In this study, the well-established Mann model is utilized to simulate wind gust fluctuations [21,22]. This model, widely accepted in structural wind engineering [23–26], is known for its spectral approach in wind generation.

2.2. Structural Modeling and Analysis

The prescribed structural model of the CAARC B building is a typical cantilever with continuous mass and stiffness, constant along its height. It cannot be modeled as a full cross-section; as with such an assumption, the prescribed modal frequencies cannot all be fulfilled at the same time. These target modes and frequencies can only be achieved by a detailed structural model, typically set up for extensive analysis during design. For the current investigation, a custom parametrizable Timoshenko beam with an FEM formulation is developed and used, referred to as ParOptBeam. Two structural models were used in this study: the high-fidelity model (HFM), which is a detailed FEM model, and the ParOptBeam model, which is a low-fidelity model (LFM).

A simplified structural model for the dynamic time history analysis of tall buildings with different structural systems is presented here. The three-dimensional (3D) structure is replaced by a simplified structural model that considers both bending and shear deformations. The 3D Timoshenko beam theory is therefore used to model the equivalent structural system. Ref. [27] presented a simplified methodology for the analysis of the framed structure shear wall interaction problem. It can be seen that the Timoshenko beam model can effectively model various structural systems used for tall buildings in practice.

2.2.1. Timoshenko Beam Model

The Timoshenko beam theory is adopted to model the simplified structural system to analyze the actual 3D building. A 3D prismatic homogeneous isotropic beam element is used for the element formulation. The element considers both the shear deformation and rotational inertia (related to bending). Since the shear deformations are taken into account, the planes that were initially perpendicular to the neutral axis are no longer perpendicular to it after deformation.

A consistent mass formulation is used. Hence, both the stiffness and mass matrix depend on the relative importance of the shear deformations (relative to the contribution from bending). This ratio is given by

$$\Phi = \frac{12EI}{G(A/\alpha)L^2} \tag{2}$$

where E and G are the moduli of elasticity and shear of the material, respectively, I is the moment of inertia of the cross-section, A/α is the effective shear area, and L is the length of the discrete beam element. In 3D, the properties of cross-sections can change in the two principle directions; hence, two relative shear importance factors may be defined. The torsional stiffness is computed from the polar inertia of the equivalent cross-section. A Rayleigh damping model is assumed. For a dynamic problem, the governing equation of motion (EOM) is shown in (3).

The numerical structural models of the version described are set up such that the eigenfrequencies can be met. Furthermore, in the initial description of CAARC B, linear mode shapes (with linear referring to the distribution along the height and not referring to geometric/material linearity) are mentioned, as these would be typically achieved by the target study at that time involving a HFFB measurement. These distributions along the height are only possible if the model is an infinitely rigid beam along the height, and deformations are only possible at the mount (clamped edge). This could be seen as a rigid body deformation along the height permitted by springs at the fixity. These mode shapes are, however, not representative (nor realistic) of typical aeroelastic models used in wind tunnels, nor of high-rise structures modeled using computational methods. Realistically, these have mode shapes and sway deformations that lie somewhere between the theoretical limits of pure shear buildings and pure bending cantilevers and are definitely not linear. The latter would correspond to the Bernoulli/Timoshenko beam formulation, whereas the former can only be achieved with a dedicated MDoF shear model. Current modeling techniques and implementations permit the modification of the relative importance of shear. Also, the addition of point mass and stiffness entries to model the possible effect of outriggers is enabled.

2.2.2. Implementation Details of the Structural Model in ParOptBeam

The Timoshenko beam model realized in the current work (ParOptBeam [28]) is capable of carrying out a 3D analysis of tall buildings subjected to wind forces. A linear (geometric and material) model is used for the structure, and the time domain dynamic problem is solved with a *generalized* $-\alpha$ time integration scheme. Mode shapes (two sway modes and torsion) are uncoupled.

Analysis in the Modal Coordinates in the Framework

The structural solver is capable of computing time-dependent responses in the modal coordinates as well. Such computations help to reduce the time required for analysis, as one needs to solve a smaller set of equations (which comprise a series of scalar solves). This can be seen as useful if the number of DoFs of the system is very large, and if the system is mainly influenced by a few of the modes. The capabilities are implemented for the structural model. The responses computed in transformed coordinates are thereafter transferred back to the initial system of reference.

The Capability of Stepping

In many designs in practice, it so happens that the stiffness values are reduced along the height of the structure towards the top. This is realized by different zones (intervals) having changing prescribed stiffness and mass values. Constant stiffness (material and geometric) and mass (density) are assumed within an interval.

Modeling of Outriggers

In practice, outriggers are generally designed in-between changes of mass and stiffness. These are mostly used to help reduce the bending moment at the core. The mass and stiffness contributions of outriggers are added to the corresponding matrices.

Variations in the model regarding mass and stiffness considered are shown in Figure 2 and are as follows:

- Continuous—constant distribution over height;
- With intervals—varying (but constant on intervals) distribution of these over height (three intervals considered);
- With outriggers—additional inclusion of outriggers (at the two changes between intervals).

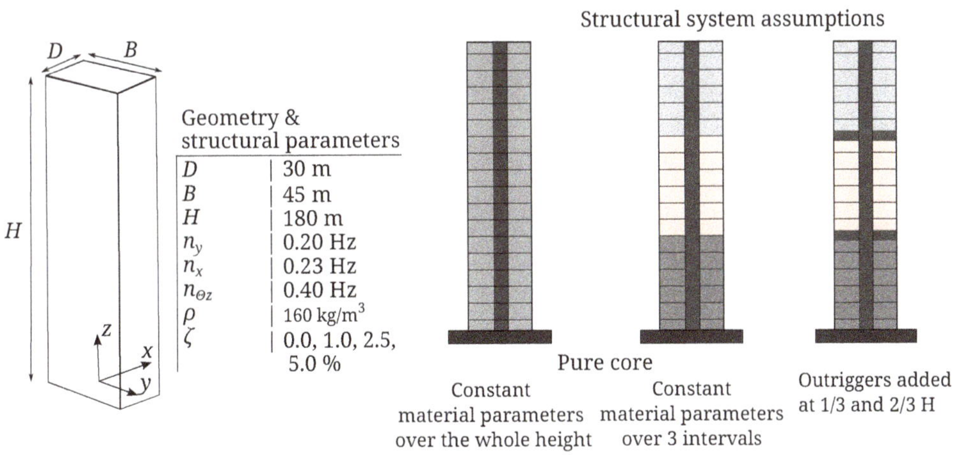

Figure 2. Main assumptions and parameters of the structural model.

2.3. Coupling of Fluid and Structure

The wind flow results in nodal forces on the surface mesh of the structure, typical for the numerical approach (CFD) at hand. These fluid forces need to be transferred to the structural model (modeled using computational structural mechanics—CSM methods) in a conservative way, referring to the conservation of forces in an integral sense. Thus, the resulting forces on the surface mesh on the fluid solver side need to be the same as the acting forces on the structure during the CSM simulation. These data transfer operations can be generically seen as a mapping operation. It needs to be noted that the deformations of the structural model are not sent back to the fluid solver; as such, the wind flow CWE model acts as if we were only able to see the rigid structure throughout the total duration of the simulation.

This manner of assessing and transferring fluid forces is typical to either HFPI techniques (on a rigid model) or CFD simulations and leads to a one-way coupled fluid–structure interaction (FSI) approach, which means that only the wind flow affects the structure (due to arising forces), but structural deformations (kinematics) do not have an influence on the geometry of the flow domain considered.

The main goal is to identify the minimum amount of considered structural detailing (represented here by the minimum amount of finite elements in discretization and/or number of mode shapes used for modal reduction), which leads only to a minimal degree of variation from the full resolution model. The minimum (or acceptable) level of refinement is established based on intended applications. The EOM in spatial coordinates is displayed in Equation (3), where $x(t)$ denotes the vector of the DoFs of the generalized displacement, with time derivatives also being marked. The matrices $[M]$, $[B]$, and $[K]$ represent mass (consistent), stiffness, and damping (Rayleigh), respectively. The corresponding vector of external forces is represented by $\{F(t)\}$. The sizes of the matrices and vectors are shown using the number of DoFs n.

Generalized discrete EOM.

$$[M]_{n\times n}\{\ddot{x}(t)\}_{n\times 1}+[B]_{n\times n}\{\dot{x}(t)\}_{n\times 1}+[K]_{n\times n}\{x(t)\}_{n\times 1}=\{F(t)\}_{n\times 1} \tag{3}$$

Transformation into modal coordinates. The eigenvalue problem for the conservative (undamped) structure is

$$[K]_{n\times n}[\Phi]_{n\times n}=[M]_{n\times n}[\Phi]_{n\times n}[\Lambda]_{n\times n} \tag{4}$$

where $[\Phi]_{n\times n} = [\{\phi_1\}_{n\times 1},\{\phi_2\}_{n\times 1},\ldots,\{\phi_n\}_{n\times 1}]_n^T$ is the modal matrix, consisting of n modal vectors; $[\Lambda]_{n\times n} = [diag\{\lambda_1,\lambda_2,\ldots,\lambda_n\}]_{n\times n}$ is the modal stiffness matrix, with $\lambda_i = \omega_i^2$ for $i = 1, n$, consisting of n eigenvalues on the main diagonal. This assumption should be well accepted for civil engineering structures, which are typically highly underdamped.

Mass-normalization is used for obtaining the modal matrix $[\Phi]_{n\times n}$, such that

$$\begin{aligned}
[\Phi]_{n\times n}^T[M]_{n\times n}[\Phi]_{n\times n} &= [I] = [\overline{M}]_{n\times n}\\
[\Phi]_{n\times n}^T[K]_{n\times n}[\Phi]_{n\times n} &= [\Lambda] = [\overline{K}]_{n\times n}\\
[\Phi]_{n\times n}^T[B]_{n\times n}[\Phi]_{n\times n} &= [\mathrm{B}] = [\overline{B}]_{n\times n}
\end{aligned} \tag{5}$$

where $[\mathrm{B}]_{n\times n} = [diag\{\beta_1,\beta_2,\ldots,\beta_n\}]_{n\times n}$ is the modal damping matrix, with $\beta_i = 2\zeta_i\omega_i$ for $i = 1, n$, ζ_i being the damping ratio for the i-th mode.

The kinematics spatial coordinates and respective time derivatives are transformed into modal coordinates using

$$\{x(t)\}_{n\times 1}=[\Phi]_{n\times n}\{\overline{x}(t)\}_{n\times 1} \tag{6}$$

and pre-mutiplying the resulting equation with $[\Phi]^T_{n\times n}$, leading to

$$[\overline{M}]_{n\times n}\{\ddot{\overline{x}}(t)\}_{n\times 1} + [\overline{B}]_{n\times n}\{\dot{\overline{x}}(t)\}_{n\times 1} + [\overline{K}]_{n\times n}\{\overline{x}(t)\}_{n\times 1} = \{\overline{F}(t)\}_{n\times 1} \tag{7}$$

where $[\overline{M}]_{n\times n}$, $[\overline{B}]_{n\times n}$ and $[\overline{K}]_{n\times n}$ are the modal mass, stiffness and damping matrices (all diagonal), respectively.

It will be assumed that the following condition holds:

$$[\overline{B}]_{n\times n}[\overline{M}]^{-1}_{n\times n}[\overline{K}]_{n\times n} = [\overline{K}]_{n\times n}[\overline{M}]^{-1}_{n\times n}[\overline{B}]_{n\times n} \tag{8}$$

ensuring the fact that the damped system will possess the same mode shapes as its undamped counterpart. Consequently, the eigenvectors will be real and the equation will contain n uncoupled systems of equations, representing an SDoF-analogy for each modal component and leading to a scalar system for each, individually.

Modal reduction.

Systematically decreasing the involved modal components, one chooses $m \leq n$ modes to represent a subset $[\Phi]_{n\times m} = [\{\phi_1\}_{n\times 1}, \{\phi_2\}_{n\times 1}, \ldots, \{\phi_n\}_{n\times 1}]^T_m$ for $i = 1, m$ of the full modal matrix. Carrying out the transformation considering only this selection of modes, the following equation holds:

$$[\overline{M}]_{m\times m}\{\ddot{\overline{x}}(t)\}_{m\times 1} + [\overline{B}]_{m\times m}\{\dot{\overline{x}}(t)\}_{m\times 1} + [\overline{K}]_{m\times m}\{\overline{x}(t)\}_{m\times 1} = \{\overline{F}(t)\}_{m\times 1} \tag{9}$$

representing the EOM in reduced modal space. Consequently, the response in the initial coordinate system is made up only by the contribution (mathematically a linear superposition) of a reduced subset of (decoupled) modes.

$$\{x(s,t)\} = \sum_{j=1}^{\infty} \Phi_j(s)\overline{x}_j(t) \rightarrow \ldots \rightarrow \{x(t)\}_{n\times 1} = [\Phi]_{n\times m}\{\overline{x}(t)\}_{m\times 1} \tag{10}$$

It should be noted that transferring from continuous space to finite space is implicitly already a model-order reduction, where the number of modes is reduced from ∞ to n, further reduced here to m.

For large systems, determining all mode shapes and frequencies is practically prohibitive. Analytical solutions are not viable for such cases. Thus, numerical methods (such as the Rayleigh–Ritz) exist, and they approximate a certain subset of eigenforms and values, mostly starting at the lowest frequency (in civil/structural engineering applications). Similarly, experimental mode shape analysis only enables the determination of a limited number of lower-frequency modes. The result is, from either procedure, that one only has a limited subset of eigenforms corresponding to the lower eigenfrequencies. The current study uses a parametric FEM formulation of the Timoshenko beam, such that all eigenforms and values are assessed for a (relatively) highly detailed spatial model. The spatial and modal resolution is systematically reduced such that the effect of a lower number of FE elements or the mode truncation can be properly measured.

3. Numerical Study

The time history of the forces acting on the structure is recorded level-wise. Figure 3 represents how the structure model is split into 60 regions along the height, such that a series of force components (six per floor: three forces and three moments in a body-attached coordinate system) are made available with the intention of being transferable to the structural model for structural analysis. This defines the highest spatial resolution, meaning that at the highest detailing level, 60 floors along the height are taken into account, with six DoFs at each floor and corresponding force/moment components. Due to the time settings (total time, time step) of the computational wind flow simulation, the time resolution is set to 50 steps/s, leading to 30,000 of total time instances for 10 min, which defines the temporal resolution. The total simulation time of 10 min is deemed sufficient to capture naturally turbulent wind loading effects on structures at full scale.

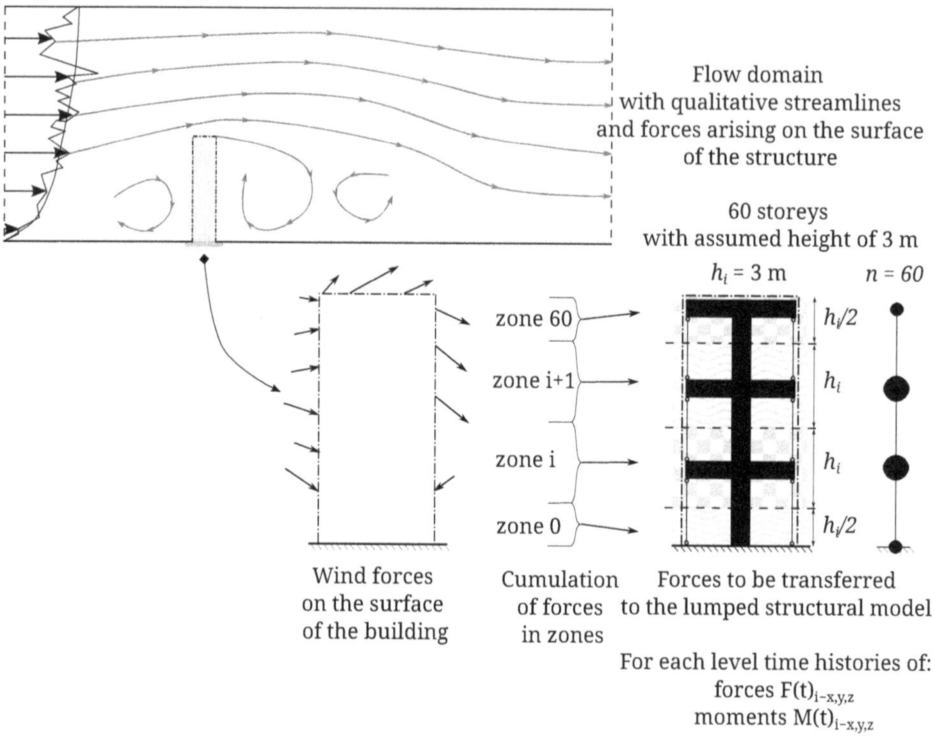

Figure 3. Transfer of wind forces to the structure.

The reduction of the forces from each point on the surface mesh of the structure is similar in manner to what would be available from HFPI experimental wind tunnel techniques. Forces and moments are reduced to a center point, which represents the geometrical center of each fictitious floor of the considered structural model. Thus, the resulting nodal forces (on mesh nodes in flow-attached coordinates) are part of the solution of the CFD solving procedure. These values are transferred onto the structural model, which is reduced to the centers of floors and transformed into the body-attached coordinate system (corresponding to the principal axes of inertia) of the building.

3.1. Numerical Study I: Comparison of a Detailed FEM Model for Structures with Low-Fidelity ParOptBeam Model

In this study, a detailed CFD model for wind is coupled to a detailed FEM model and compared with the parametric optimizable beam (ParOptBeam) model. The same wind forces from the CFD model are used for both models.

The high-fidelity model (HFM) for the case study was developed to conduct a detailed time series analysis, focusing on the wind effects on the CAARC building. GiD was used during the preprocessing stage to set up the model, which included defining the geometry, assigning structural elements, and generating a finite element mesh. The material properties were considered linear-elastic, as non-linear behavior was deemed excessive for assessing operational performance. The modeling of the structural elements is summarized as slabs and walls, which were modeled as thick shell elements, beams and columns as Timoshenko beam elements, and supports, which were modeled as pinned fixities at the base.

Geometrical non-linearities, which impact the amplification of structural responses in tall and slender buildings, were also incorporated into the HFM. Although these effects can be ignored when displacements are minimal, it is advisable to account for them.

To account for geometrical non-linearities, the beam and shell elements were modeled using the co-rotating theory [29], and the Newton–Raphson method was applied in the analysis. The co-rotating approach references elements relative to a local coordinate system rather than a global one, where total displacements and rotations are considered. This method decomposes the element's overall motion into two components: rigid-body motion and deformable motion. The inclusion of shear deformations is particularly significant for deeper beams and thicker shell elements. Thick shells may be ideal for applications like foundation plates, while thin elements may be more suitable for slabs. Incorporating shear deformation provides a more realistic approach to the HFM.

To assess wind effects on a tall building, it is crucial to examine the first three modes, as illustrated in Figure 4. This focus is due to the fact that wind forces in tall buildings are not commonly excited by higher-order modes since wind forces do not exert force in one direction up to a certain height and then shift to another direction for the remainder of the structure.

(a) (b) (c)

Figure 4. Mode shapes of the high-fidelity model: (**a**) 1st mode bending; (**b**) 2nd mode bending; (**c**) 1st mode in torsion.

A modal analysis was conducted on the HFM to explore the building's dynamic behavior. The results of the eigenvalue analysis are depicted in Figure 4. The first and second eigen-mode is characterized mainly by a dominant bending effect, with minimal torsional impact, representing a nearly pure bending scenario. In contrast, the third eigen-modes exhibit translation in the y-direction combined with torsional rotation along the building's Z-axis.

Low-Fidelity Model

After creating the HFM of the building, a low-fidelity model (LFM) consisting of beam elements can be established by considering the characteristics of the building. Dynamic simulations with the LFM require less computational time to complete. To create the LFM, it is necessary to ensure that this model has attributes similar to those of the detailed building model, such as eigen-frequencies and mode shapes. Timoshenko beam theory was used, as elaborated in Section 3.

The top-floor accelerations are presented for two angles of the wind in Tables 1 and 2. Time statistics of the two models are compared in the table. Mean, standard deviation (std), mean + 3 sd, maximum, and the 99% values are compared for the two models for two angles of the incoming wind. The base reactions are presented in Tables 3 and 4. It can be seen that the approximate model can capture the base reactions and the top-floor acceleration with reasonable accuracy. The mean of the top-floor acceleration is almost zero, and extreme statistics are of importance. Further, the time series of accelerations and base reactions are presented in Figures 5–7. The computations were performed on the SuperMUC-NG system equipped with an Intel Xeon Platinum 8174 processor using 8 nodes and each node has 48 cores per node (Intel Corporation, Santa Clara, CA, USA). The computational time for both analyses was compared, and the high-fidelity model required approximately 13,824 CPU hours, whereas the low-fidelity model completed the analysis in approximately 144 CPU hours. The LF model demonstrated a significant computational saving, using only 1% of the resources required by the HF model. Table 5 compiles a few of the QoIs that are interesting to each load case, and the % of the difference is presented from HFM. In both acceleration and base reactions, the low-fidelity model was able to capture the responses accurately to less than 3%. In almost all cases, the LFM underpredicts the responses. However, this difference is expected and is considerably justified considering the computational saving it presents.

Table 1. Comparison of HFM and LFM—acceleration at the top for 2.5% damping and 0° angle of wind.

	Mean	Std	Mean + 3Std	Max	99th Percentile
HFM					
A_x [m/s^2]	1.67×10^{-4}	1.32×10^{-1}	3.96×10^{-1}	4.34×10^{-1}	3.04×10^{-1}
A_y [m/s^2]	-1.42×10^{-4}	1.17×10^{-1}	3.52×10^{-1}	3.73×10^{-1}	2.73×10^{-1}
A_rz [rad/s^2]	-2.34×10^{-7}	1.30×10^{-3}	3.90×10^{-3}	4.96×10^{-3}	3.22×10^{-3}
LFM					
A_x [m/s^2]	1.60×10^{-4}	1.41×10^{-1}	4.25×10^{-1}	4.21×10^{-1}	3.32×10^{-1}
A_y [m/s^2]	-2.04×10^{-5}	1.14×10^{-1}	3.43×10^{-1}	3.65×10^{-1}	2.69×10^{-1}
A_rz [rad/s^2]	-5.15×10^{-7}	1.30×10^{-3}	3.90×10^{-3}	4.39×10^{-3}	3.23×10^{-3}

Table 2. Comparison of HFM and LFM—acceleration at the top for 2.5% damping and 90° angle of wind.

	Mean	Std	Mean + 3Std	Max	99th Percentile
HFM					
Ax [m/s^2]	-1.75×10^{-4}	1.84×10^{-1}	5.51×10^{-1}	5.41×10^{-1}	4.22×10^{-1}
Ay [m/s^2]	3.39×10^{-5}	6.40×10^{-2}	1.92×10^{-1}	2.24×10^{-1}	1.53×10^{-1}
Arz [rad/s^2]	-4.28×10^{-7}	2.44×10^{-3}	7.33×10^{-3}	9.06×10^{-3}	5.92×10^{-3}
LFM					
Ax [m/s^2]	-1.17×10^{-4}	1.85×10^{-1}	5.56×10^{-1}	5.31×10^{-1}	4.10×10^{-1}
Ay [m/s^2]	2.40×10^{-5}	5.17×10^{-2}	1.55×10^{-1}	1.50×10^{-1}	1.20×10^{-1}
Arz [rad/s^2]	-8.56×10^{-7}	2.45×10^{-3}	7.35×10^{-3}	8.66×10^{-3}	5.95×10^{-3}

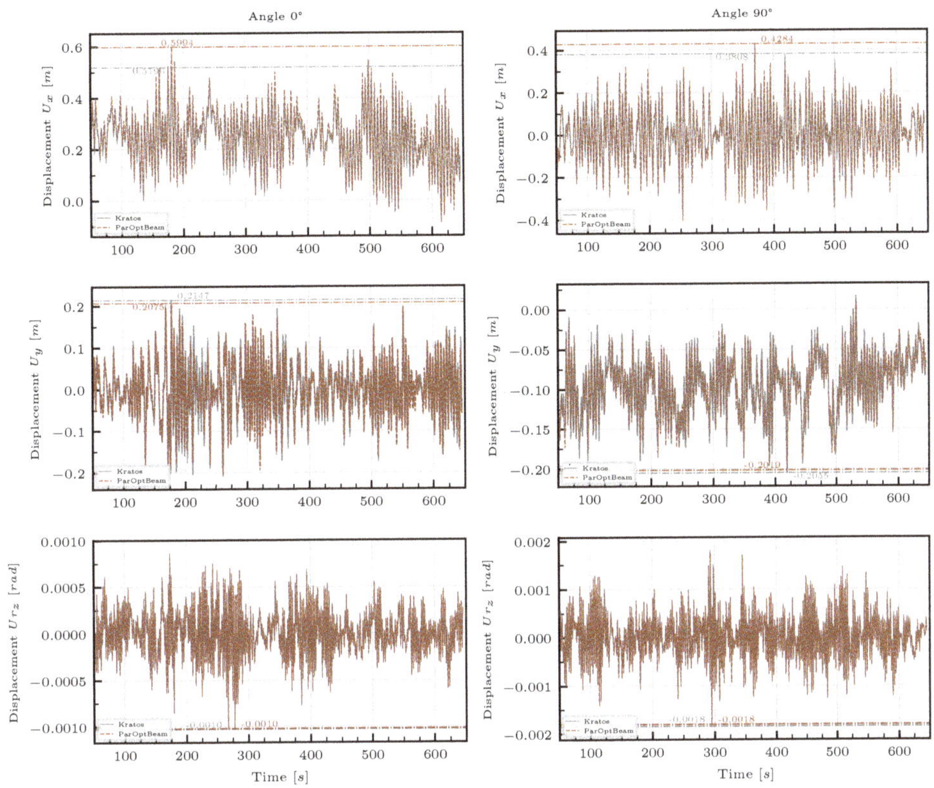

Figure 5. Top-floor displacement—comparison of HFM and LFM for two angles of wind and damping ratio ζ = 2.5%.

Table 3. Comparison of HFM and LFM—base reactions for 2.5% damping and 0° angle of wind.

	Mean	Std	Mean + 3Std	Max	99th Percentile
HFM					
Fx [MN]	8.94×10^{0}	4.11×10^{0}	2.13×10^{1}	2.49×10^{1}	1.92×10^{1}
Fy [MN]	-5.46×10^{-2}	3.67×10^{0}	1.09×10^{1}	1.41×10^{1}	8.78×10^{0}
Fz [MN]	-3.71×10^{2}	4.59×10^{1}	-2.34×10^{2}	-1.73×10^{-1}	8.47×10^{1}
Mx [MNm]	3.25×10^{-1}	3.26×10^{2}	9.77×10^{2}	9.70×10^{2}	8.14×10^{2}
My [MNm]	8.72×10^{2}	3.60×10^{2}	1.95×10^{3}	1.85×10^{3}	1.66×10^{3}
Mz [MNm]	2.84×10^{-1}	1.62×10^{1}	4.90×10^{1}	5.63×10^{1}	4.01×10^{1}
LFM					
Fx [MN]	8.85×10^{0}	3.21×10^{0}	1.85×10^{1}	1.79×10^{1}	1.57×10^{1}
Fy [MN]	1.21×10^{-2}	2.81×10^{0}	8.45×10^{0}	8.12×10^{0}	6.59×10^{0}
Fz [MN]	-3.67×10^{2}	4.60×10^{1}	-2.29×10^{2}	0.00×10^{0}	-8.09×10^{1}
Mx [MNm]	9.93×10^{-1}	3.11×10^{2}	9.34×10^{2}	9.70×10^{2}	7.71×10^{2}
My [MNm]	8.53×10^{2}	3.60×10^{2}	1.93×10^{3}	1.93×10^{3}	1.64×10^{3}
Mz [MNm]	3.27×10^{-1}	1.34×10^{1}	4.06×10^{1}	4.55×10^{1}	3.24×10^{1}

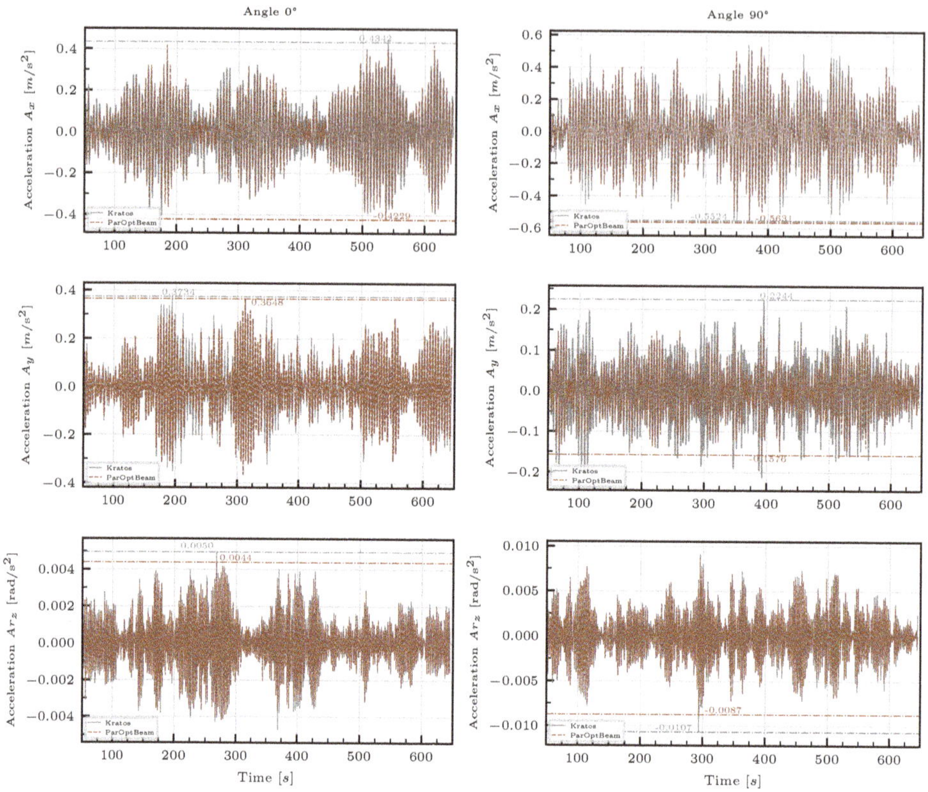

Figure 6. Top-floor acceleration—comparison of HFM and LFM for two angles of wind and damping ratio ζ = 2.5%.

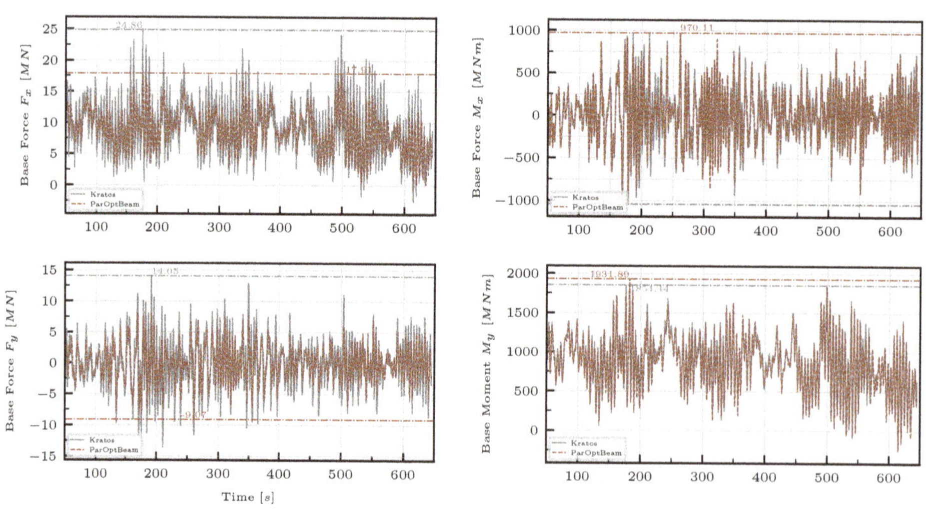

Figure 7. Base reaction responses—comparison of HFM and LFM for angle 0 and damping ratio ζ = 2.5%.

Table 4. Comparison of HFM and LFM—base reaction for for 2.5% damping 90° wind.

	Mean	Std	Mean + 3Std	Max	99th Percentile
HFM					
Fx [MN]	3.04×10^{-2}	4.01×10^{0}	1.21×10^{1}	1.73×10^{1}	5.04×10^{0}
Fy [MN]	-4.66×10^{0}	1.80×10^{0}	7.25×10^{-1}	6.88×10^{-1}	-2.46×10^{0}
Fz [MN]	-3.71×10^{2}	4.59×10^{1}	-2.34×10^{2}	-1.73×10^{-1}	-3.78×10^{2}
Mx [MNm]	4.70×10^{2}	1.83×10^{2}	1.02×10^{3}	9.73×10^{2}	7.16×10^{2}
My [MNm]	1.96×10^{0}	4.22×10^{2}	1.27×10^{3}	1.34×10^{3}	5.52×10^{2}
Mz [MNm]	7.72×10^{-1}	2.60×10^{1}	7.88×10^{1}	1.10×10^{2}	3.37×10^{1}
LFM					
Fx [MN]	9.17×10^{-4}	3.45×10^{0}	1.03×10^{1}	1.08×10^{1}	4.55×10^{0}
Fy [MN]	-4.64×10^{0}	1.63×10^{0}	2.50×10^{-1}	2.95×10^{-2}	-2.64×10^{0}
Fz [MN]	-3.67×10^{2}	4.60×10^{1}	-2.29×10^{2}	0.00×10^{0}	-3.76×10^{2}
Mx [MNm]	4.58×10^{2}	1.77×10^{2}	9.88×10^{2}	9.59×10^{2}	6.97×10^{2}
My [MNm]	1.88×10^{0}	4.16×10^{2}	1.25×10^{3}	1.35×10^{3}	5.44×10^{2}
Mz [MNm]	8.72×10^{-1}	2.42×10^{1}	7.36×10^{1}	9.54×10^{1}	3.15×10^{1}

Table 5. Comparison of the HFM and LFM—summary.

Case	Quantity	HFM	LFM	Difference in % from HFM
Table 1	Max A_y [m/s^2]	0.373	0.365	2.14
Table 2	Max A_x [m/s^2]	0.541	0.531	1.84
Table 3	Mean F_x [MN]	8.94	8.85	1.01
Table 3	Mean M_y [MNm]	872	853	2.18
Table 4	Mean F_x [MN]	−4.66	−4.64	0.43
Table 2	Mean M_y [MNm]	470	458	2.55

3.2. Numerical Study II: Comparison of Responses with Reduced Degrees of Freedom

The low-fidelity model, which was elaborated and proven to be useful in the numerical Study I, is used here in this study for further parametric studies. A systematic parametric study is carried out in order to evaluate various modeling choices for the low-fidelity model elaborated above. One of the major points looked into was how many DoFs need to be present for the accurate prediction of the structural responses. The number of nodes or elements is systematically reduced in this study. The top-floor displacements, acceleration, and base reactions are the quantities of interest. Sixty elements are considered to comprise the highest resolution model. This resolution was used in the previous study to validate with a high-fidelity model. The major focus of the section is to identify the effect of the number of elements used for FEM modeling and the number of mode shapes considered in modal analysis for a reduction in various quantities of interest.

Outline of the Parametric Study

The parametric study is thought out such that the effect of variations in

- Load scenario:
  - Wind flow angle (0°, 90°)
  - Presence of turbulence (smooth exponential inlet profile or with fluctuations in time and space)
- Structural modeling: type of considered models
  - Damping characteristics
  - Mass and stiffness distribution along the height
  - Detailing the level of the model represented by spatial resolution, achieved by

  * Varying the number of finite elements used (in physical space - spatial reduction)
  * Systematic modal reduction (in modal space)

are considered.

The measured quantities of interest are

- Kinematics
  - At the top: considering the occurrence (time) and magnitude of maximum
    * Displacement
    * Acceleration
- Base moments: at the bottom (fixity) considering the occurrence (time) and magnitude of maximum
- Base shear at the bottom (fixity).

Numerous structural parameters and modeling parameters (like the number of elements considered) are varied, and the results are analyzed and discussed. The simulations are carried out for 10-min winds, and the structural response time history is evaluated. In many cases, the chosen representative values for structural responses are those of the absolute maximum. These are reported for comparative purposes.

The time series of responses, either displacement or base reactions for the base scenario, are presented in Figures 8 and 9 for cases with and without incoming turbulence in the wind. The following figures compare the top-floor accelerations and base reactions with and without damping. Also, the variation in damping is presented. It can be seen that the response reduction is prominent in the case of a structure with damping. It is uncommon for the structures to have zero damping, as shown in the figure. However, this is presented for a reference.

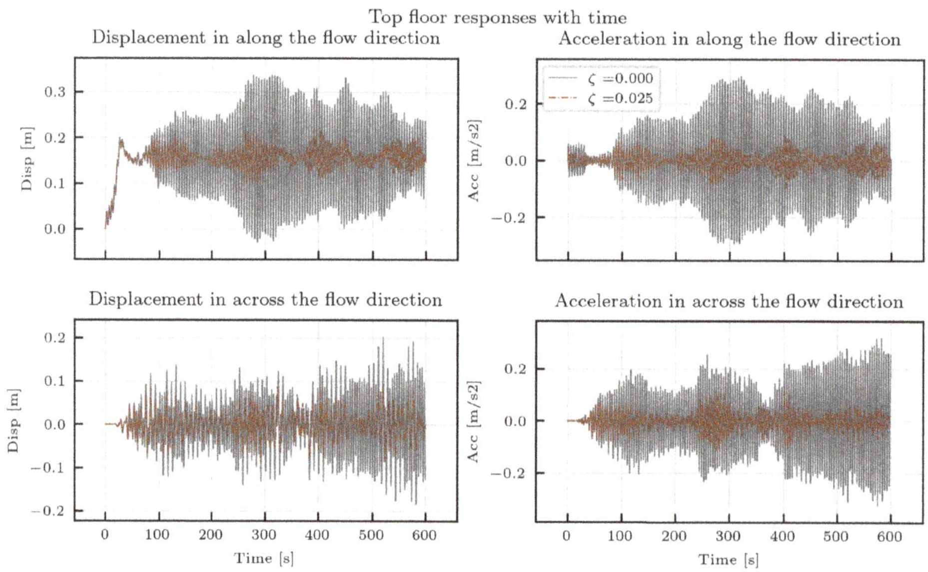

Figure 8. Top-floor response (acceleration and displacement) time history with no turbulence in the wind. Two damping ratios, ζ are plotted.

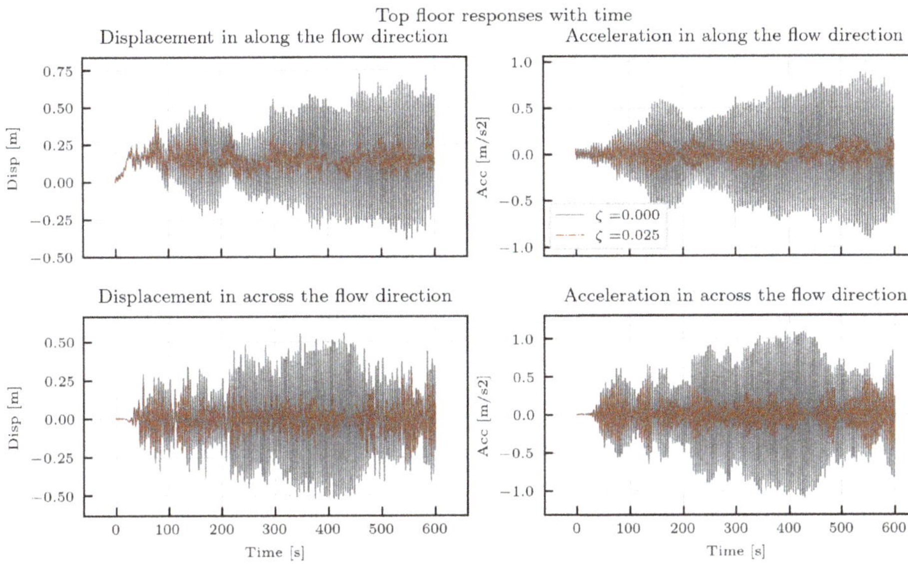

Figure 9. Top-floor response (acceleration and displacement) time history with no turbulence in the wind. Two damping ratio ζ are plotted.

For the spatial evaluation from 15 elements onward (for some cases even starting from 3 elements), maximum convergency is observed for top-floor displacements and accelerations. This is tabulated in Table 6. Similar observation can be made for Table 7. The tabulated and compared quantities are the maximum and the RMS of displacements/accelerations at the top.

Table 6. Spatial reduction summary for no turbulence, damping ratio ζ = 0.01, continuous, 0° angle of wind.

No. of Elements	$Disp_{max}$	$Disp_{RMS}$	Acc_{max}	Acc_{RMS}
1	0.1565	0.1081	0.2323	0.0210
2	0.2056	0.1452	0.0745	0.0278
3	0.2364	0.1572	0.1576	0.0524
15	0.2238	0.1588	0.1319	0.0429
30	0.2265	0.1593	0.1386	0.0463
60	0.2234	0.1591	0.1307	0.0446

Table 7. Spatial reduction summary for with turbulence, damping ζ = 0.01, continuous, 0° angle of wind.

No. of Elements	$Disp_{max}$	$Disp_{rms}$	Acc_{max}	Acc_{rms}
1	0.5884	0.2233	0.4389	0.1855
2	0.6804	0.2922	0.5340	0.2292
3	0.8332	0.3266	0.8698	0.3220
15	0.6785	0.3099	0.5926	0.2549
30	0.8567	0.3201	0.9044	0.3015
60	0.7511	0.3061	0.6936	0.2497

This part of this study focuses on the evaluation of an extensive parametric study, showing a clear summary of the comprehensive data available. The values on the vertical (y) axis are normalized with the values of the highest level of spatial resolution. The effects of

- Wind flow conditions
  - Angle of attack
  - Presence of natural incoming turbulence
- Structural model

are plotted for the top-floor displacement in all the above cases in Figures 10–13. It can be observed that for the cases with and without turbulence, the quantity of interest shows similar characteristics. This may be due to the normalization with respect to the highest resolution. As commented earlier, the resolution of 15 spatial elements is accurate in representing the maximum and RMS of displacement at the top. The amount of structural damping plays an insignificant role compared to the spatial resolution of modeling.

A similar study was conducted on the model results and is plotted in Figures 14–17. Interestingly, it can be seen that in the modal domain, the number of modes required to capture the mean and RMS of acceleration are very low. As seen in Table 8 even with three modes, the responses can be captured with a 99% accuracy. This is similar to what is being used for wind tunnels with approximate modeling. However, it is important to note that the accuracy of the results depends on the accuracy of these mode shapes. It is important to obtain these shapes and eigen-frequencies correctly, as seen in Study I. It is also seen that from the table, the mean can be approximated by the lower mode considerations somewhat accurately. However, the RMS needs more modes. This observation is valid for more accelerations compared to displacements. Some additional figures that support this observation is shown in additional figures in Appendix A.

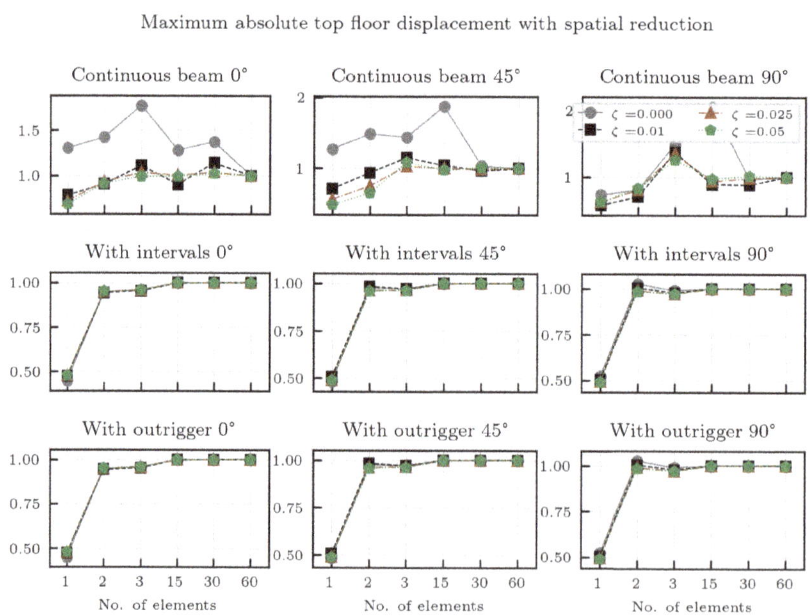

Figure 10. Spatial reduction—Maximum top-floor displacement as a ratio of the finest resolution model against the number of elements in the beam model with turbulence in the incoming wind. Four damping ratios ζ are plotted. The title of each sub-figure is on top.

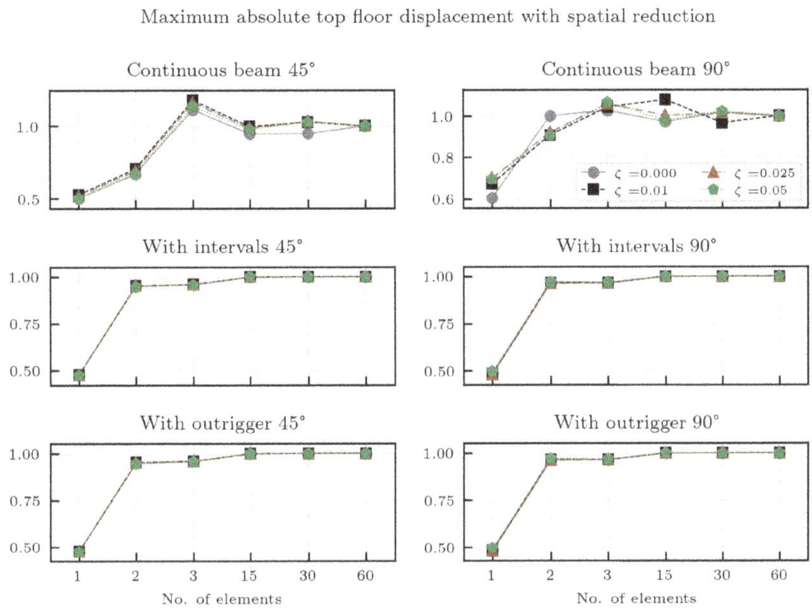

Figure 11. Spatial reduction—Maximum top-floor displacement as a ratio of the finest resolution model against the number of elements in the beam model with no turbulence in the incoming wind. Four damping ratios ζ are plotted. The title of each sub-figure is on top.

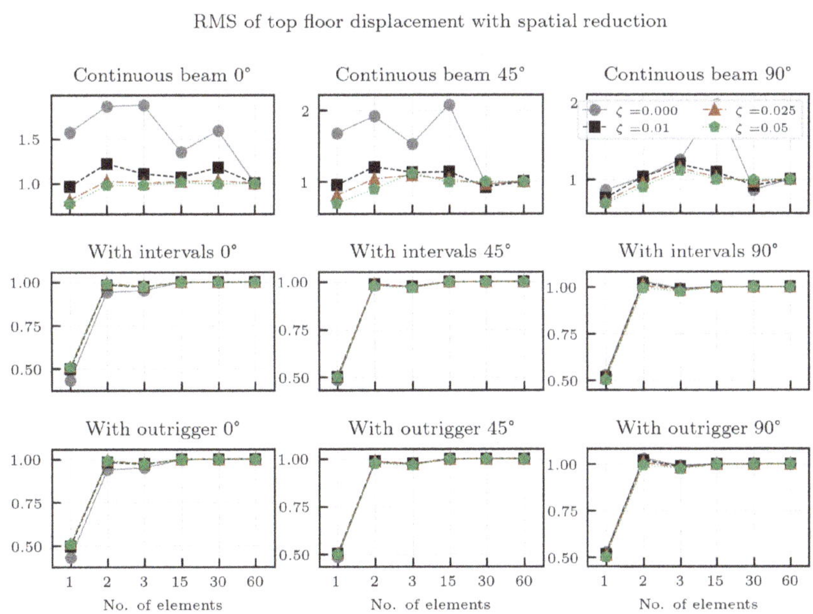

Figure 12. Spatial reduction—RMS of top-floor displacement as a ratio of the finest resolution model against the number of elements in the beam model with turbulence in the incoming wind. Four damping ratios ζ are plotted. The title of each sub-figure is on top.

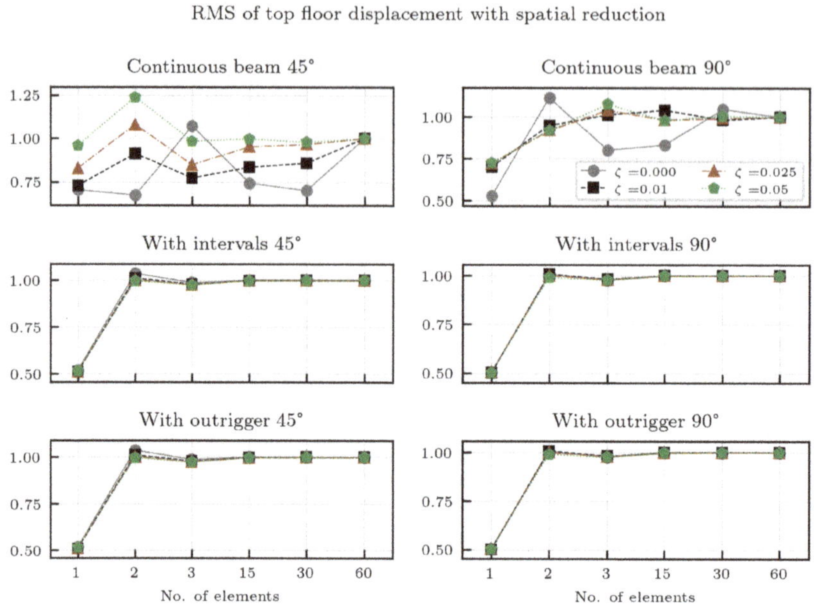

Figure 13. Spatial reduction—RMS of top-floor displacement as a ratio of the finest resolution model against the number of elements in the beam model with no turbulence in the incoming wind. Four damping ratios ζ are plotted. The title of each sub-figure is on top.

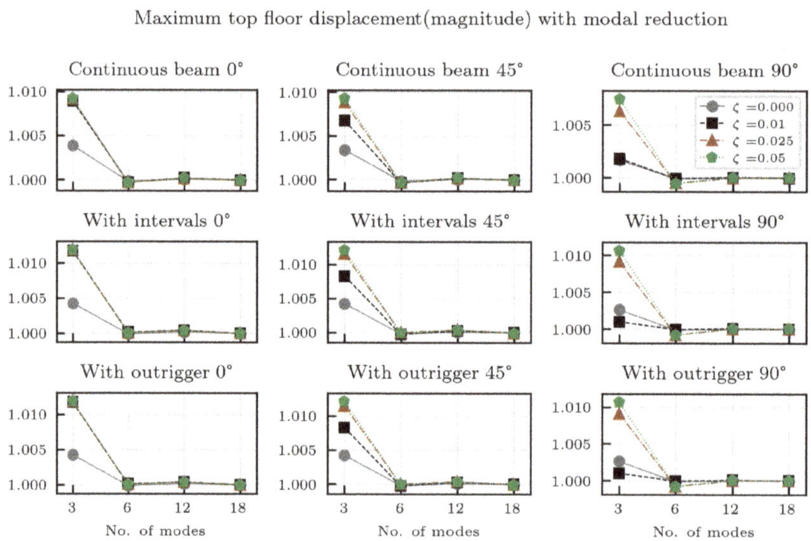

Figure 14. Modal reduction—Maximum top-floor displacement as a ratio of the finest resolution model against the number of modes considered with turbulence in the incoming wind. Four damping ratios ζ are plotted. The title of each sub-figure is on top.

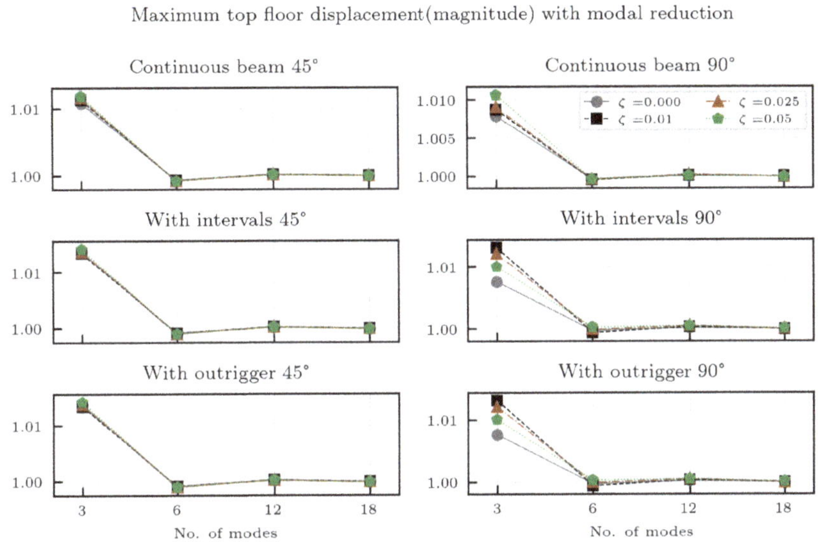

Figure 15. Modal reduction—Maximum top-floor displacement as a ratio of the finest resolution model against the number of modes considered with no turbulence in the incoming wind. Four damping ratios ζ are plotted. The title of each sub-figure is on top.

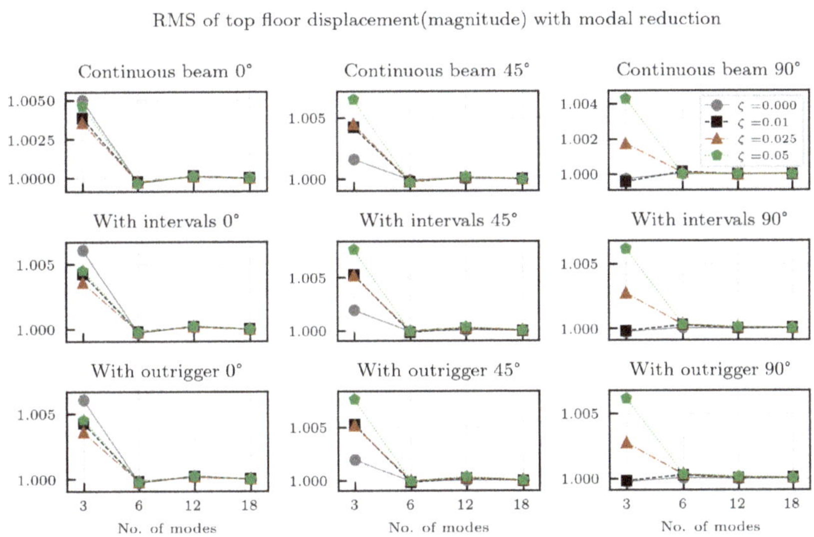

Figure 16. Modal reduction—RMS of top-floor displacement as a ratio of the finest resolution model against the number of modes considered with turbulence in the incoming wind. Four damping ratios ζ are plotted. The title of each sub-figure is on top.

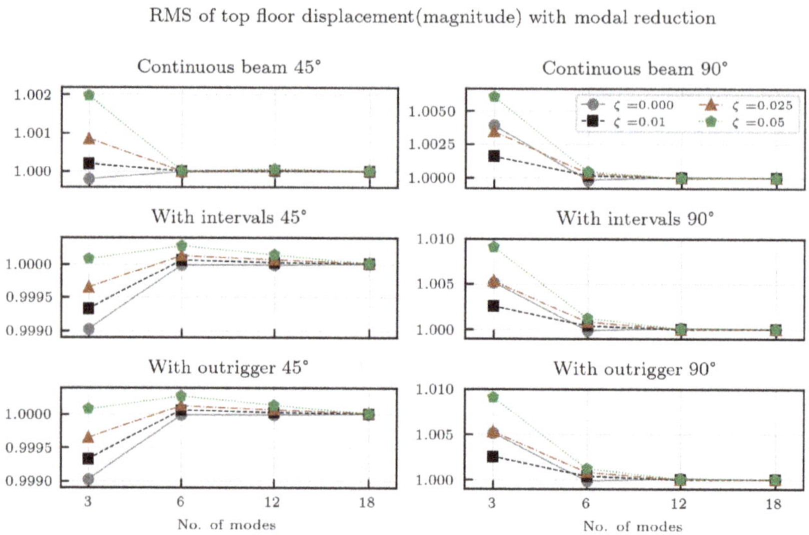

Figure 17. Modal reduction—RMS of top-floor acceleration as a ratio of the finest resolution model against the number of modes considered with no turbulence in the incoming wind. Four damping ratios ζ are plotted. The title of each sub-figure is on top.

Table 8. Model reduction summary for no turbulence, damping ζ = 0.025, continuous, 90° angle of wind.

No. of Modes	$Disp_{max}$	$Disp_{RMS}$	Acc_{max}	$Disp_{RMS}$
3	0.1997	0.1249	0.1722	0.0549
6	0.1978	0.1232	0.1723	0.0549
12	0.1979	0.1233	0.1724	0.0549
18	0.1979	0.1233	0.1724	0.0549
90	0.1979	0.1233	0.1724	0.0549
180	0.1979	0.1233	0.1724	0.0549

4. Conclusions

This study emphasizes the significance of approximate structural modeling in the early design phases of high-rise buildings with a focus on natural wind load conditions. The specific conclusions from the work are

- These early-phase structural models often exhibit inherent limitations due to simplifying assumptions. These limitations can greatly influence the accuracy of the measured responses to dynamic loads, especially wind loads. As a designer, one needs to be aware of the limitations of respective modeling assumptions.
- The parametric study reveals that low-fidelity models are a good representation of high-fidelity models, provided that the frequencies of the modes are matched.
- This study demonstrates that reduced structural models, specifically lumped-mass models with three lumped masses, can effectively capture the behavior of high-rise structures under wind loads, provided that the frequencies and mode shapes are matched.
- These models can approximate the behavior of complete systems with a certain degree of accuracy, quantified as around 11% for displacements and 20% for accelerations. Moreover, these LF models are fast and easy to evaluate.

These models are particularly beneficial in the initial design stages, where rapid, cost-effective, and moderately accurate evaluations are required. They are also useful for computationally intensive tasks, like optimization, uncertainty quantification, optimization under uncertainties, and the design of damping devices to reduce the structural acceleration responses with supplemental damping.

Author Contributions: Conceptualization, M.P., R.W. and F.W.; Methodology, A.K. and M.P.; Software, M.P.; Validation, A.K.; Formal analysis, A.K.; Investigation, A.K. and M.P.; Data curation, A.K.; Writing—original draft, A.K.; Writing—review & editing, M.P., R.W. and F.W.; Visualization, A.K.; Supervision, K.-U.B. and R.W.; Project administration, K.-U.B. and F.W.; Funding acquisition, K.-U.B. and F.W. All authors have read and agreed to the published version of the manuscript.

Funding: This research received no external funding.

Data Availability Statement: The original contributions presented in this study are included in the article. Further inquiries can be directed to the corresponding author.

Acknowledgments: We would like to thank Enis Alite and Muhamed Horo for providing helpful discussions and contributing to this work.

Conflicts of Interest: Author Felix Weber was employed by the company Maurer Switzerland GmbH. The remaining authors declare that the research was conducted in the absence of any commercial or financial relationships that could be construed as a potential conflicts of interest.

Appendix A. Additional Figures

Here, a few additional figures to numerical Study II are presented. They indicate similar observations to Section 3.2.

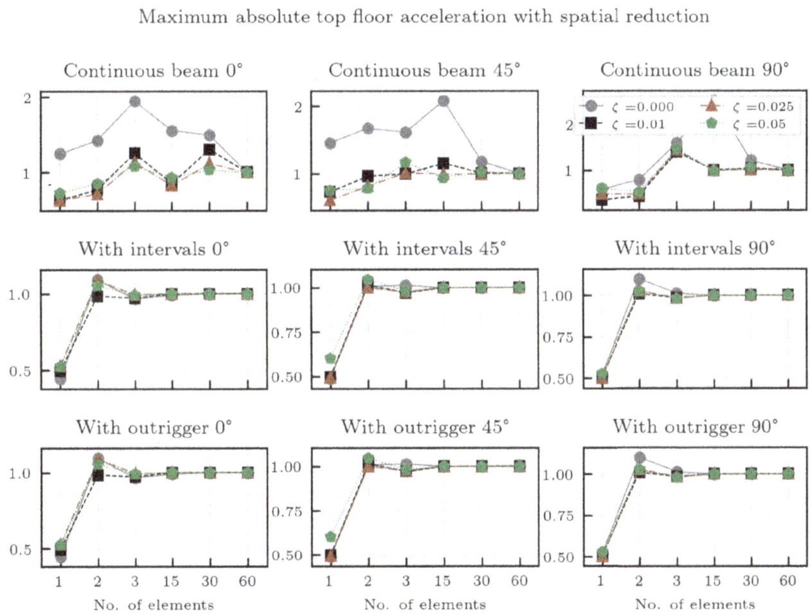

Figure A1. Spatial reduction—maximum top-floor acceleration with turbulence case.

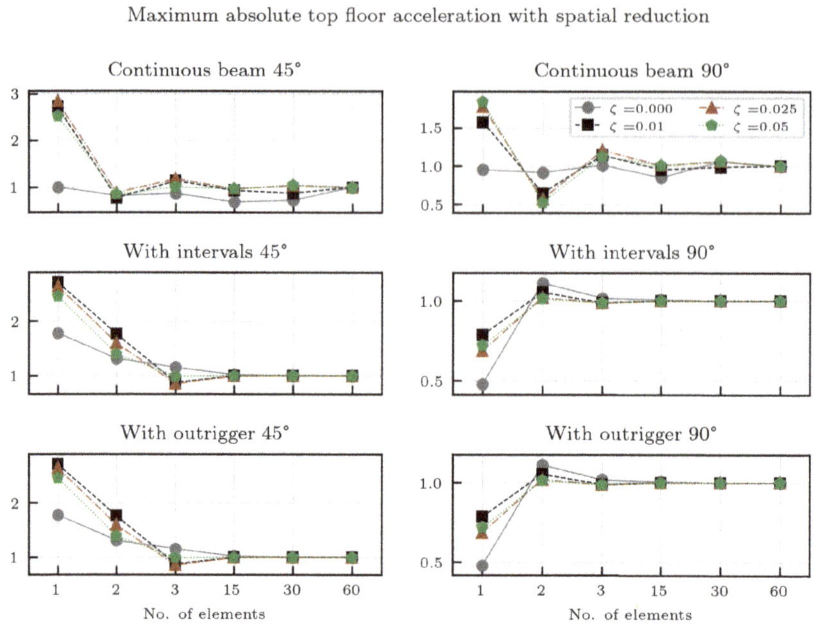

Figure A2. Spatial reduction—maximum top-floor acceleration with no turbulence case.

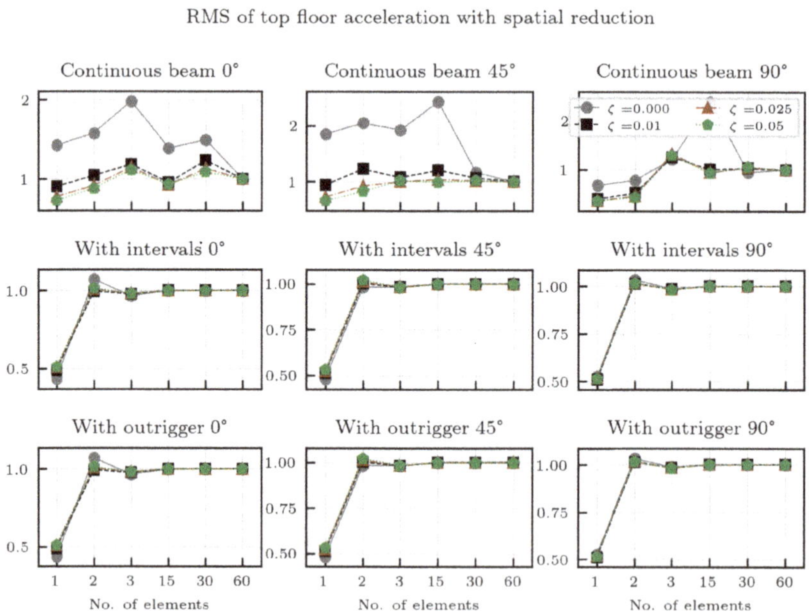

Figure A3. Spatial reduction—RMS of top-floor acceleration with turbulence case.

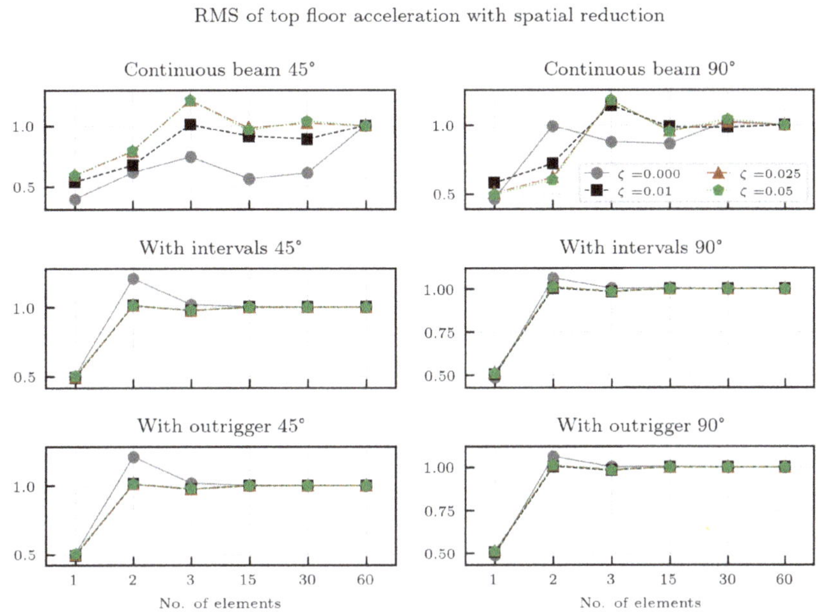

Figure A4. Spatial reduction—RMS of top-floor acceleration with no turbulence case.

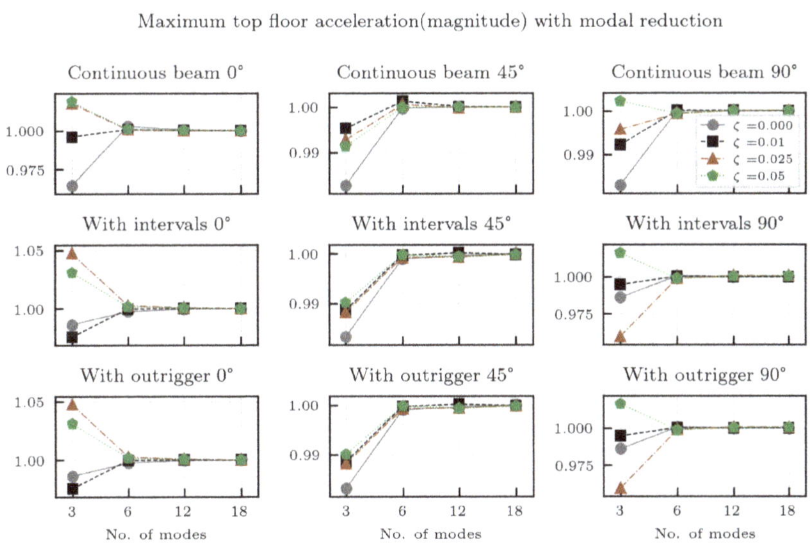

Figure A5. Modal reduction—maximum top-floor acceleration with turbulence case.

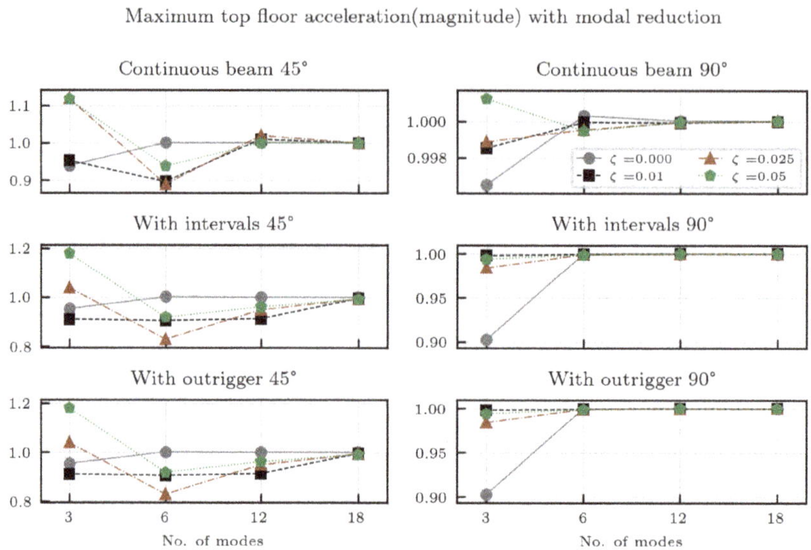

Figure A6. Modal reduction—maximum top-floor acceleration no turbulence case.

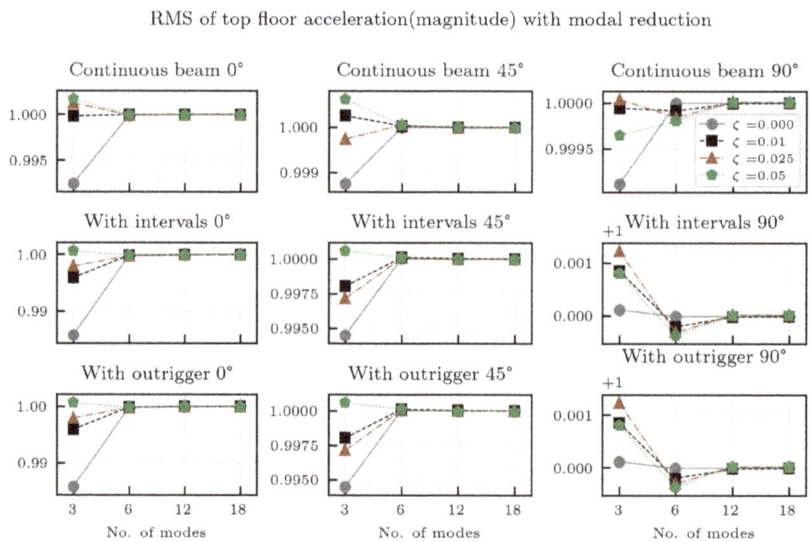

Figure A7. Modal reduction—RMS of top-floor acceleration with turbulence case.

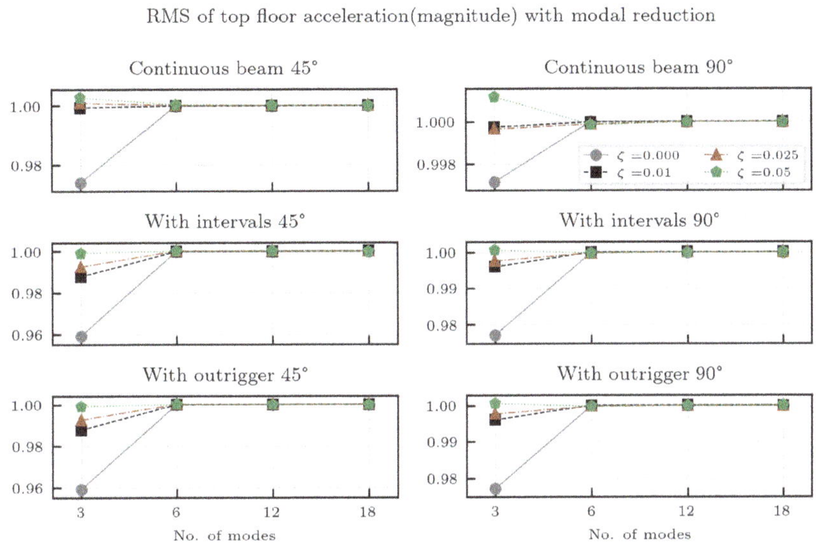

Figure A8. Modal reduction—RMS of top-floor acceleration with no turbulence case.

References

1. Blocken, B. 50 years of Computational Wind Engineering: Past, present and future. *J. Wind. Eng. Ind. Aerodyn.* **2014**, *129*, 69–102. [CrossRef]
2. Chen, Z.; Huang, H.; Tse, K.; Xu, Y.; Li, C.Y. Characteristics of unsteady aerodynamic forces on an aeroelastic prism: A comparative study. *J. Wind. Eng. Ind. Aerodyn.* **2020**, *205*, 104325. [CrossRef]
3. Wang, L.; Liang, S.; Huang, G.; Song, J.; Zou, L. Investigation on the unstability of vortex induced resonance of high-rise buildings. *J. Wind. Eng. Ind. Aerodyn.* **2018**, *175*, 17–31. [CrossRef]
4. Wang, L.; Fan, X.Y.; Liang, S.G.; Song, J.; Wang, Z.K. Improved expression for across-wind aerodynamic damping ratios of super high-rise buildings. *J. Wind. Eng. Ind. Aerodyn.* **2018**, *176*, 263–272. [CrossRef]
5. Dongmei, H.; Ledong, Z.; Quanshun, D.; Xue, Z.; Wei, C. Aeroelastic and aerodynamic interference effects on a high-rise building. *J. Fluids Struct.* **2017**, *69*, 355–381. [CrossRef]
6. Elshaer, A.; Aboshosha, H.; Bitsuamlak, G.; El Damatty, A.; Dagnew, A. LES evaluation of wind-induced responses for an isolated and a surrounded tall building. *Eng. Struct.* **2016**, *115*, 179–195. [CrossRef]
7. Huang, S.; Li, Q.; Xu, S. Numerical evaluation of wind effects on a tall steel building by CFD. *J. Constr. Steel Res.* **2007**, *63*, 612–627. [CrossRef]
8. Wijesooriya, K.; Mohotti, D.; Chauhan, K.; da Costa, D.D. Numerical investigation of scale resolved turbulence models (LES, ELES and DDES) in the assessment of wind effects on supertall structures. *J. Build. Eng.* **2019**, *25*, 100842. [CrossRef]
9. Mohotti, D.; Wijesooriya, K.; da Costa, D.D. Comparison of Reynolds Averaging Navier-Stokes (RANS) turbulent models in predicting wind pressure on tall buildings. *J. Build. Eng.* **2019**, *21*, 1–17. [CrossRef]
10. International Association for Wind Engineering. Benchmark Buildings for an International HFBB Comparison. 2009. Available online: https://www.iawe.org/committees/HFBB-spec.pdf (accessed on 2 November 2024).
11. Holmes, J.; Tse, T. International high-frequency base balance benchmark study. In Proceedings of the 2012 World Congress on Advances in Civil, Environmental, and Materials Research (ACEM' 12), Seoul, Republic of Korea, 26–29 August 2012; International Association of Structural Engineering and Mechanics: Daejeon, Republic of Korea, 2012.
12. Kratos Multiphysics Development Team Kratos Multiphysics—Open-Source Framework for Multiphysics Simulations. 2024. Available online: https://github.com/KratosMultiphysics/Kratos (accessed on 2 November 2024)
13. Dadvand, P.; Rossi, R.; Gil, M.; Martorell, X.; Cotela, J.; Juanpere, E.; Idelsohn, S.R.; Oñate, E. Migration of a generic multi-physics framework to HPC environments. *Comput. Fluids* **2013**, *80*, 301–309. [CrossRef]
14. Dadvand, P.; Rossi, R.; Oñate, E. An object-oriented environment for developing finite element codes for multi-disciplinary applications. *Arch. Comput. Methods Eng.* **2010**, *17*, 253–297. [CrossRef]
15. Cotela-Dalmau, J.; Rossi, R.; Oñate, E. A FIC-based stabilized finite element formulation for turbulent flows. *Comput. Methods Appl. Mech. Eng.* **2017**, *315*, 607–631. [CrossRef]

16. Pentek, M.; Winterstein, A.; Vogl, M.; Kupás, P.; Bletzinger, K.U.; Wüchner, R. A multiply-partitioned methodology for fully-coupled computational wind-structure interaction simulation considering the inclusion of arbitrary added mass dampers. *J. Wind. Eng. Ind. Aerodyn.* **2018**, *177*, 117–135. [CrossRef]
17. Kareem, A.; Tamura, A. *Advanced Structural Wind Engineering*; Springer: Berlin/Heidelberg, Germany, 2015.
18. Tabor, G.R.; Baba-Ahmadi, M. Inlet conditions for large eddy simulation: A review. *Comput. Fluids* **2010**, *39*, 553–567. [CrossRef]
19. Sanderse, B.; Van der Pijl, S.; Koren, B. Review of computational fluid dynamics for wind turbine wake aerodynamics. *Wind Energy* **2011**, *14*, 799–819. [CrossRef]
20. Wu, X. Inflow turbulence generation methods. *Annu. Rev. Fluid Mech.* **2017**, *49*, 23–49. [CrossRef]
21. Mann, J. The spatial structure of neutral atmospheric surface-layer turbulence. *J. Fluid Mech.* **1994**, *273*, 141–168. [CrossRef]
22. Mann, J. Wind field simulation. *Probabilistic Eng. Mech.* **1998**, *13*, 269–282. [CrossRef]
23. IEC. *Wind Turbines–Part 1: Design Requirements*; International Electrotechnical Commission: Geneva, Switzerland, 2005.
24. Michalski, A.; Kermel, P.D.; Haug, E.; Löhner, R.; Wüchner, R.; Bletzinger, K.U. Validation of the computational fluid-structure interaction simulation at real-scale tests of a flexible 29m umbrella in natural wind flow. *J. Wind. Eng. Ind. Aerodyn.* **2011**, *99*, 400–413. [CrossRef]
25. Andre, M.; Mier-Torrecilla, M.; Wüchner, R. Numerical simulation of wind loads on a parabolic trough solar collector using lattice Boltzmann and finite element methods. *J. Wind. Eng. Ind. Aerodyn.* **2015**, *146*, 185–194. [CrossRef]
26. Dong, L.; Lio, W.H.; Simley, E. On turbulence models and lidar measurements for wind turbine control. *Wind. Energy Sci.* **2021**, *6*, 1491–1500. [CrossRef]
27. Malekinejad, M.; Rahgozar, R. An analytical model for dynamic response analysis of tubular tall buildings. *Struct. Des. Tall Spec. Build.* **2014**, *23*, 67–80. [CrossRef]
28. Pentek, M.; Kodakkal, A.; Contributors. ParOptBeam: A Parametric Optimizable Beam Model for Structural Analysis. 2025. Available online: https://github.com/mpentek/ParOptBeam (accessed on 10 January 2025).
29. Krenk, S. *Non-Linear Modeling and Analysis of Solids and Structures*; Cambridge University Press: Cambridge, UK, 2009.

applied sciences

MDPI

Article

Tuning of a Viscous Inerter Damper: How to Achieve Resonant Damping without a Damper Resonance

Jan Høgsberg

Department of Civil and Mechanical Engineering, Technical University of Denmark, Koppels Allé, Building 404, DK-2800 Kongens Lyngby, Denmark; jahog@dtu.dk

Abstract: Inerter dampers are effectively employed to mitigate and dampen structural vibrations in slender or high-rise buildings. The simple viscous inerter damper, with a viscous dashpot placed in series with an inerter, is designed to create resonant vibration damping, although the damper itself is without an internal resonance. The apparent resonant behavior is instead obtained by increasing the damper inertance until the two lowest modes of the considered building model interact, whereafter the viscous coefficient is adjusted until the desired response mitigation is achieved. The present modal interaction tuning requires that the reduced-order single-mode dynamic model of the building includes both inertia and flexibility from the (other) modes otherwise discarded by the model reduction. While the inertia correction adjusts the modal mass of the inerter damper, the corresponding flexibility introduces the apparent damper stiffness that creates the desired damper resonance. Thus, the accurate representation of other modes is essential for the design and resonant tuning of the simple viscous inerter damper. The resonant damper performance by the non-resonant viscous inerter damper is illustrated by a numerical example with a 20-story building model, for which the desired resonant modal interaction requires an inertance of almost ten times the entire translational building mass.

Keywords: structural dynamics; vibration damping; vibration control; damper tuning; inerter; tuned inerter damper; residual mode correction

1. Introduction

In slender civil engineering structures, external dampers are used for response mitigation, the improvement of comfort, and earthquake protection (among others). The present paper proposes a simple viscous inerter damper, tuned to act resonantly for a building structure model, without an inherent damper resonance. Thus, the tuning of this apparent resonant damper is based on classic results for tuned mass dampers (TMDs), for which the book by Den Hartog [1] is considered a fundamental reference, introducing the so-called fixed-point tuning principle. For applications to civil engineering buildings, the fixed-point tuning principle results in different expressions for optimal stiffness and damping in the case of either harmonic steady-state response or white-noise stochastic excitation. An early and comprehensive summary of tuning expressions for various types of loading and response of a single-degree-of-freedom (sdof) structure model has been provided by [2], while the extension to full multi-dof (mdof) structures has mainly aimed at the numerical optimization of TMDs to narrow-band earthquake and/or wind loading [3]. In [4], the TMD is compared to a tuned liquid column damper (TLCD) for building vibrations, while recently, [5] considered the TMD inside Taipei 101 as a special case study. Recent review papers about the TMD damping of buildings structures are given in [6–8], emphasizing the wide application of TMDs for slender high-rise structures. For the present damper, optimal tuning is obtained by maximizing modal damping instead of mitigating response measures. For the TMD, this equal modal damping principle has been proposed in [9] and more recently extended to mdof structures in [10]. For the maximum damping approach, the tuning principle becomes independent of the particular loading and response type and thus results in a rather robust tuning principle.

1.1. Inerter Dampers

A recent extension of the classic TMD is the introduction of the so-called inverter, a two-terminal inertia element [11] that is most commonly realized by converting translational deflection into rotation [12], whereby gearing systems can be used to realize extraordinary large effective damper masses [13,14]. Because the absorber theory dictates that damper performance increases with damper mass, the mass gearing property has accelerated research and development of inerter-based devices.

The basic inerter-type dampers mostly combine the inerter with a spring and dashpot to create a fundamental resonant damper. The inerter and spring create a damper resonance that is commonly targeted to the dominant vibration mode of the structure. The fundamental analysis of tuned inerter dampers (TIDs) is introduced in [15], in which the damper is tuned to optimal response reduction, compared to the performance of the classic TMD, and investigated for mass ratios of up to 50% of the fundamental modal mass. In [16], tuning expressions for inerter dampers are carefully derived for various damper configurations and extended in [17] to maximize the base-isolation performance when the sdof structure is exposed to stochastic loading.

Analytical tuning expressions for TIDs are presented in [18,19], where the modal coupling with other modes is explicitly accounted for by additional (artificial) flexibility and inertia terms. This inclusion of residual mode correction enables the derivation of accurate explicit tuning formulae for inerter dampers, an approach also followed in the present paper. An exact numerical optimization of the two fundamental TIDs is conducted in [20,21] for an sdof structure exposed to stochastic base excitation.

Tuned inerter dampers have been proposed for several applications in high-rise building technology, such as for seismic retrofit [22,23] and the optimization of multiple TIDs for the vibration control of an mdof structure model under stochastic base excitation. Furthermore, inerter dampers have been extended by the introduction of negative-stiffness springs [24,25], which improves the TID's damping capability, and by the installation of hysteretic dampers in [26].

A particular property of inerter elements is their theoretical ability to connect, and thus transfer force, between floors separated by several stories. An example is provided by [27], in which the inerter bridges four stories to create a larger relative damper deflection for the wind-induced vibration control of a high-rise building. The ultimate realization defines the so-called ground inerter damper, for which one of the inerter terminals is attached to the building support level. In [28], the inerter damper is tuned for optimal vibration damping, while in [29], the grounded concept is applied for seismic isolation.

The simplest case of grounded dampers are those acting as base-isolation systems. In [30], tuning expressions are derived optimal base-isolation applications of inerter dampers. As in [17], the optimally tuned inerter dampers aim at minimizing the transmissibility from base to structure. Inerter dampers are proposed for base-isolation of storage tanks by [31], while simple inerter systems (with hysteretic damping) are proposed for seismic isolation of building models in [32].

1.2. Hybrid Mass and Inerter Dampers

Because of the working similarity of TMDs and inerter-based TIDs, the merger of the two damper concepts has been considered extensively in the research literature, especially in connection with the damping and protection of building-type structures. The fundamental extension of the TMD by simply connecting an inerter between the auxiliary TMD mass and another building floor is introduced in [33] and referred to as the tuned mass damper inerter (TMDI). In [12], the TMDI is investigated in detail for various applications: vibration suppression, damper mass reduction, and the harvesting of energy. Subsequently, the TMDI has been considered in [34–36] for installation in building structures, focusing on harmonic excitation, earthquake loading, and wind-induced vibrations. As for the basic inerter dampers, the TMDI has been shown to possess superior base-isolation abilities with limited damper stroke in [37,38]. Furthermore, the TMDI also allows for the bridging of stories, as considered in [35,39] for wind-induced vibrations. In the detailed study in [39], it was shown that for a realistic bridging of stories, corresponding to the space occupied by a pendulum absorber, the classic TMD outperforms the TMDI.

More advanced inerter augmentations of the classic TMD have been proposed in the literature. The basic TMDI in [39] was improved by the extending the inerter by a spring and the TMD dashpot [14]. This more advanced TMD inerter device was shown in [14] to give better performance when tuned precisely. The inclusion of various inerter-type devices in the classic TMD is considered [16], whereby double-resonance absorbers are created. It is shown that these augmented absorbers create a very wide and flat plateau in the frequency response curves at the targeted structural frequency. Closed-form solutions for the tuning of such double-resonant TMD inerter absorbers were derived in [40] for the most common combinations of inerter, spring, and dashpot. Reviews of inerter-based dampers for structural vibration damping are given by [41,42].

1.3. Outline of Paper

The aim of the present paper is to design a simple viscous inerter damper, for which the inertance is chosen so that the first two structural vibration modes interact, creating an apparent resonant absorber characteristic, with a damper without an internal resonance. Section 2 introduces the mdof building model and the theory for attaching a general inerter damper. The system reduction from a full mdof model to a single-dof model for the targeted vibration mode includes the influence from the other modes by the residual stiffness and inertance terms, of the same form as in [18].

In Section 3, the viscous inerter damper is introduced, with a viscous dashpot placed in series with the inerter. This model is referred to as the C2 configuration in [17]. This viscous inerter damper is without a spring and thus is in itself non-resonant. However, because of the residual mode flexibility term, the characteristic equation for the reduced-order single-mode model is of quartic order in the complex natural frequency, whereby the equal modal damping from [9,18,43] can be applied to create resonant vibration damping. The residual mode correction coefficients are determined from the two limiting natural frequencies for vanishing (free) and infinite (clamped) viscous damper coefficients, following the general approach for vibration absorbers in [44]. However, in the present case, the damper is not resonant in itself, which is opposite to the absorbers in [44], which implies that the inertance of the present viscous inerter damper is calibrated by a simple iterative procedure to create the desired modal interaction between the first two modes of the building model. Section 3 concludes by tuning the viscous dashpot so that optimal frequency response curves are obtained for the response velocity amplitude of the structure.

The performance of the resonantly tuned viscous inerter damper is illustrated in Section 4 for the 20-floor building model in [14,39]. The damper is placed between ground and bottom floor, similar to a base-isolation system. The iterative tuning procedure gives a mass ratio of 10.5% of mode 1 modal mass, which is safely within what is considered applicable for inerter dampers in the literature. A root locus analysis, in which the complex-valued natural frequencies are traced in the complex plane for an increasing viscous coefficient, shows that the mass ratio tuning almost creates the desired bifurcation point condition. The subsequent frequency response analysis demonstrates that the novel resonant tuning of the simple viscous inerter damper creates a flat plateau for the top-floor velocity amplitude. The numerical example also investigates the influence of placing a spring in parallel to the dashpot, with a stiffness that is sufficiently small to not interfere too much with the sensitive resonant tuning but large enough to re-center the viscous inerter damper in practice (avoid drift).

The final conclusions in Section 5 highlight the conceptual novelty of the proposed damper configuration, which is simple and robust and creates the desired resonant type vibration mitigation, although the damper is without an internal resonance (no spring included).

2. Governing Equations of Motion

Figure 1a shows a simple shear frame building model with $n = 20$ stories. The story mass is assumed to be concentrated (lumped) at each floor, where it is denoted as m_f, while k_f is the corresponding inter-story stiffness, as depicted in Figure 1b. The governing equation of motion in the time domain is then written in the common form as

$$\mathbf{M}\ddot{\mathbf{q}}(t) + \mathbf{C}\dot{\mathbf{q}}(t) + \mathbf{K}\mathbf{q}(t) + \mathbf{b}f(t) = \mathbf{f}_{ext}(t) \tag{1}$$

where t is time and $\dot{(\;)} = d(\;)/dt$ represents the time derivative. For the present building model, the 20 floor displacements are collected in the displacement vector

$$\mathbf{q}^T = [q_1, q_2, \ldots, q_{20}] \tag{2}$$

thereby constituting the governing degrees of freedom (dofs) for the present building model, which corresponds to the model used by Weber et al. [14,39]. In (1), $\mathbf{M}$ is the mass matrix, $\mathbf{C}$ is the damping matrix, and $\mathbf{K}$ is the stiffness matrix. The construction of the building model is presented in [14,39] and summarized in Section 4. The external loading is represented by the vector process $\mathbf{f}_{ext}(t)$, which in Figure 1a will be assumed to be evenly distributed across the building floors. Finally, $f(t)$ represents the force exerted by the inerter damper onto the structure, with the particular location of the damper determined by the connectivity vector $\mathbf{b}$. For the location between the ground and first floor in Figure 1, the connectivity vector is given as

$$\mathbf{b}^T = [1, 0, \ldots, 0] \tag{3}$$

with entries 2 to 20 all zero.

In the following damper design and tuning analysis, the behavior of the damper is conveniently represented in the frequency domain, in which time derivatives are included by 90-degree phase shifts. Thus, the time-based equation of motion (1) is conveniently converted into the frequency domain by introducing the steady-state frequency representations

$$\mathbf{q}(t) = \mathbf{q}\exp(i\omega t) \quad , \quad f(t) = f\exp(i\omega t) \quad , \quad \mathbf{f}_{ext}(t) = \mathbf{f}_{ext}\exp(i\omega t) \tag{4}$$

where the frequency variables $\mathbf{q}$, f, and $\mathbf{f}_{ext}$ represent steady-state amplitudes and thus are written without the time argument. By substitution of (4) into (1), the structural equation of motion can be expressed as a frequency relation in the following form

$$\left(-\omega^2\mathbf{M} + i\omega\mathbf{C} + \mathbf{K}\right)\mathbf{q} + \mathbf{b}f = \mathbf{f}_{ext} \quad (5)$$

where the damper force f is defined in the following sections based on the inerter damper considered in the present paper. The frequency solution enables tuning based on both frequency response curves or pole-placement techniques. As shown for the classic tuned mass damper (TMD) in [9] and for general control methods in [43], a balanced pole-placement method will also provide effective response amplitude mitigation in frequency response curves. Thus, the pole-placement technique is applied in Section 3 for inerter damper tuning.

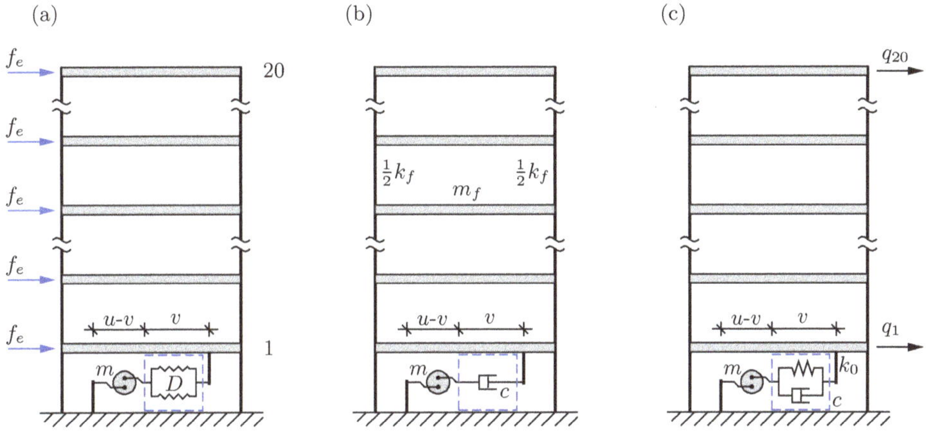

Figure 1. Shear frame building model with an inerter damper (with inertance m) between the ground and the first floor. (**a**) General inerter damper with damper function $D(\omega)$, (**b**) viscous inerter damper with viscous coefficient c, and (**c**) viscoelastic inerter damper with viscous coefficient c and parallel spring stiffness k_0 introduced to avoid drift in dashpot deflection.

2.1. Inerter Damper Model

Inerter dampers can generally be constructed as an inerter (with inertance m) placed in series with an arbitrary damper element, which contains a viscous dashpot c in combination with springs and potentially additional inerters. This dynamic combination, placed in series with the front-end inerter m, is in the present (general) inerter damper model denoted $D(\omega)$; see Figure 2. This general inerter damper model is also used in Figure 1a when acting inside the building model. Furthermore, the figure shows two examples in (b) and (c), where the pure viscous inerter damper in Figure 1b is represented by $D(\omega) = i\omega c$, with c denoting the tunable viscous dashpot coefficient. In Figure 1c, the viscous dashpot is extended by a spring k_0 to (in principle) form a tuned inerter damper (TID) by the damper function $D(\omega) = i\omega c + k_0$. However, in the present case, the pure viscous damper function in Figure 1b is designed to operate resonantly by creating modal interaction between the two fundamental modes of the building structure, while k_0 in (c) is only introduced in the numerical example in Section 4 to avoid the drift of the damper stroke during operation.

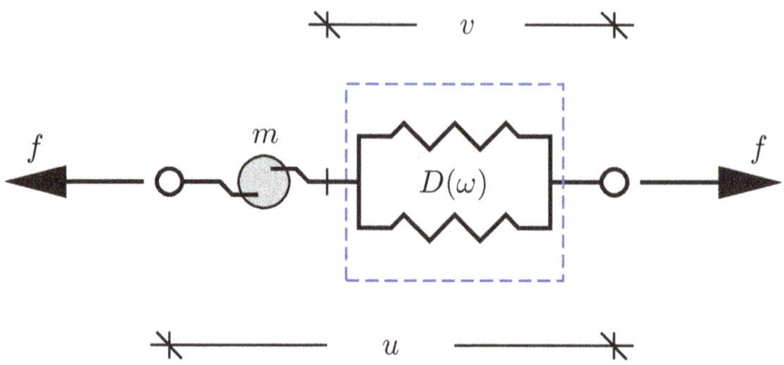

Figure 2. General inerter damper with front-end inertance m in series with general damper element $D(\omega)$.

In the present section, the general inerter damper is assumed to be attached to the flexible dynamic structure, as shown for the building model in Figure 1a. The governing equation for the general inerter damper in Figure 2 is written in flexibility format as

$$\left(\frac{1}{-\omega^2 m} + \frac{1}{D(\omega)}\right) f = u \tag{6}$$

where the displacement u in Figure 2 represents the full stroke across the inerter damper. For the building model in Figure 1, this entire stroke u is determined by the connectivity vector as

$$u = \mathbf{b}^T \mathbf{q} \tag{7}$$

which, therefore, by construction, is the energy conjugate displacement to the damper force f.

For the general inerter damper, the undamped case is retrieved by $D \to 0$, in which case $f = 0$ in (6). On the other hand, for $D \to \infty$, the damper link, inside the blue-dashed box in Figure 2, will become fully rigid, thus ideally transferring the force, whereby $f = -\omega^2 m u$ will add pure inertia to the structure. These two limiting cases will be important in the dynamic system reduction applied next.

2.2. Limiting Eigenvalue Problems

A typical damper design aims at optimizing the damping in a single vibration mode, which dominates the structural response of the structure. For high-rise buildings and other slender structures, such as wind turbines, towers, and masts, the dominant mode is likely the first (or fundamental) vibration mode. An effective system reduction procedure must isolate the targeted vibration mode, referred to below as mode number s, while still accounting for the spillover from all the other modes. For that, the present subsection introduces the limiting eigenvalue problems associated with $D \to 0$ and ∞, from which the corresponding natural frequencies are used to calibrate the modal spill-over to mode s from the non-targeted modes.

In Section 2.3, the modal expansion is based on the undamped vibration modes associated with $D = 0$ and thereby $f = 0$ from (6). In this free damper limit, the frequency equation of motion (5) simplifies to the generalized eigenvalue problem

$$\left(-\omega_j^2 \mathbf{M} + \mathbf{K}\right) \mathbf{u}_j = \mathbf{0} \tag{8}$$

for vanishing damping and external loading. In (8), the angular natural frequency ω_j is determined from the eigenvalue, while $\mathbf{u}_j$ is the associated mode shape vector for vibration mode j. Below, this undamped limit for $D \to 0$ is denoted as the free condition, whereby (8) is referred to as the free eigenvalue problem.

In the other limit $D \to \infty$—referred to below as the clamped limit—the damper element (inside the dashed-blue box) in Figure 2 will fully lock, whereby the generalized eigenvalue problem will change into

$$\left(-\hat{\omega}_j^2(\mathbf{M}+m\mathbf{b}\mathbf{b}^T)+\mathbf{K}\right)\hat{\mathbf{u}}_j = \mathbf{0} \tag{9}$$

where m will increase the system's inertia at the damper location. The hat symbol $\hat{()}$ depicts solutions or parameters associated with the clamped limit, e.g., the clamped natural frequency $\hat{\omega}_j$ and clamped mode shape vector $\hat{\mathbf{u}}_j$. For the target mode $j = s$, it is assumed that the addition of inertia at the damper location will result in the clamped natural frequency being smaller than the associated free natural frequency, i.e., $\hat{\omega}_s < \omega_s$.

In the next section, the free mode shapes are conveniently used in the modal expansion, as these mode shapes do not depend on the front-end inertance m of the inerter damper in Figure 2. Therefore, the clamped natural frequencies are used below to determine the influence from other modes, which in the present damper is of great importance, as the resonant damper characteristics are derived from modal interaction and not by an explicit damper resonant. The derivations are presented in the most compact form when both the free and clamped mode shapes are normalized to unit damper deflections, enforced by the relations

$$\mathbf{b}^T\mathbf{u}_j = 1 \quad , \quad \mathbf{b}^T\hat{\mathbf{u}}_j = 1 \tag{10}$$

For the indirect location of the inerter damper in Figure 1, the modal mass will be much larger than for a TMDI placed at the top floor, as in [14,39].

2.3. Modal Representation

As mentioned in the previous section, the free mode shapes are conveniently used in the modal expansion, as they are damper-independent because $D \to 0$ implies $f = 0$. The modal representation of the structural displacement vector is therefore expressed as

$$\mathbf{q} = \sum_j \mathbf{u}_j r_j \tag{11}$$

The normalization in (10) to unit modal deflection across the inerter damper then simplifies the damper deflection u in (7),

$$u = \sum_j \mathbf{b}^T\mathbf{u}_j r_j = \sum_j r_j \tag{12}$$

which is simply the sum of modal coordinates. The governing (frequency-domain) modal equation of motion is obtained by substitution of (11) into (5), which, upon pre-multiplication by $\mathbf{u}_j^T$, can be written as

$$(-\omega^2 m_j + k_j)r_j + f = f_j \tag{13}$$

This introduces the modal mass, stiffness, and load as

$$m_j = \mathbf{u}_j^T\mathbf{M}\mathbf{u}_j \quad , \quad k_j = \mathbf{u}_j^T\mathbf{M}\mathbf{u}_j \quad , \quad f_j = \mathbf{u}_j^T\mathbf{f}_{ext} \tag{14}$$

while f appears un-scaled because of the unit-normalization of the free mode shape in (10).

In the free limit, where $f = 0$ is realized by $D \to 0$ in (6), the solution to (13) must define the free natural frequency ω_j from the corresponding eigenvalue problem in (8). For

$f_j = 0$ and $f = 0$, the modal equation of motion (13) determines the free natural frequency $\omega = \omega_j$ as

$$\omega_j = \sqrt{\frac{k_j}{m_j}} \tag{15}$$

which corresponds to the well-known modal stiffness-to-mass ratio.

2.4. Residual Mode Correction

The damper force relation in (6) must be modified to give the damper force f as a function of (not u, but) the modal coordinate r_s for the target mode s. This modal damper force relation will inherently contain spill-over from other modes ($j \neq s$), simply because the placement of a local damper, presently on the first floor, violates the modal orthogonality conditions. Substitution of the modal damper deflection representation (12) into the damper force relation (6) gives

$$\left(\frac{1}{-\omega^2 m} + \frac{1}{D(\omega)}\right) f = r_s + \sum_{j \neq s} r_j = r_s - \underbrace{\sum_{j \neq s} \frac{1}{-\omega^2 m_j + k_j}}_{1/R(\omega)} f = r_s - \frac{f}{R(\omega)} \tag{16}$$

On the right-hand side of the first equality, the sum of r_j is separated into r_s for the target mode and a summation over all other (residual) modes $j \neq s$. The modal coordinates for these residual modes are replaced by the modal Equation (13) for $f_j = 0$, which must be accurate for isolated tuning to mode $j = s$. Finally, in (16), all dynamic residual mode terms have f as a common factor, whereby they can be used in a general frequency-dependent residual mode function $R(\omega)$, to be determined below.

When collecting all terms proportional to the damper force f on the left-hand side, the inerter Equation (16) can be reduced to

$$\left(\frac{1}{-\omega^2 m} + \frac{1}{D(\omega)} + \frac{1}{R(\omega)}\right) f = r_s \tag{17}$$

which constitutes the governing damper equation for the target mode $j = s$. The closed-loop equations for mode s are thus composed of the modal Equation (13) for $j = s$ and the inerter damper force Equation (17).

As argued in [18,44–46], the influence from residual modes may be split into two separate components, representing stiffness and inertia. For the residual mode function $R(\omega)$ in (16), this separation can be expressed as

$$\frac{1}{R(\omega)} \simeq \frac{1}{-\omega^2 m_s'} + \frac{1}{k_s'} \tag{18}$$

with m_s' and k_s' representing artificial inertia and stiffness components that account for the interaction with other modes ($j \neq s$). For the present inerter damper, the substitution of this $1/R$ into the force relation (17) gives

$$\left(\frac{1}{-\omega^2 \bar{m}_s} + \frac{1}{D(\omega)} + \frac{1}{k_s'}\right) f = r_s \tag{19}$$

which directly defines the combined inertance $\bar{m}_s$ by the reciprocal relation

$$\frac{1}{\bar{m}_s} = \frac{1}{m} + \frac{1}{m_s'} \tag{20}$$

In the case of vanishing influence from residual modes, the reciprocal inertia correction will vanish ($1/m'_s \rightarrow 0$), whereby the combined inertance recovers the front-end inertance, i.e., $\bar{m}_s \rightarrow m$. The modal damper model is shown in Figure 3, representing the modal damper relation (19) with combined inertance $\bar{m}_s$ and residual mode stiffness k'_s. In the upcoming Section 3, these two system parameters are determined for a pure viscous inerter damper with a damper function $D = i\omega c$.

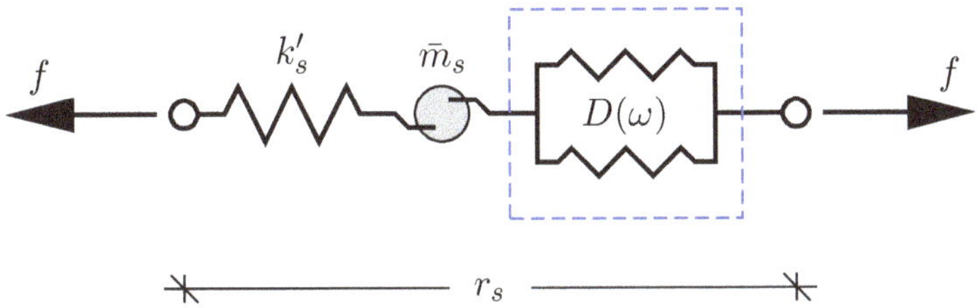

Figure 3. Model of mode s viscous inerter damper, with combined inertance $\bar{m}_s$ and additional spring stiffness k'_s to account for residual-mode flexibility.

For target mode $j = s$, the frequency response amplitude is conveniently derived from a frequency response function (FRF), which is obtained by eliminating f between (13) and (17). In normalized form, the FRF can then be expressed as

$$\frac{r_s k_s}{f_s} = \frac{[-\rho^2\bar{\mu}_s + \eta_s(\rho) - \rho^2\bar{\mu}_s\eta_s(\rho)/\kappa'_s]}{(-\rho^2+1)[-\rho^2\bar{\mu}_s + \eta_s(\rho) - \rho^2\bar{\mu}_s\eta_s(\rho)/\kappa'_s] - \rho^2\bar{\mu}_s\eta_s(\rho)} \tag{21}$$

when introducing

$$\rho = \frac{\omega}{\omega_s} \quad , \quad \bar{\mu}_j = \frac{\bar{m}_j}{m_s} \quad , \quad \eta(\rho) = \frac{D(\omega)}{k_s} \quad , \quad \kappa'_j = \frac{k'_j}{k_s} \tag{22}$$

as the frequency ratio (ρ), the effective mass ratio ($\bar{\mu}_j$), the normalized damper function (η), and the stiffness correction ratio (κ'_j). The normalization of the combined mass ratio $\bar{\mu}_s$ in (22) implies that (20) can also be written in the non-dimensional form

$$\frac{1}{\bar{\mu}_s} = \frac{1}{\mu} + \frac{1}{\mu'_s} \tag{23}$$

where the front-end and correction mass ratios

$$\mu = \frac{m}{m_s} \quad , \quad \mu'_j = \frac{m'_j}{m_s} \tag{24}$$

follow from the same definition as $\bar{\mu}_j$ in (22). The non-dimensional parameters and ratios defined in (22) and (24) are introduced in (21) and (23) for the target mode $j = s$.

3. Viscous Inerter Damper

The simplest damper function possible is the pure viscous dashpot, with

$$D(\omega) = i\omega c \tag{25}$$

in Figure 4. This simple series viscous inerter damper is shown attached to the present building model in Figure 1b, while in [16,17], it is referred to as a C2-type damper. By substitution of the viscous damper function in (25), the damper force relation (6) is given as

$$\left(\frac{1}{-\omega^2 m} + \frac{1}{i\omega c}\right) f = u \tag{26}$$

The present section will aim at a resonant design of this damper, where the extra resonance needed to create the resonance is not contained in the damper model, but instead obtained by controlled modal interaction between the two fundamental modes of the building structure.

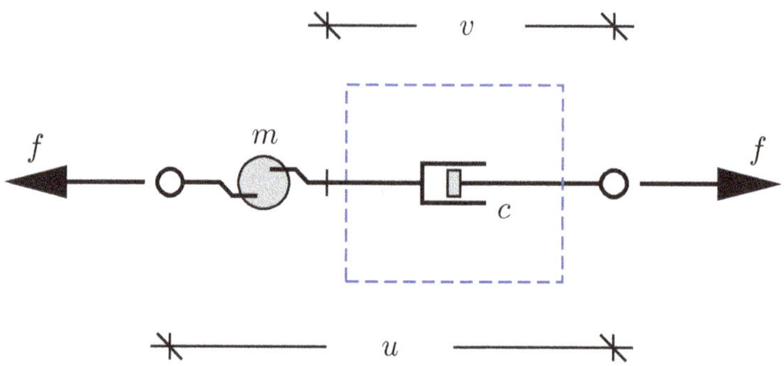

Figure 4. Inerter damper with front-end inertance m in series with viscous dashpot, represented by the damper function $D(\omega) = i\omega c$.

3.1. Single-Mode Model

For the pure viscous inerter damper in Figure 4, the non-dimensional damper function η in (22) can be expressed as

$$\eta = \frac{i\omega c}{k_s} = i\rho\beta \tag{27}$$

which introduces the non-dimensional damper ratio

$$\beta = \frac{c}{\sqrt{k_s m_s}} \tag{28}$$

Note that c is normalized by the modal mass and stiffness (and not the damper mass and stiffness). The damper function $\eta = i\rho\beta$ is then substituted into the modal frequency response function (21), which upon division of both numerator and denominator by $i\rho\beta$ can be written as

$$\frac{r_s k_s}{f_s} = \frac{\left[i\rho\dfrac{\bar{\mu}_s}{\beta} + 1 - \dfrac{\rho^2\bar{\mu}_s}{\kappa_s'}\right]}{(-\rho^2+1)\left[i\rho\dfrac{\bar{\mu}_s}{\beta} + 1 - \dfrac{\rho^2\bar{\mu}_s}{\kappa_s'}\right] - \rho^2\bar{\mu}_s} \tag{29}$$

with the damping parameter β appearing in its reciprocal form. This FRF describes a single-dof structure model with modal coordinate r_s and the augmented damper model in Figure 3 attached. This simplified structural model is shown in Figure 5, in which m_s' (contained in $\bar{m}_s$) and k_s' represent the interaction with the other modes ($j \neq s$). Thus, these correction parameters introduce the modifications needed to secure accurate modal results when reducing the system from the full mdof building mode in Figure 1b to the 2-dof (modal) model in Figure 5.

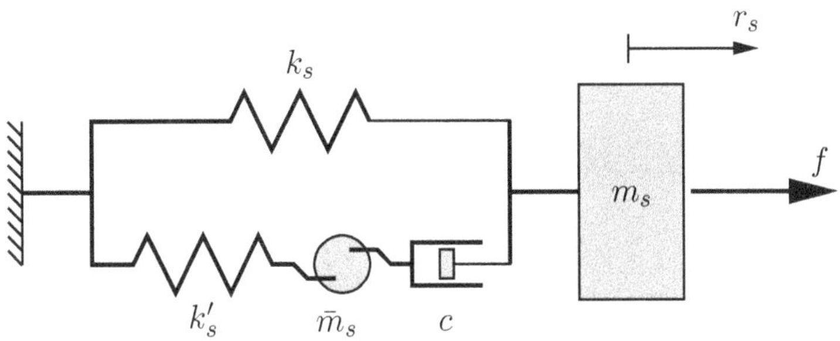

Figure 5. Resulting 2-dof model for mode $j = s$, combining the structural model with the modal coordinate r_s and the viscous inerter damper with a combined inertance $\bar{m}_s$ and additional spring stiffness k'_s accounting for the interaction with residual modes ($j \neq s$).

The correction coefficients m'_s (or $\bar{m}_s$) and k'_s are determined in the next section so that the system poles (complex natural frequencies) approach the correct (real-valued) natural frequencies in the clamped limit. The characteristic equation, which governs the system poles, is identified from the denominator of the FRF in (21) as

$$(-\rho^2 + 1)\left(1 - \rho^2 \frac{\bar{\mu}_s}{\kappa'_s}\right) - \rho^2 \bar{\mu}_s + i\rho \frac{\bar{\mu}_s}{\beta}(-\rho^2 + 1) = 0 \quad (30)$$

It is worth noticing that finite values of κ'_s will imply that this equation has two complex conjugate solution pairs. For the viscous inerter damper with a small front-end inertance, the damper will act as a viscous damper, whereby the stiffness correction term must vanish for the characteristic equation to only have a single pair of complex conjugate poles. However, in the present design, the front-end inertance is increased until coupling with an adjacent vibration mode allows two system poles to interact and thereby create apparent resonant damper properties.

Figure 6 shows the trajectories of the complex system roots for the full 20-dof building model when attaching a pure viscous inerter damper. The equations of motion and the corresponding eigenvalue problem that governs the complex poles are presented in detail in Section 4. The blue circles "Fine with blue font" on the real frequency axis represent the free natural frequency $\rho_j = \omega_j/\omega_1$. Thus, for mode $j = s = 1$, the blue circle is by definition located at $\rho = 1$ when $\eta = 0$. On the other hand, the blue squares represent the clamped natural frequencies at $\rho = \hat{\omega}_j/\omega_1 = \hat{\rho}_j$, which the complex trajectories reach when $\eta \to \infty$.

In the top subplot in Figure 6a, the front-end mass $m = 0.067m_1$ chosen is slightly too small, as the damper trajectory approaches its clamped limit (blue square) at around $\rho \simeq 1.3$. For this inertance, the mode 1 locus is very small with limited attainable damping because of the very limited modal interaction between the mode 1 locus and the damper locus. In the next subplot (b), the front-end mass increases to $m = 0.10m_1$, which clearly implies a great interaction with the mode 1 locus. It is seen that the two loci almost create a bifurcation point, which in the following is the tuning condition for the front-end inertance m. In the final subplot (c), the inertance is increased to $m = 0.15m_1$, whereby the damper branch terminates at $\rho \simeq 0.75$, with little modal interaction and thus limited attainable damping for the semi-circular mode 1 locus. This sequence of plots in Figure 6 indicates that the optimal tuning must aim at creating a bifurcation point between the two clamped modes $\hat{\rho}_1$ and $\hat{\rho}_2$, with the free root at $\rho_1 = 1$ as the intermediate root that places a backbone curve that approaches the imaginary axis along the unit circle. Thus, the present damper shows that resonant damping, as tuned for a TMD or TID in [9,18], can be realized without having a stiffness–inertance damper element.

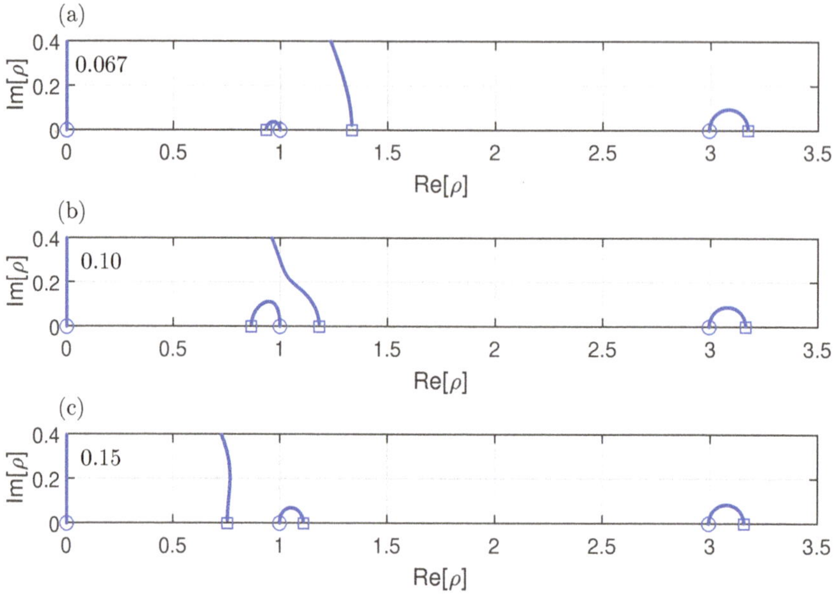

Figure 6. Root loci for $\mu = 0.067$ (**a**), 0.10 (**b**) and 0.15 (**c**). Free natural frequencies ω_j by circles and clamped natural frequencies $\hat{\omega}_j$ by squares.

3.2. Correction Coefficients

For the damper ratio $\beta \to 0$, the solution to the characteristic Equation (30) becomes $\rho = 0$ and $\rho = 1$. In Figure 6, the undamped pole (or normalized free natural) frequencies are depicted by circles, which are located at the origin, at $\rho = 1$ for mode 1 and then at $\rho = \omega_2/\omega_1 \simeq 3$ for the next vibration mode $j = 2$.

The aim of the tuning procedure is to place the first two clamped natural frequencies ($\hat{\rho}_1$ and $\hat{\rho}_2$) so that they create apparent resonant absorber characteristics around the first free natural frequency $\rho = 1$. Thus, the two solutions to the characteristic equation in (30) for $s = 1$ must be

$$\hat{\rho}_1 = \frac{\hat{\omega}_1}{\omega_1} \quad , \quad \hat{\rho}_2 = \frac{\hat{\omega}_2}{\omega_1} \tag{31}$$

when $\beta \to \infty$.

In the clamped limit $\beta \to \infty$, the last term (proportional to $i\rho$) in (30) vanishes, whereby the characteristic equation reduces to

$$\rho^4 - \rho^2\left(1 + \kappa_1' + \frac{\kappa_1'}{\bar{\mu}_1}\right) + \frac{\kappa_1'}{\bar{\mu}_1} = 0 \tag{32}$$

with subscript $s = 1$ indicating that the damper design targets the first vibration mode of the building model. For a given value of the front-end inertance m, the clamped natural frequencies are determined by solving the clamped eigenvalue problem (9). Thus, the solutions $\rho = \hat{\rho}_1$ and $\hat{\rho}_2$ to (32) are known from (9).

For the present normalization of the clamped-limit characteristic Equation (32), the product of the roots $\hat{\rho}_1$ and $\hat{\rho}_2$ is equal to the constant term, while the sum is the linear coefficient with the opposite sign. These relations can be used to solve for the modal damper parameters as a function of the clamped natural frequencies. The expressions can be written as

$$\bar{\mu}_1 = \left(1 - \frac{1}{\hat{\rho}_2^2}\right)\left(\frac{1}{\hat{\rho}_1^2} - 1\right) \tag{33}$$

and

$$\kappa_1' = (\hat{\rho}_2^2 - 1)(1 - \hat{\rho}_1^2) \tag{34}$$

which are similar to the corresponding expressions for resonant inerter-based absorbers in [47]. When introducing $\bar{\mu}_1$ from (33), the correction mass ratio μ_1' is isolated in (23) as

$$\frac{1}{\mu_1'} = \frac{1}{\bar{\mu}_1} - \frac{1}{\mu} = \frac{1}{\left(1 - \frac{1}{\hat{\rho}_2^2}\right)\left(\frac{1}{\hat{\rho}_1^2} - 1\right)} - \frac{1}{\mu} \tag{35}$$

where the front-end mass ratio is $\mu = m/m_1$ is defined in (24).

3.3. Optimal Mass Ratio

As seen from the sequence of subplots in Figure 6, the optimal front-end inertance m is associated with the two branches meeting at a bifurcation point, similar to the tuning concept for a classic tuned mass or inerter damper [9,18]. This bifurcation point condition requires the two clamped roots $\rho = \hat{\rho}_1$ and $\hat{\rho}_2$ (squares) to be inverse points to the free root at $\rho = 1$ (intermediate circle), which mathematically can be expressed by the simple relation

$$\hat{\rho}_1 \hat{\rho}_2 = 1 \tag{36}$$

When this relation is squared, the product $\hat{\rho}_1^2 \hat{\rho}_2^2$ is actually the product of the two roots to the clamped characteristic Equation (32), whereby $\hat{\rho}_1^2 \hat{\rho}_2^2$ must equal the constant term in (32). This provides the optimality condition

$$\hat{\rho}_1^2 \hat{\rho}_2^2 = \frac{\kappa_1'}{\bar{\mu}_1} = 1 \qquad \Rightarrow \qquad \bar{\mu}_1 = \kappa_1' \tag{37}$$

which secures that the two complex roots meet at a bifurcation point.

The optimality criterion (37) is used to find the optimal front-end mass ratio μ by the relation

$$\frac{1}{\mu} = \frac{1}{\kappa_1'} - \frac{1}{\mu_1'} \tag{38}$$

where $\bar{\mu}_1$ has been eliminated by its definition in (23). If the residual mode ratios κ_1' and μ_1' are determined with sufficient accuracy in (34) and (35), the expression (38) can be used directly to determine the front-end inertance m that secures optimal resonant damper behavior. However, in general, the inertance $m = \mu m_1$ from (38) will differ from the m used initially to solve the clamped eigenvalue problem in (9). Thus, the tuning of m is an inherently iterative process, conveniently solved by the stepwise procedure in Table 1.

Table 1. Tuning procedure for optimal front-end inertance m.

(1)	Choose m and determine $\mu = m/m_1$
(2)	Solve (9) and determined $\hat{\rho}_1^2$ and $\hat{\rho}_2^2$ from (31)
(3)	Determine κ_1' from (34) and μ_1' from (35)
(4)	Update μ from (38) and $m = \mu m_1$
(5)	Repeat (2) to (4) until μ has converged

3.4. Optimal Damping

When the optimal front-end inertance is determined by the iterative procedure in Table 1, the optimal viscous coefficient c is chosen to optimize the attainable damping. In theory, the maximum damping is obtained exactly at the bifurcation point. However, as demonstrated in [9,18,43], the bifurcation point damper tuning results in interaction between the two modes associated with the first vibration form, which yields undesirable response amplification in FRF curves. Thus, for the present damper, the ratio β_{bif} at the bifurcation point is multiplied by $\sqrt{2}$ to achieve a flat plateau in the frequency response curves. The FRF curves for optimal and bifurcation point damper ratios are compared in the numerical example in Section 4.5.

For the optimal mass ratio tuning $\bar{\mu}_1/\kappa_1' = 1$ in (37), the characteristic Equation (30) is readily reduced to

$$\rho^4 - \rho^2(2+\bar{\mu}_s) + 1 + i\rho\frac{\bar{\mu}_s}{\beta}(-\rho^2+1) = 0 \tag{39}$$

This characteristic equation secures that the two roots are inverse points to unity when $\beta \to \infty$, i.e., at the clamped damper limit. When reducing β, the complex poles move along semi-circular paths, still as inverse points to the unit circle. At the particular value $\beta = \beta_{bif}$, the two roots meet at a bifurcation point, after which they split and (approximately) trace the unit circle, with the pole approaching the real axis being the less damped and thus the pole governing the structural dynamics. When locating the poles with sufficient distance to the bifurcation point to avoid the dynamic interference between the two modes, the desired characteristic equation can be written as

$$\rho^4 - \rho^2\left(2+8\zeta_*^2\right) + 1 + i\rho(4\zeta_*)(-\rho^2+1) = 0 \tag{40}$$

as derived in [9,18,43]. In this equation, ζ_* represents the attainable damping ratio introduced by the inerter damper to the target mode, which in this case is the fundamental mode $s = 1$.

The optimal $\beta = \beta_{opt}$ and the corresponding attainable damping ratio ζ_* are determined by one-to-one comparison between the actual characteristic Equation (39) for the present viscous inerter damper and the desired characteristic Equation (40). It follows directly from a comparison of the quadratic terms that the attainable damping ratio is

$$\zeta_* = \sqrt{\tfrac{1}{8}\bar{\mu}_1} \tag{41}$$

which indicates that the damping ratio scales with the square root of the combined mass ratio $\bar{\mu}_1$, in practice evaluated by (33). Subsequently, the optimal damper ratio β_{opt} is derived by comparison of the factors to the $(-\rho^2+1)$ parenthesis. This gives the optimal damper ratio

$$\beta_{opt} = \sqrt{\tfrac{1}{2}\bar{\mu}_1} = 2\zeta_* \tag{42}$$

as twice the attainable damping ratio. The tuning expressions (41) and (42) canonize the combined mass ratio $\bar{\mu}_1$ as the overall governing damper parameter that determines both tuning and performance.

4. Damping of the Building Model

The present section numerically investigates the proposed damper tuning, in which the viscous inerter damper is tuned to create modal coupling between the two fundamental modes of the 20 story model from [14,39], shown in Figure 1. In the equation of motion for the 20-floor model structure in [14,39], the mass matrix is diagonalized, while the stiffness matrix is a tri-band matrix based on assumed linear spring connections. Thus, the mass and stiffness matrix are composed as

$$\mathbf{M} = m_f\begin{bmatrix} 1 & & & \\ & 1 & & \\ & & \ddots & \\ & & & 1 \end{bmatrix}, \qquad \mathbf{K} = k_f\begin{bmatrix} 2 & -1 & & \\ -1 & 1 & -1 & \\ & -1 & \ddots & -1 \\ & & -1 & 1 \end{bmatrix} \tag{43}$$

In the subsequent simulations, the structural damping matrix is omitted, whereby all damping is introduced by the viscous inerter damper.

4.1. Modal Parameters

The analytical expression for the natural frequency of this particular matrix structure can be derived from the finite difference discretization of a string [48], which gives

$$\omega_j = 2\Omega_f \sin\left(\frac{\pi}{2}\frac{2j-1}{2n+1}\right) \tag{44}$$

as shown in [49]. In this expression, the reference angular frequency $\Omega_f = \sqrt{k_f/m_f}$ is the natural frequency of a single-story building, i.e., for $n = 1$. The floor mass is chosen as $m_f = 5000\,\mathrm{t} = 5 \times 10^6$ kg, whereafter the floor stiffness is determined from (44), which gives $k_f = 659 \times 10^6\,\mathrm{N/m}$ when choosing $\omega_1/(2\pi) = 0.14\,\mathrm{Hz}$ from [14,39]. The natural frequencies for the first 10 modes of the 20-dof building model are shown in the first row of Table 2. The second row presents the free modal mass for mode shapes normalized to unity deflection across the damper connection, i.e., $u_j = 1$. For the present damper location, this means a unit first-floor deflection for each of the mode shapes. Therefore, the present modal masses are substantially larger than the common modal mass values obtained by normalizing the maximum mode shape deflection to unity. Furthermore, the modal mass decreases for increasing mode number, as the first floor deflection becomes relatively larger for higher mode shapes.

Table 2. Normalized natural frequencies and modal masses for free (top row) and clamped (bottom row) conditions. Frequencies are normalized by the free mode 1 natural frequency $\omega_1 = 2\pi \times 0.14\,\mathrm{Hz} = 0.88\,\mathrm{rad/s}$, and modal masses are normalized by the free mode 1 modal mass $m_1 = 8746 \times 10^6$ kg. The front-end inertance $m = 919.4 \times 10^6$ kg is obtained by the design procedure in Table 1.

j	1	2	3	4	5	6	7	8	9	10
ρ_j	1.00	2.99	4.97	6.92	8.82	10.68	12.47	14.19	15.83	17.37
m_j/m_1	1.000	0.113	0.042	0.022	0.015	0.011	0.008	0.007	0.006	0.006
$\hat{\rho}_j$	0.853	1.17	3.16	5.23	7.27	9.26	11.2	13.1	14.8	16.5
$\hat{m}_j/m_1$	0.185	0.259	9.37	29.6	61.6	106.87	168.35	250.1	358.3	502.1

4.2. Tuning Parameters

The first overall task in the damper tuning is to determine the front-end damper inertance m using the iterative procedure in Table 1. In the present case, a first guess is chosen as $m = 0.05m_1$, i.e., 5% mode 1 modal mass. The iterative procedure is run for 10 iterations, with a relative change of less than 10^{-12} for $\bar{\mu}_1 = 0.1051$ in the final iteration. The final inertance is then obtained as $m = 919.4 \times 10^6$ kg, which results in the governing tuning parameters in Table 3.

Table 3. Output from tuning procedure in Table 1.

μ	$\bar{\mu}_1$	μ_1'	κ_1'	β_{opt}
0.1051	0.1025	4.1440	0.1025	0.2264

For the optimal front-end inertance m, the clamped natural frequencies $\hat{\omega}_j$ are obtained by solving the generalized eigenvalue problem (9). These frequencies are presented as $\hat{\rho}_j = \hat{\omega}_j/\omega_1$ for target mode $s = 1$ in the bottom half of Table 2. It is seen that $\hat{\rho}_1\hat{\rho}_2 = 1$, as required by the inverse point relation in (37a) and enforced by step (4) of the iterative tuning procedure in Table 1.

It is seen in Table 3 that the effective mass ratio $\bar{\mu}_1 = 0.1025$ is approximately equal to the actual mass ratio $\mu = 0.1050$, which indicates that the correction by $\mu_1' = 4.1440$ in (23) is seemingly small. Thus, when using the common mass ratio in the damper tuning, i.e., $\bar{\mu}_1 = \mu$, only moderate detuning is expected. In Table 3, the correction stiffness ratio $\kappa_1' = \bar{\mu}_1$, as required by (37), while the optimal damper ratio $\beta = \beta_{opt}$ from (42) secures a flat plateau in the frequency response curves. It is noted that the attainable damping is half of the damper ratio, which for the present case yields $\zeta_* = \frac{1}{2}\beta = 0.113$ for damping of the first vibration mode ($s = 1$).

4.3. Coupled Equations of Motion

As presented in Sections 4.4 and 4.5, the damping and mitigation performance of the viscous inerter damper is assessed by root locus and frequency response analyses. For that purpose, the present section introduces the coupled equations for the damped structure in Figure 1. The equations also include the stiffness k_0 in Figure 7, which vanishes ($k_0 = 0$) in the ideal viscous inerter damper but is included in the numerical example to avoid drift between inerter and dashpot.

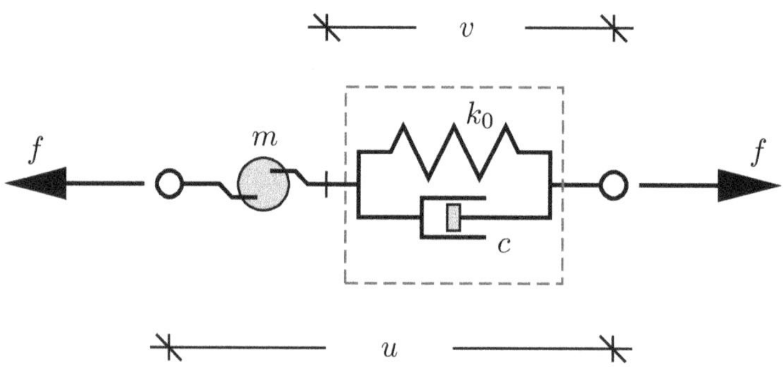

Figure 7. Inerter damper with front-end inertance m, in series with visco-elastic dashpot: $D(\omega) = k_0 + i\omega c$.

A complex frequency—or pole—is derived from the damped eigenvalue problem, governed by the coupled equations (5) for the structure and (26) for the viscous inerter damper. It is convenient to eliminate the force $f = -\omega^2 y$, with $y = u - v$ being the deformation across the inerter. Thereby, the governing Equations (5) and (26) can be written in coupled form as

$$\left(-\omega^2\bar{\mathbf{M}} + i\omega\bar{\mathbf{C}} + \bar{\mathbf{K}}\right)\bar{\mathbf{q}} = \bar{\mathbf{f}} \tag{45}$$

introducing the augmented matrices

$$\bar{\mathbf{M}} = \begin{bmatrix} \mathbf{M} & \mathbf{0} \\ \mathbf{0}^T & m \end{bmatrix}, \quad \bar{\mathbf{C}} = \begin{bmatrix} \mathbf{C} + c\mathbf{b}\mathbf{b}^T & -c\mathbf{b} \\ -c\mathbf{b}^T & c \end{bmatrix}, \quad \bar{\mathbf{K}} = \begin{bmatrix} \mathbf{K} + k_0\mathbf{b}\mathbf{b}^T & -k_0\mathbf{b} \\ -k_0\mathbf{b}^T & k_0 \end{bmatrix} \tag{46}$$

for the augmented displacement and load vector

$$\bar{\mathbf{q}} = \begin{bmatrix} \mathbf{q} \\ y \end{bmatrix}, \quad \bar{\mathbf{f}} = \begin{bmatrix} \mathbf{f}_{ext} \\ 0 \end{bmatrix} \tag{47}$$

where the latter assumes that the inerter damper is unexposed to external forcing. These augmented matrices correspond to those derived for a classic TMD, with y representing the motion of the auxiliary TMD mass.

The complex system poles (or natural frequencies) are governed by the coupled equation of motion (45) without external loading ($\bar{\mathbf{f}} = \mathbf{0}$). This eigenvalue problem is conveniently solved in the first-order state space format

$$(\mathbf{A} - i\omega\mathbf{I})\mathbf{z} = \mathbf{0} \tag{48}$$

where the system matrix and state-space vector are introduced as

$$\mathbf{A} = \begin{bmatrix} \mathbf{0} & \mathbf{I} \\ -\mathbf{M}^{-1}\mathbf{K} & -\mathbf{M}^{-1}\mathbf{C} \end{bmatrix} \quad , \quad \mathbf{z} = \begin{bmatrix} \bar{\mathbf{q}} \\ i\omega\bar{\mathbf{q}} \end{bmatrix} \tag{49}$$

It is noted that in (45) to (49), the entry $\mathbf{0}$ is the zero vector or matrix and $\mathbf{I}$ is the identity matrix that fit the given dimensions.

In Section 4.4, the first-order eigenvalue problem (48) is solved for a varying viscous coefficient c to trace the complex roots in a root locus diagram, as already applied in Figure 6. In Section 4.5, the frequency response curves are obtained by solving the frequency-domain equation of motion (45) for optimal tuning and varying (real-valued) angular frequency ω. These analyses will illustrate the ability of the viscous inerter damper to realize apparent resonant damping, with k_0 being sufficiently small to not interfere while still large enough to avoid any drift across the viscous dashpot.

4.4. Root Locus Analysis

As shown by the sequence of root locus diagrams in Figure 6, the optimal tuning of the viscous inerter damper is derived from a modal interaction that creates a bifurcation point in the complex frequency plane (in this case the normalized ρ-plane). For the optimal front-end inertance, obtained by the design procedure in Table 1 and summarized in Table 3, the complex poles $\rho = \omega/\omega_1$ are obtained by solving the eigenvalue problem (48) from $c \to 0$ to $c \to \infty$. Figure 8 shows the root locus diagram for the optimally tuned viscous inerter damper. By comparison with the almost optimal loci in Figure 6b for $\mu = 0.1000$, the optimal tuning curve in Figure 8 for $\mu = 0.1051$ almost realizes the desired bifurcation point. The circles represent the free natural frequencies, while the squares are the clamped natural frequencies. The tuning procedure in Table 1 secures that the product of the first two normalized roots is unity, according to the inverse point relation in (37).

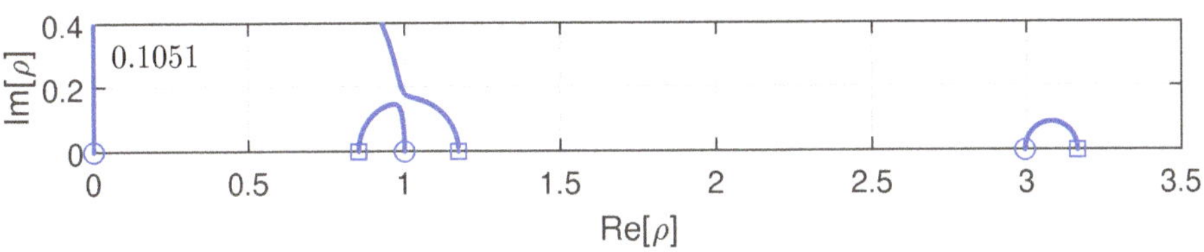

Figure 8. Root loci for inertance tuning to mode $s = 1$. The front-end mass ratio $\mu = 0.1051$ is obtained by the tuning procedure in Table 1.

Figure 9a shows a detailed plot of the semi-circular loci for the mode 1 roots. The circle represents the real-valued root for $\beta \to 0$. The other free limit root is located at the origin, as depicted by the circle at the origin in Figure 8. For increasing β, the root from $\rho = 1$ moves into the complex plane, approximately along the backbone unit circle, toward the bifurcation point. The other root leaves the origin along the imaginary axis, meets its conjugate pole decending from $\rho = i\infty$, whereby it follows the backbone unit circle toward the bifurcation point. In the present case, the root from $\rho = 1$ branches off before the ideal bifurcation point and traces the semi-circular path towards the smaller of the two clamped natural frequencies (blue square), while oppositely, the root from the imaginary axis follows the semi-circular path toward the larger clamped frequency.

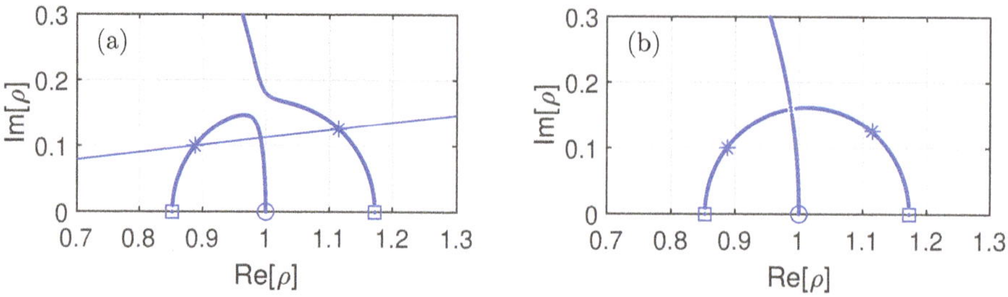

Figure 9. (**a**) Root loci for mode 1 with asterisk ($*$) depicting the roots for the optimal $\beta = \beta_{opt}$ by (42). (**b**) Ideal root locus diagram from the quartic polynomial in (39).

The ideal root locus diagram is obtained by solving the quartic polynomial in (39) with optimal $\bar{\mu}_1$ and plotted in Figure 9b. It is seen that for the idealized sdof model, without actual influence from other modes, the two semi-circular branches do meet exactly at a bifurcation point, where the damper parameter is $\beta_{bif} = \beta_{opt}/\sqrt{2} = \sqrt{\frac{1}{4}\bar{\mu}_1}$. Thus, the inability of the root loci for the 20-story shearframe building model in Figure 9b to exactly meet at a bifurcation point indicates the approximations applied in the representation of residual mode effects in Section 3.2.

In the root locus diagram in Figure 9, the asterisks represent the roots for optimal β from the design formula (42). The inclined straight line intersects the origin, whereby its inclination approximates the modal damping ratio. It is seen that both optimal roots are placed on the same line, which demonstrates a balanced damper design, where both roots associated with the target mode 1 are equally damped. The two damping ratios ζ_{1-} and ζ_{1+} at the two asterisks in Figure 9 are presented in Table 4 for $k_0 = 0$. It is seen that the two damping ratios for mode 1 and $k_0 = 0$ are identical (0.1126) to the fourth decimal and practically equal to $\zeta_* = 0.1132$, which validates the accuracy of the proposed inertance tuning procedure in Table 1.

Table 4. Damping ratios ζ_{1-} and ζ_{1+} for the two mode 1 poles for optimal tuning from Table 3. Prediction of maximum attainable damping: $\zeta_* = \frac{1}{2}\beta_{opt} = 0.1132$.

μ	$k_0 = 0$	$k_0 = 0.05\,\omega_1^2 m$	$k_0 = 0.1\,\omega_1^2 m$
ζ_{1-}	0.1126	0.1159	0.1190
ζ_{1+}	0.1126	0.1106	0.1084

As indicated in Figure 1c, the pure viscous inerter damper is in practice implemented with a spring in parallel to the dashpot to avoid drift in the damper stroke. It is important to note that in the present analysis, the stiffness k_0 of the spring is not chosen to create an additional resonance with the front-end inertance m and thus kept sufficiently small to avoid dynamic interaction with mode 1. As indicated in Table 4, the spring stiffness k_0 is chosen as 0%, 5%, and 10% of the stiffness needed for the isolated damper frequency $\omega_0 = \sqrt{k_0/m}$ to equal the free mode 1 frequency ω_1. Figure 10 shows the root loci for mode 1 for the three different choices of damper stiffness: $k_0 = 0$ (blue), $k_0 = 0.05\,\omega_1^2 m$ (magenta) and $k_0 = 0.1\,\omega_1^2 m$ (red). Thus, the blue trajectories are identical to those in

Figure 9. For finite values of k_0, the damper locus initiates from a small real-valued root (magenta and red circles) and not the origin (blue circle). For increasing values of β and sufficiently small values of k_0, the damper root approaches the imaginary axis and follows a path that is qualitatively similar to that explained for the pure viscous inerter damper in Figure 9. It is seen from the local mode 1 trajectories that increasing k_0 implies a slight (and further) deterioration of the bifurcation point property. To assess the reduction in attainable damping, the two damping ratios for optimal β are also presented in Table 4. It is seen that the damping ratio ζ_{1-} for the lower frequency branch increases with k_0, while conversely, the higher-branch damping ratio ζ_{1+} is reduced. Thus, the attainable damping (in this case the smaller ζ_{1+}) is reduced by 2% and 4% compared with the equal damping ratios for the pure viscous inerter damper (for $k_0 = 0$). Thus, including the low-stiffness spring k_0 has a limited effect on the overall damper performance, while it secures the mechanical centering of the damper during operation. In practice, it is therefore recommended to install a k_0-spring in parallel with the dashpot, which by itself would create a damper resonance (with m) a decade below the targeted natural frequency.

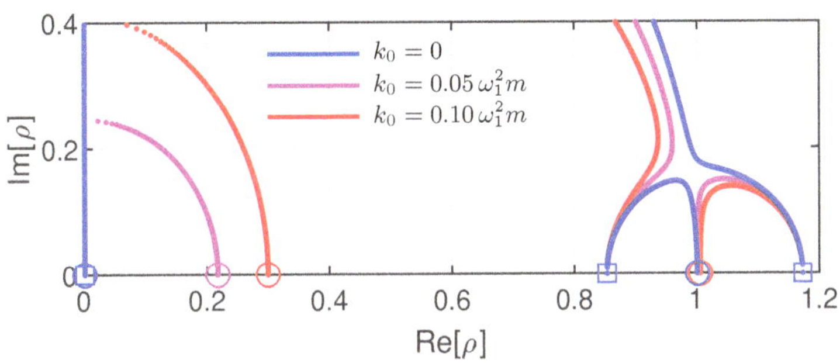

Figure 10. Root loci for mode 1 for $k_0 = 0$ (blue), $k_0 = 0.05\,\omega_1^2 m$ (magenta) and $k_0 = 0.1\,\omega_1^2 m$ (red).

4.5. Frequency Response Analysis

The present resonant tuning of the viscous inerter damper hinges on the creation of a bifurcation point in the complex root diagram; see Figures 8–10 in the previous section. Thus, the initial validation in Section 4.4 investigates the root loci for the target mode 1 for the pure viscous inerter damper ($k_0 = 0$) and for sufficiently small values of the restoring spring's stiffness k_0. However, in structural dynamics, the tuning of resonant dampers is most commonly assessed by the reduction in frequency amplitudes, i.e., by the inspection of the frequency response curves (or FRFs).

The ideal frequency response curves for the present resonant damping device are obtained from the sdof FRF in (21), in which the optimal damper ratio $\beta = \beta_{opt}$ is contained in $\eta = i\rho\beta$. These sdof FRF curves are shown in Figure 11. From the frequency displacement curve in Figure 11a it is seen that the equal damping calibration used in Section 3.3 to determine the optimal mass ratio will not provide a flat plateau in the FRF for the displacement variable. Instead, the curve is slightly inclined, with a larger low-frequency peak. However, the equal damping calibration yields a flat plateau for the structure velocity amplitude in Figure 11a. Although not shown, the acceleration amplitude will then have an opposite inclination to that for the displacement in (a), whereby the present tuning (with a flat velocity amplitude curve) might be a decent compromise between minimizing displacement to optimize load-carrying capacity and minimizing acceleration to optimize comfort.

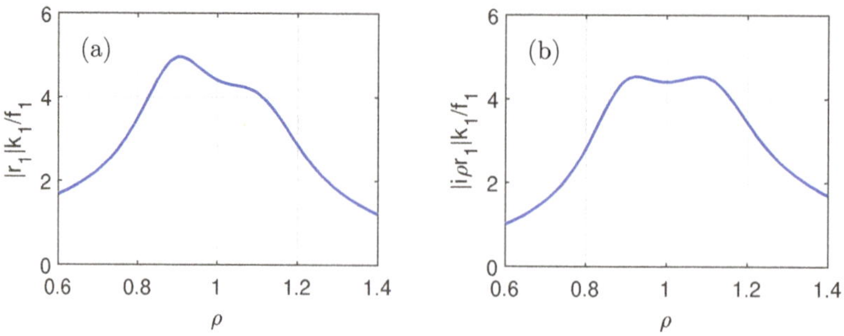

Figure 11. Frequency response functions for idealized sdof model from (21): (**a**) modal displacement amplitude and (**b**) modal velocity amplitude.

Figure 12 shows the frequency response curves for the full 20-story building model with the viscous inerter damper placed between the ground and first floor, as shown in Figure 1. The curves are obtained by solving the frequency equation of motion in (45) with evenly distributed forcing by

$$\mathbf{f}_{ext} = f_{ext}[1, 1, \ldots, 1]^T$$

where the scalar f_{ext} denotes the loading intensity. Since the response curves in Figure 12 are normalized by the corresponding static deflection $\mathbf{q}_0 = \mathbf{K}^{-1}\mathbf{f}_{ext}$, the actual value of f_{ext} is irrelevant in the present linear analysis.

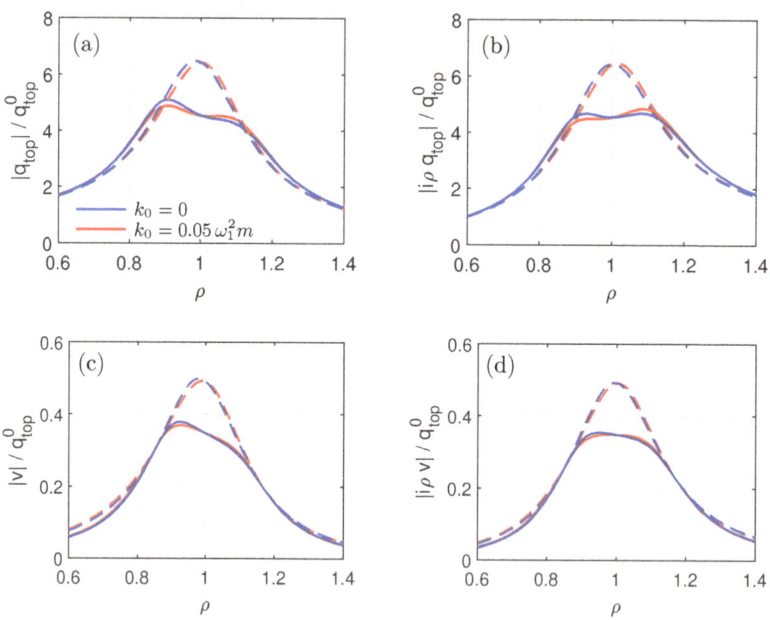

Figure 12. Frequency response functions for mode 1: (**a**,**b**) Top-floor motion q_{top}, (**c**,**d**) damper motion v, (**a**,**c**) displacement amplitude, and (**b**,**d**) velocity amplitude.

Figure 12a shows the top floor displacement amplitude, while (b) shows the corresponding velocity amplitude. The two bottom subfigures (c) and (d) show the displacement and velocity amplitudes for the damper stroke $v = u - y$ across the damper dashpot. The blue curves are for vanishing damper stiffness ($k_0 = 0$), while the red curves represent

$k_0 = 0.1\,\omega_1^2 m$ (red), as in the previous root locus diagrams. The solid-line curves represent the optimal damper parameter $\beta = \beta_{opt}$ from (42), while the dashed curves are obtained for the bifurcation-point damping at $\beta = \beta_{opt}/\sqrt{2}$. The response amplification associated with the dashed curves (for maximum damping) explains why a slightly increased damper ratio β (compared to β_{bif}) is typically chosen in a damper tuning that aims at minimizing response amplitudes.

It is seen by comparing the blue-solid top-row curves in Figure 12a,b with the ideal sdof-curves in Figure 11 that the curves for the full 20-story model are practically identical to the idealized curves, with a slight inclination for the displacement amplitude and an almost perfectly flat curve around resonance for the velocity amplitude. This comparison shows that the present tuning procedure, relying on the iterative scheme in Table 1, is very accurate for the present case with the inerter damper placed between ground and the first floor of the 20-story building model in Figure 1.

5. Conclusions

In the tuning and design of inerter-based vibration absorbers, the inclusion of correction for residual modes is important, as inerter dampers have the ability to impose sufficiently large changes in the expansion mode shape(s) to influence the tuning accuracy. In the present case, the very simple viscous inerter damper is introduced, for which the residual-mode inertia correction adjusts the effective mass ratio, while the corresponding stiffness corrections enable resonant tuning by modal interaction between the fundamental structural vibration modes. An iterative tuning procedure is derived for the optimal inertance based on the requirement of inverse clamped roots relative to the free (undamped) natural frequency. The present tuning—in its current form—directly hinges on the balanced pole placement calibration that secures equal modal damping, which means it must be altered if alternative tuning expressions or principles are to be used.

A 20-story building model is used as an example. The root locus and frequency response analysis show that the resonant damping characteristics are almost perfectly realized by the proposed tuning method. The bifurcation point is not entirely obtained, and a small inclination—relative to the optimal case—is noted for the frequency response curves. However, the resonant tuning of the non-resonant viscous inerter damper is a novel concept that (to the author's knowledge) has not been explored before.

For the present civil engineering application, the optimal mass ratio of $\mu = 0.1051$ is not extraordinarily large. However, it results in an extremely large inertance $m = 919.4 \times 10^6$ kg because the modal mass $m_1 = 8746 \times 10^6$ kg for the first vibration mode is almost 100 times larger than the total translational mass of the building itself ($20 \times 5 \times 10^6$ kg $= 100 \times 10^6$ kg). Note that the modal mass in (14) is much larger than the actual building mass because the normalization of the mode shape in (10) enforces a unit deflection at the bottom floor and not at the top floor, where the deflection is otherwise largest for mode $j = 1$. However, the inerters can be constructed with enormous gearing ratios, and although the present inerter can be placed on the ground, its size seems enormous, and it may require some development to make the concept feasible. However, if it can be constructed, the large mass ratio implies large attainable damping, which may reduce the construction material needed for reinforcement against vibrational loading. Finally, the robustness of the proposed damper tuning should be investigated in further detail, especially addressing the damper's sensitivity to changes in the inertance m, which for the present damper configuration must be chosen precisely to achieve the desired resonant damper characteristics.

Funding: This research received no external funding

Institutional Review Board Statement: Not applicable

Informed Consent Statement: Not applicable

Data Availability Statement: The original contributions presented in the study are included in the article, further inquiries can be directed to the corresponding author.

Conflicts of Interest: The authors declare no conflict of interest.

References

1. Den Hartog, J. *Mechanical Vibrations*, 4th ed.; McGraw-Hill: New York, NY, USA, 1956.
2. Warburton, G.B. Optimum absorber parameters for various combinations of response and excitation parameters. *Earthq. Eng. Struct. Dyn.* **1982**, *10*, 381–401. [CrossRef]
3. Rana, R.; Soong, T.T. Parametric study and simplified design of tuned mass dampers. *Eng. Struct.* **1998**, *20*, 193–204. [CrossRef]
4. Chang, C. Mass dampers and their optimal designs for building vibration control. *Eng. Struct.* **1999**, *21*, 454–463. [CrossRef]
5. Tuan, A.Y.; Shang, G. Vibration control in a 101-story building using a tuned mass damper. *J. Appl. Sci. Eng.* **2014**, *17*, 141–156.
6. Elias, S.; Matsagar, V. Research developments in vibration control of structures using passive tuned mass dampers. *Annu. Rev. Control* **2017**, *44*, 129–156. [CrossRef]
7. Elias, S.; Matsagar, V. Wind response control of tall buildings with a tuned mass damper. *J. Build. Eng.* **2018**, *15*, 51–60. [CrossRef]
8. Jafari, M.; Alipour, A. Methodologies to mitigate wind-induced vibration of tall buildings: A state-of-the-art review. *J. Build. Eng.* **2021**, *33*, 101582. [CrossRef]
9. Krenk, S. Frequency analysis of the tuned mass damper. *J. Appl. Mech.* **2005**, *72*, 936–942. [CrossRef]
10. Krenk, S.; Høgsberg, J. Tuned mass absorber on a flexible structure. *J. Sound Vib.* **2014**, *333*, 1577–1595. [CrossRef]
11. Smith, M. Synthesis of mechanical networks: The inerter. *IEEE Trans. Automat. Contr.* **2002**, *47*, 1648–1662. [CrossRef]
12. Marian, L.; Giaralis, A. The tuned mass-damper-inerter for harmonic vibrations suppression, attached mass reduction, and energy harvesting. *Smart Struct. Syst.* **2017**, *19*, 665–678.
13. Papageorgiou, C.; Smith, M.C. Laboratory experimental testing of inerters. In Proceedings of the 44th IEEE Conference on Decision and Control, Seville, Spain, 15 December 2005; pp. 3351–3356.
14. Weber, F.; Borchsenius, F.; Distl, J.; Braun, C. Performance of numerically optimized tuned mass damper with inerter (TMDI). *Appl. Sci.* **2022**, *12*, 6204. [CrossRef]
15. Lazar, I.; Neild, S.; Wagg, D. Using an inerter-based device for structural vibration suppression. *Earthq. Eng. Struct. Dyn.* **2014**, *43*, 1129–1147. [CrossRef]
16. Hu, Y.; Chen, M.Z. Performance evaluation for inerter-based dynamic vibration absorbers. *Int. J. Mech. Sci.* **2015**, *99*, 297–307. [CrossRef]
17. Hu, Y.; Chen, M.; Shu, Z.; Huang, L. Analysis and optimisation for inerter-based isolators via fixed-point theory and algebraic solution. *J. Sound Vib.* **2015**, *346*, 17–36. [CrossRef]
18. Krenk, S.; Høgsberg, J. Tuned resonant mass or inerter-based absorbers: Unified calibration with quasi-dynamic flexibility and inertia correction. *Proc. Math. Phys. Eng. Sci.* **2016**, *472*, 20150718. [CrossRef]
19. Krenk, S. Resonant inerter based vibration absorbers on flexible structures. *J. Frankl. Inst.* **2019**, *356*, 7704–7730. [CrossRef]
20. Pan, C.; Zhang, R.; Luo, H.; Li, C.; Shen, H. Demand-based optimal design of oscillator with parallel-layout viscous inerter damper. *Struct. Control Health Monit.* **2018**, *25*, e2051. [CrossRef]
21. Zhang, R.; Zhao, Z.; Pan, C.; Ikago, K.; Xue, S. Damping enhancement principle of inerter system. *Struct. Control Health Monit.* **2020**, *27*, e2523. [CrossRef]
22. Li, Y.; Lombardi, L.; De Luca, F.; Farbiarz, Y.; Blandon, J.J.; Lara, L.A.; Rendon, J.F.; Jiang, J.Z.; Neild, S. Optimal design of inerter-integrated vibration absorbers for seismic retrofitting of a high-rise building in Colombia. *J. Phys. Conf. Ser.* **2019**, *1264*, 012031. [CrossRef]
23. Shen, W.; Niyitangamahoro, A.; Feng, Z.; Zhu, H. Tuned inerter dampers for civil structures subjected to earthquake ground motions: Optimum design and seismic performance. *Eng. Struct.* **2019**, *198*, 109470. [CrossRef]
24. Barredo, E.; Rojas, G.; Mayén, J.; Flores-Hernández, A. Innovative negative-stiffness inerter-based mechanical networks. *Int. J. Mech. Sci.* **2021**, *205*, 106597. [CrossRef]
25. Chowdhury, S.; Banerjee, A.; Adhikari, S. The optimal design of negative stiffness inerter passive dampers for structures. *Int. J. Mech. Sci.* **2023**, *258*, 108551. [CrossRef]
26. Zhao, Z.; Zhang, R.; Jiang, Y.; De Domenico, D.; Pan, C. Displacement-dependent damping inerter system for seismic response control. *Appl. Sci.* **2019**, *10*, 257. [CrossRef]
27. Wang, Q.; Qiao, H.; De Domenico, D.; Zhu, Z.; Xie, Z. Wind-induced response control of high-rise buildings using inerter-based vibration absorbers. *Appl. Sci.* **2019**, *9*, 5045. [CrossRef]
28. Chen, B.; Zhang, Z.; Hua, X. Equal modal damping-based optimal design of a grounded tuned mass-damper-inerter for flexible structures. *Struct. Contr. Health Monit.* **2022**, *29*, e3106. [CrossRef]

29. De Angelis, M.; Giaralis, A.; Petrini, F.; Pietrosanti, D. Optimal tuning and assessment of inertial dampers with grounded inerter for vibration control of seismically excited base-isolated systems. *Eng. Struct.* **2019**, *196*, 109250. [CrossRef]
30. Sun, H.; Zuo, L.; Wang, X.; Peng, J.; Wang, W. Exact H_2 optimal solutions to inerter-based isolation systems for building structures. *Struct. Contr. Health Monit.* **2019**, *26*, e2357. [CrossRef]
31. Jiang, Y.; Zhao, Z.; Zhang, R.; De Domenico, D.; Pan, C. Optimal design based on analytical solution for storage tank with inerter isolation system. *Soil Dyn. Earthq. Eng.* **2020**, *129*, 105924. [CrossRef]
32. Jangid, R.S. The role of a simple inerter in seismic base isolation. *Appl. Sci.* **2024**, *14*, 1056. [CrossRef]
33. Marian, L.; Giaralis, A. Optimal design of a novel tuned mass-damper–inerter (TMDI) passive vibration control configuration for stochastically support-excited structural systems. *Probabilistic Eng. Mech.* **2014**, *38*, 156–164. [CrossRef]
34. Giaralis, A.; Marian, L. Use of inerter devices for weight reduction of tuned mass-dampers for seismic protection of multi-story building: The Tuned Mass-Damper-Interter (TMDI). In Proceedings of the SPIE Smart Structures and Materials + Nondestructive Evaluation and Health Monitoring, Las Vegas, NV, USA, 20–24 March 2016.
35. Giaralis, A.; Petrini, F. Wind-induced vibration mitigation in tall buildings using the tuned mass-damper-inerter. *J. Struct. Eng.* **2017**, *143*. 04017127. [CrossRef]
36. Giaralis, A.; Taflanidis, A. Optimal tuned mass-damper-inerter (TMDI) design for seismically excited MDOF structures with model uncertainties based on reliability criteria. *Struct. Contr. Health Monit.* **2018**, *25*, e2082. [CrossRef]
37. De Domenico, D.; Ricciardi, G. An enhanced base isolation system equipped with optimal tuned mass damper inerter (TMDI). *Earthq. Eng. Struct. Dyn.* **2018**, *47*, 1169–1192. [CrossRef]
38. De Domenico, D.; Impollonia, N.; Ricciardi, G. Soil-dependent optimum design of a new passive vibration control system combining seismic base isolation with tuned inerter damper. *Soil Dyn. Earthq. Eng.* **2018**, *105*, 37–53. [CrossRef]
39. Weber, F.; Huber, P.; Borchsenius, F.; Braun, C. Performance of TMDI for tall building damping. *Actuators* **2020**, *9*, 139. [CrossRef]
40. Barredo, E.; Blanco, A.; Colín, J.; Penagos, V.; Abúndez, A.; Vela, L.; Meza, V.; Cruz, R.; Mayén, J. Closed-form solutions for the optimal design of inerter-based dynamic vibration absorbers. *Int. J. Mech. Sci.* **2018**, *144*, 41–53. [CrossRef]
41. Liu, C.; Chen, L.; Lee, H.P.; Yang, Y.; Zhang, X. A review of the inerter and inerter-based vibration isolation: Theory, devices, and applications. *J. Frankl. Inst.* **2022**, *359*, 7677–7707. [CrossRef]
42. Ma, R.; Bi, K.; Hao, H. Inerter-based structural vibration control: A state-of-the-art review. *Eng. Struct.* **2021**, *243*, 112655. [CrossRef]
43. Krenk, S.; Høgsberg, J. Equal modal damping design for a family of resonant vibration control formats. *J. Vib. Control* **2013**, *19*, 1294–1315. [CrossRef]
44. Høgsberg, J.; Lossouarn, B.; Deü, J.F. Tuning of vibration absorbers by an effective modal coupling factor. *Int. J. Mech. Sci.* **2024**, *268*, 109009. [CrossRef]
45. Høgsberg, J.; Krenk, S. Calibration of piezoelectric RL shunts with explicit residual mode correction. *J. Sound Vib.* **2017**, *386*, 65–81. [CrossRef]
46. Raze, G.; Dietrich, J.; Kerschen, G. Tuning and performance comparison of multiresonant piezoelectric shunts. *J. Intell. Mater. Syst. Struct.* **2022**, *33*, 2470–2491. [CrossRef]
47. Høgsberg, J. Consistent frequency-matching calibration procedure for electromechanical shunt absorbers. *J. Vib. Control* **2020**, *26*, 1133–1144. [CrossRef]
48. Géradin, M.; Rixen, D. *Mechanical Vibrations: Theory and Application to Structural Dynamics*, 2nd ed.; John Wiley: Chichester, UK, 1997.
49. Høgsberg, J.; Krenk, S. Linear control strategies for damping of flexible structures. *J. Sound Vib.* **2006**, *293*, 59–77. [CrossRef]

applied sciences

MDPI

Article

Probabilistic Modeling of Congested Traffic Scenarios on Long-Span Bridges

Xuejing Wang [1,2], Xin Ruan [1], Joan R. Casas [2,*] and Mingyang Zhang [3]

1 Department of Bridge Engineering, Tongji University, Shanghai 200092, China; 1710011@tongji.edu.cn (X.W.); ruanxin@tongji.edu.cn (X.R.)
2 Department of Civil and Environmental Engineering, Universitat Politècnica de Catalunya (UPC-BarcelonaTech), Jordi Girona 1-3, North Campus, 08034 Barcelona, Spain
3 College of Civil and Transportation Engineering, Shenzhen University, Shenzhen 518060, China; zhang601@szu.edu.cn
* Correspondence: joan.ramon.casas@upc.edu; Tel.: +34-934016513

Abstract: This paper aims to extend a previously developed probabilistic model for simulating extreme response scenarios to include congested traffic flow on long-span bridges, addressing the challenge of accurately modeling traffic loads under changing conditions. While the model was initially designed for free-flow traffic, this study demonstrates how it can be adapted for congested conditions, with the objective of improving the accuracy of traffic load models. To overcome the limitation of traditional Weigh-in-Motion (WIM) systems in capturing congested traffic, congested flow characteristics were inferred from available free-flow data. The cellular automata (CA) method was applied to generate realistic congested traffic scenarios, which were used as input for the probabilistic model. Key simulation parameters, such as cell length and vehicle weight distribution, were adjusted to reflect congested conditions. The results validate the model's flexibility, showing how, with the adaptation of some parameters, it can simulate both free-flow and congested traffic patterns effectively. This research provides a basis for improving traffic load models used in the design and assessment of long-span bridges, addressing the current limitations in existing codes and standards.

Keywords: traffic simulation; congested traffic; extreme response scenario; long-span bridges; traffic load models

1. Introduction

The aging of bridge infrastructure is a global issue. In the US, bridges are, on average, 50 years old, with over 46,000 in poor condition, requiring major rehabilitation or replacement [1,2]. North American bridges, largely built between 1960 and 1980, are also showing significant signs of wear [3,4], while European bridges constructed between 1970 and 2001 face similar challenges [5,6]. Meanwhile, the increasing freight demands, with the USDOT projecting a 40% rise in freight volume by 2045, are further stressing these aging infrastructures, which were not designed to handle such heavy traffic [7]. In China, the rapid growth of the transportation sector, with a freight traffic increase of 6% annually, has exacerbated vehicle overloading issues, pushing traffic loads beyond design capacities [8–15]. Overloading has become a leading cause of bridge failures worldwide, as revealed by surveys of over 4500 bridge failures from 2010 to 2016 [16–20]. This situation highlights the urgent need for structural reassessments to ensure the safety of existing bridge infrastructure under modern traffic demands [2,6].

In this context, the issue of traffic loading on long-span bridges becomes particularly pronounced. As a critical loading traffic situation for long-span bridges, congested flow traffic is important in the design and assessment of bridge structural safety. Most of the current load codes, which are primarily developed for short- and medium-span bridges, are not suitable for the design and assessment of long-span bridges [11,21,22]. Additionally, the increasing volume of traffic often leads to severe congestion on long-span bridges, posing threats to both human safety and structural performance. For short-span bridges, maximum traffic load effects are typically governed by one single heavy vehicle or several heavy vehicles side-by-side. However, for long-span bridges, as the span length increases, the number of vehicles that can be accommodated on the bridge deck also significantly increases, thereby complicating vehicle load management and safety assessments [11,23,24]. Numerous studies have indicated that traffic congestion constitutes the most critical scenario for long-span bridges [25–27]. However, the majority of WIM systems that are utilized globally to gather traffic data are unreliable when congestion appears [28]. In the research of Micu et al. [28], the free-flow traffic of WIM data are used to modeling congested flow traffic by collapsing the inter-vehicle gaps to minimum values.

Traditional congested traffic modeling methods can simulate the congestion on bridge decks very well [29–31]. However, not all congestion scenarios can cause structural effects to reach extreme values. Based on common sense, congested traffic with many stationary vehicles on the span is the critical situation to find the maximum traffic load on the bridge. However, this cannot be the case for maximum traffic effects, as these will depend on the shape of the influence line of the considered load effect. The congested traffic situation also applies traffic load in the areas of the influence line with an effect opposite to the one under analysis. It should be noted that to find the worst loading scenarios for a specific load effect, the term "long-span bridge" is not always equivalent to a "long-loaded length", where the latter is defined as the part of the bridge that should be loaded to find the worst load effect. As an example, regarding the maximum tensile force in the main cable of a suspension bridge, the worst traffic condition will be congestion over the whole length (in this case "long-span" and "long-loaded length" are equivalent). However, in looking to the maximum positive bending moment at the mid-span of a cable-stayed bridge and taking into account the shape of the influence line, as shown in Figure 1, scenarios where vehicles are only present near the mid-span may lead to the worst effect. In this case, a long span is not equivalent to a long-loaded length from the point of view of the worst loading effect.

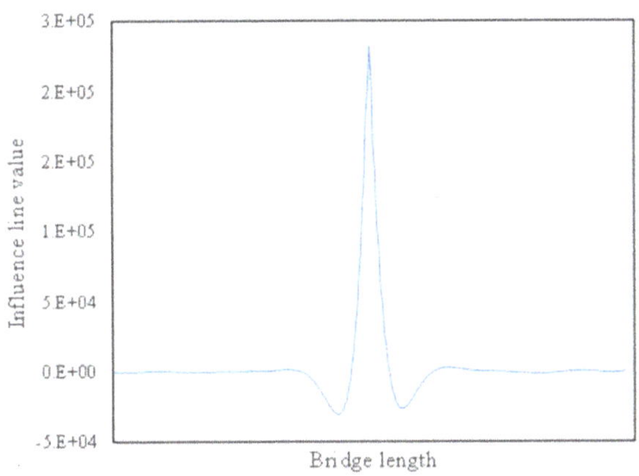

Figure 1. Influence line of bending moment at mid-span of cable-stayed bridge.

The filtering effect of the influence line may convert a long-span bridge into a short-loaded length bridge from the point of view of deriving the maximum load effects. In this sense, it is relevant to cite one of the conclusions from the thesis of Lipari [32]: "The queue of vehicles at a standstill, widely used in previous research, is not always the most critical congested state for long-span bridge loading (Section 2). This is due to the fact that full stop queues involve only one realization of vehicles per congestion event, which may actually reduce the probability of finding an extreme loading scenario". Therefore, there exists a research gap for the simulation of extreme loading scenarios of congested traffic on long-span bridges.

To address this research gap, this study investigated and simulated extreme response scenarios of congested traffic on long-span bridges using an extreme scenario simulation probabilistic model previously developed by the authors for free-flowing traffic [33]. In fact, effectively simulating extreme scenarios of congested traffic first requires a substantial number of extreme scenario samples, which need to be selected from a larger pool of normal congested traffic scenarios. As WIM (Weigh-in-Motion) traffic data are not representative of congested traffic scenarios, currently, congestion traffic data are typically obtained through bridge-mounted camera systems. Due to the slow flow speed of congested traffic, this method of data acquisition often results in low efficiency and limited samples. According to the research of Micu et al. [28], congested traffic can be obtained from jammed traffic images. In this study, the inherent correlation between congested and free-flow traffic at the same location was explored, and the characteristics of congested flow from those of free flow were deduced, allowing the relevant information contained in the WIM data to be fully explored. Then, the congested flow was modeled using the cellular automata (CA) approach, providing the simulated samples of the previous proposed extreme response scenario [33].

The framework of this paper is as follows: Section 2 provides a detailed review of the methodologies employed to develop the probabilistic model for simulating extreme response scenarios, outlining its theoretical foundations; Section 3 deals with an exploration of how the CA method is utilized to simulate congested traffic, detailing the process and the adaptations made for accurate representations of traffic congestion; Section 4 analyzes the application of the extreme response scenario method to model the impact of congested traffic loading on infrastructure, particularly focusing on the process and results of simulating these conditions; Section 5 provides a focused discussion on the key findings of the study, emphasizing the practical implications of the probabilistic model for evaluating extreme response scenarios on long-span bridges under varying traffic conditions; and Section 6 summarizes the key findings of the research, discussing the effectiveness and adaptability of the extreme response scenario simulation method and highlighting potential areas for future research to enhance the accuracy and efficiency of traffic simulations to obtain maximum load effects in long-span bridges.

2. Review of Probabilistic Model of Extreme Response Scenarios in Long-Span Bridges

A probabilistic model was derived and fully explained in a previous publication by the authors [33]. The application to the case of free-flow traffic can also be found in [33]. The present study demonstrates how the model can also be applied to the case of congested traffic. Therefore, only the main characteristics of the model are described here as the full description can be obtained in [33].

When obtaining extreme values of load effects on long-span bridges, the possible vehicle distribution on the bridge deck can vary significantly. To simulate the vehicle distribution pattern on the bridge deck during these extreme scenarios, the bridge deck was divided into cells of a certain length. Given that the number of vehicles on the bridge deck during extreme scenarios is relatively high and that lighter vehicles contribute minimally to the extreme effect values, they were omitted in the modeling process to simplify the simulation and more clearly study the distribution characteristics of heavy vehicles. To further facilitate the simulation, it was assumed that the centroid of each heavy vehicle was located at the center of the nearest bridge deck cell in the same lane. The details of the extreme response scenarios simulation method were already explained in [33]. Moreover, the driving habits of heavy vehicles also significantly impact their distribution, where heavy vehicles tend to cluster in the outer lanes. Therefore, the distribution of heavy vehicles across different lanes should be simulated during the modeling process. Two key factors needed to be considered when modeling the extreme response scenarios on the bridge deck: heavy vehicle weights and positions within the deck. They are described in the following subsections.

2.1. Heavy Vehicle Weights and Correlations

Since the distribution levels of heavy vehicle weights vary across different lanes, parameter fitting and simulation were conducted separately for the outer lanes and the inner lanes during the extreme scenario modeling process. The vehicle weight distribution shows clear multimodal characteristics that cannot be described with typical unimodal probability distribution models [27,34]. To better fit the multimodal characteristics of the weights, the Gaussian Mixture Model (GMM), which can accurately capture multiple peak attributes, was applied in this study [35].

The clustering of consecutive heavy vehicles in the same lane is a key factor contributing to extreme load effects, posing a threat to the structural safety of bridges. To simulate this clustering effect, it was assumed that there was a certain correlation between the weights of heavy vehicles in adjacent cells. The Nataf transformation method, combined with the GMM distribution, was introduced to simulate this correlation [36]. The Nataf transformation utilizes the marginal probability density functions of the variables, the correlation coefficient matrix, and the multivariate standard normal distribution to construct a joint probability density function through the Gaussian Copula function. This process converts correlated non-normal variables into independent standard normal variables. In this method, the weight correlation over a certain length is characterized by the Pearson correlation coefficient:

$$\rho_{j,k} = \frac{\alpha_{j,k}}{\lambda_j \cdot \lambda_k} \tag{1}$$

where $\rho_{j,k}$ represents the correlation coefficient of the total weights of heavy vehicles at positions j and k, while $\alpha_{j,k}$ corresponds to the covariance between the total weights of the two heavy vehicles. λ_j and λ_k denote the standard deviations of the total weights of heavy vehicles at positions j and k, respectively. It is evident that the correlation coefficient $\rho_{j,k}$ between the total weights of two heavy vehicles weakens as the distance between them increases.

2.2. Heavy Vehicle Positions

Stochastic process models have been widely used for modeling vehicle positions, such as compound Weibull processes and compound Poisson processes. In the study of extreme response scenarios, a suitable probability distribution needs to be chosen to simulate the occupancy state of cells occupied by heavy vehicles on specific lanes. The Poisson distribution, commonly used in statistics, describes the number of expected events occurring within a specific period or area and is used to simulate the number of occurrences of a specific event within a specific area [33].

Based on the statistical results of the number of cells occupied by heavy vehicles on each lane, a simple and reasonable single-lane heavy vehicle spatial distribution model was established using the Poisson point process. In assuming that the number of cells occupied by heavy vehicles within the loading length L, the probability of t can be determined as follows:

$$P(T_L = t) = \frac{(\lambda L)^t exp(-\lambda L)}{t!}, \quad t = 0,\ 1,\ 2, \ldots \tag{2}$$

where $P(\cdot)$ represents the probability of the event occurring, and λ denotes the average occurrence rate of vehicle occupancy in the cells. For a stationary heavy vehicle distribution process, λ remains constant over the entire loading length.

In addition to the stationary process, the position distribution of heavy vehicles can also be considered a non-stationary process, where the occurrence rate λ varies along the loading length. For a given load effect, the occurrence rate of heavy vehicles occupying a cell can be represented by the Poisson parameter $\lambda(l)$. This parameter is assumed to vary continuously with the loading position, and this variation is directly related to the shape of the influence line for the effect being considered (loading vehicles near the region of the influence line's maximum value tends to yield extreme effects, thus the probability of a cell being occupied by a heavy vehicle in extreme scenarios is correlated with the shape of the influence line). Since the occurrence probability is determined by the position of the cells along the loading length, $\lambda(l)$ can be calculated using the following equation:

$$\lambda(l) = \lambda_0(l) \cdot \theta \tag{3}$$

where $\lambda_0(l)$ represents the average occurrence rate of a cell being occupied by a heavy vehicle, and θ reflects the variation in this parameter along the loading position. The observation of maximum load effects in long-span bridges has shown that the positions of the heavy vehicles are better reflected with the non-stationary Poisson process. Therefore, this process is considered in the probabilistic model of extreme response scenarios.

2.3. Simulation of Extreme Response Scenario

According to Chen et al. [37], the average total axle length of heavy vehicles in China is 15.13 m (from the first axle to the last axle). Considering the length of the front and rear trailers and the necessary following distance during driving, the bridge deck cell length in the authors' previous study was set to 20 m for free-flow traffic.

The specific simulation process was as follows: (1) The bridge deck is divided into cells of 20 m within lanes. (2) Based on WIM data and specific load effects (influence lines) of the investigated bridge, extreme scenario samples are obtained. (3) The heavy vehicle weight distribution parameters of the sample scenarios and the parameters of the non-stationary Poisson process of heavy vehicle positions are calculated. (4) For each lane, a heavy vehicle weight is assigned to each cell using the Nataf transformation and GMM fitting methods. (5) Based on the Poisson process and the non-stationary characteristics of the uneven distribution of heavy vehicles, whether each cell on the lane is occupied by a heavy vehicle or not is simulated. (6) if the cell is occupied, the heavy vehicle weight is that weight obtained from step (4); if the cell is not occupied, the weight is 0. (7) Repeat the above steps for each lane to obtain a set of extreme response scenarios. In applying the Monte Carlo method, multiple sets of simulated extreme response scenarios can be obtained.

3. Congested Traffic Simulation Based on CA Method

For a certain bridge, the site-specific traffic is dynamic, the traffic volume, percentage of heavy vehicles and other characteristics are constantly changing. Disregarding special factors such as lane maintenance, car accidents, etc., it is suggested that congested flow and free flow at the same specific site are interchangeable and somehow intrinsically related as a free-flow situation can automatically result in congestion because of an increment in the traffic volume. In this section, the intrinsic correlation between congested and free flow at the same site is discussed, and the traffic characteristics of congested flow are inferred based on free flow. The congested flow is then modeled based on the CA approach [38].

To simulate the congested traffic flow on long-span bridges, the traffic flow WIM data at the site of a real operational long-span cable-stayed bridge in China, with a span length of 1088 m, was investigated, as shown in Figure 2. From the WIM measurement of 423 days of free-flow traffic, 423 daily extreme response scenarios (i.e., the maximum-per-day event of the effect of the cable force of the investigated bridge) of free flow were obtained from WIM data as the sample for the further simulation of extreme response scenario in free-flow conditions. The point in time at which these 423 samples achieved the maxima (i.e., the response extremes achieved at the points of the day, on a 24 h basis) were further counted, as shown in Figure 3. It can be seen that most of the extreme response scenarios occurred between 20:00 pm and 7:00 am the next day, and only a small number of extreme response scenarios (86 in total) occurred during the daytime hours. Common sense dictates that trucks tend to travel at night. At the same time, almost all large cities in the vicinity of the WIM site are closed to truck traffic during daytime hours, even though there are no restrictions on truck travel times on highways where the WIM station is located. For example, in Shanghai, trucks are banned from the roads in the city area from 07:00 am to 20:00 pm daily. It can be seen that for the daily state free traffic flow, the daily extremes tend to occur more often at night (i.e., with a higher proportion of heavy vehicles and lower traffic volume), when traffic is less likely to be congested.

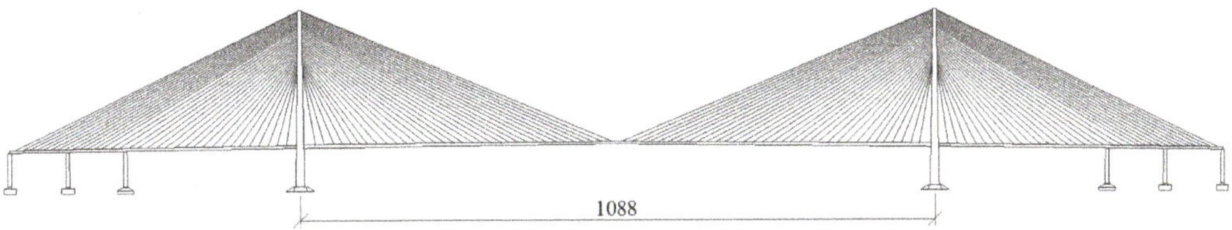

Figure 2. Longitudinal layout of the investigated bridge (m).

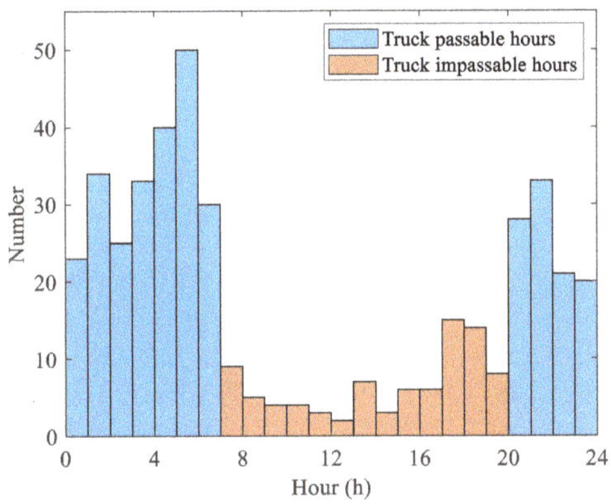

Figure 3. Histogram of occurrence times of free-flow extreme response scenarios.

The 86 extreme response scenario samples that achieved extreme response values during the daytime mentioned above were further analyzed. For these samples, the ratio of extreme traffic response to the modeled traffic load in the Chinese D60 code were calculated, and simultaneous hourly traffic volumes were counted, as plotted in Figure 4. The attributes of each extreme response scenario sample are identified with a marked shape. For example, circles indicate that the samples belong to holidays and weekends. The results reveal that extreme response scenarios with high hourly traffic flow typically have low ratios. The highest hourly traffic flow is 4083 veh/h, with a ratio of the response value to the Chinese code D60 of only 0.165. Nonetheless, higher hourly traffic volumes pose a potential risk for congestion in a free traffic flow scenario.

To examine the traffic composition in the case of higher hourly traffic volumes, the gross vehicle weight (GVW) information associated with the extreme response scenarios of hourly traffic volumes above 3000 veh/h was obtained. For each high-traffic-volume extreme response scenario, the GVW information within one hour before and after the moment of extreme values was examined, as shown in Figure 5a. For comparison, Figure 5b shows the histogram of the GVW when the extreme response values are high. It can be seen that the GVW tends to concentrate in vehicles weighing less than 4 tonnes for high-traffic-volume extreme response scenarios (Figure 5a). Meanwhile, the number of vehicles weighing between 4 and 66 tonnes are relatively rare for high-traffic-volume extreme response scenarios but very important for high extreme responses (Figure 5b). It should be noted that the histograms in Figure 5a,b correspond to the GVW distribution for extreme response scenarios with different traffic volumes. The histogram of the GVW distribution of all the vehicles of the WIM data can be seen in Figure 5c. In comparing histograms in Figure 5b,c, it can be noticed that when the value of the daily extreme response is larger, the level of the GVW distribution in the extreme response scenario is significantly higher compared to the GVW of the global WIM data. On the other hand, when the daily extreme response is obtained with a larger traffic volume, the level of GVW distribution in the extreme response scenario is significantly smaller compared to the GVW of the WIM data (compare Figure 5a and Figure 5c).

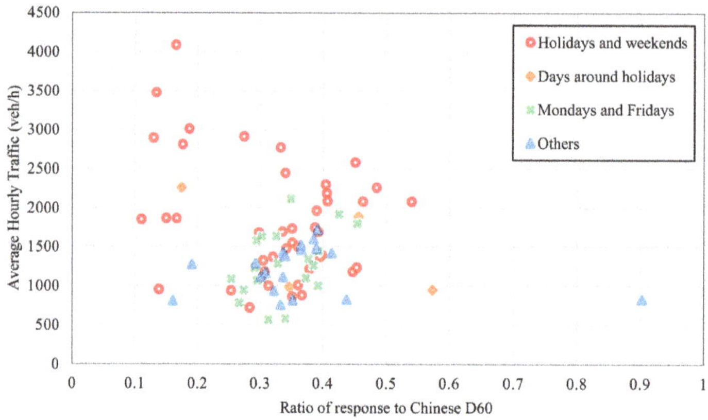

Figure 4. The average hourly traffic and ratio of response to D60 for extreme response scenarios during daytime.

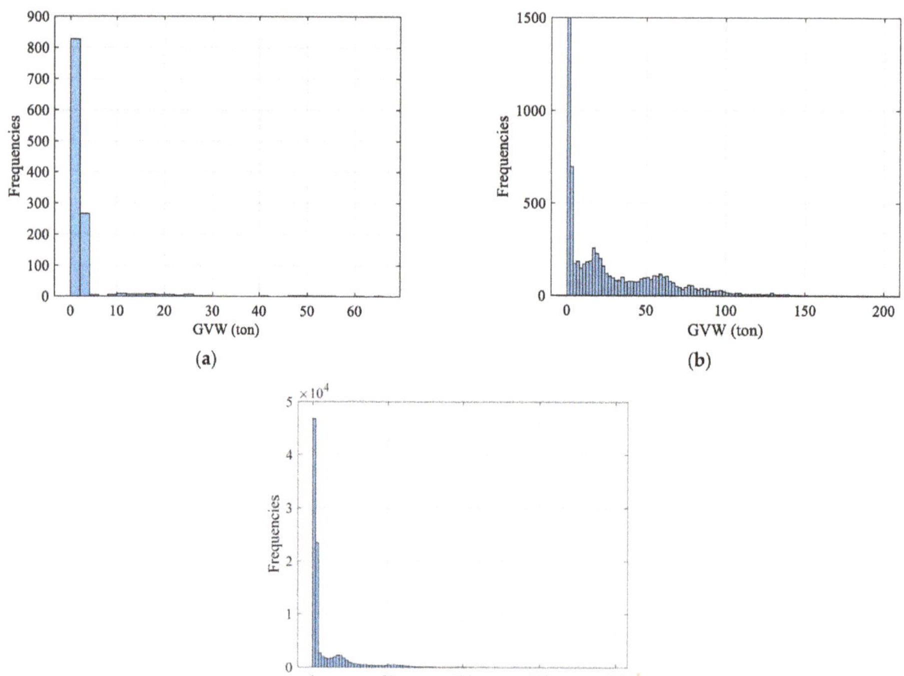

Figure 5. GVW distribution of (**a**) high hourly traffic volume extreme response scenarios; (**b**) high response value extreme response scenarios; and (**c**) WIM data.

According to this, the congestion-prone free-flow traffic is characterized by a high volume of traffic and low percentage of heavy vehicles. Therefore, in the event that traffic flow becomes congested due to excessive traffic volume, the percentage of heavy vehicles in the congested situation must be lower than the actual percentage in the original WIM data.

In order to obtain enough extreme congestion response scenarios for the comparison with free-flow extreme response scenarios, a micro-simulation model can be used to produce the congested flow traffic suitable for critical long-span bridge-loading scenarios, as presented by Enright et al. [39]. The CA method has been used in several fields. With the aid of the CA method, complex movements of vehicles, such as speed changes, lane changes, etc. [35], can be simulated successfully and the generated long-term traffic flow maximum effects can be obtained. Traffic micro-simulation based on CA is applied to obtain the detailed information associated with vehicle load, vehicle velocity, headway distance, and lane position [38]. After a vehicle is produced, its movement is controlled by the local rules that determine the driving modes (i.e., acceleration, cruising, coasting, and braking). The accuracy and efficiency of the CA method has been verified, as presented by Tian et al. [40]. Thus, the congestion traffic data used in this study were obtained using the CA model.

An input flow rate of 12,000 vehicles per hour was used for the six lanes of the bridge. A total of 24 h of congested traffic was generated. To implement the bottleneck, each vehicle's desired velocity was changed to 15–20 km/h. This arrangement lead to homogeneous congested traffic, which is characterized by a low average velocity. However, unlike fully jammed traffic, it still maintains a relatively high traffic flow [39]. The average speed histogram, traffic flow histogram for each hour, and GVW histogram of the generated congested traffic are shown in Figure 6. It can be seen that the level of vehicles with a GVW less than 10 tonnes is magnified in the simulated congested flow.

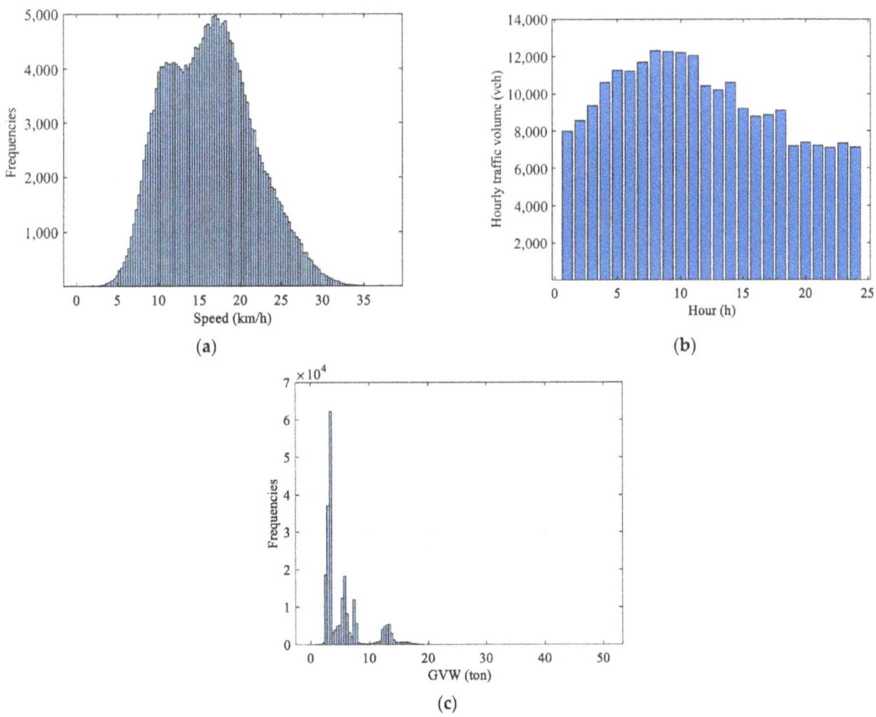

Figure 6. The (**a**) average speed histogram; (**b**) traffic flow histogram for each hour; and (**c**) GVW histogram of the generated congested traffic.

4. Extreme Response Scenario of Congested Traffic Loading

The extreme response scenarios of congested flow are calculated for five representative load effects of the investigated bridge: the axial force of the longest cable (AFC, effect 1); the vertical deflection at the mid-span of the main span (VDMS, effect 2); the longitudinal displacement at the top of the pylon (LDTP, effect 3); the bending moment at the mid-span of the main span (BMMS, effect 4); and the bending moment at the bottom of the pylon (BMBP, effect 5). For each effect, the extreme response scenario is obtained every second, and the top 100 extreme response scenarios with the largest response are selected as simulation samples for further validation of the probabilistic model proposed by the authors [33] under congested flow traffic.

For the extreme response scenarios of the congestion flow, the probability of occurrence of vehicles in each cell on the outer lanes was counted, as shown in Figure 7. It can be seen that in the extreme response scenarios of congestion, the distribution of vehicles along the longitudinal direction of the bridge, which varies according to the load effect, is not uniform, and the vehicles tend to concentrate in the zones of the influence lines where the maximum appears depending on the load effect. Therefore, the simulation of the location of vehicles on the bridge deck can still be performed using a non-stationary Poisson distribution as in the free-flow case [33].

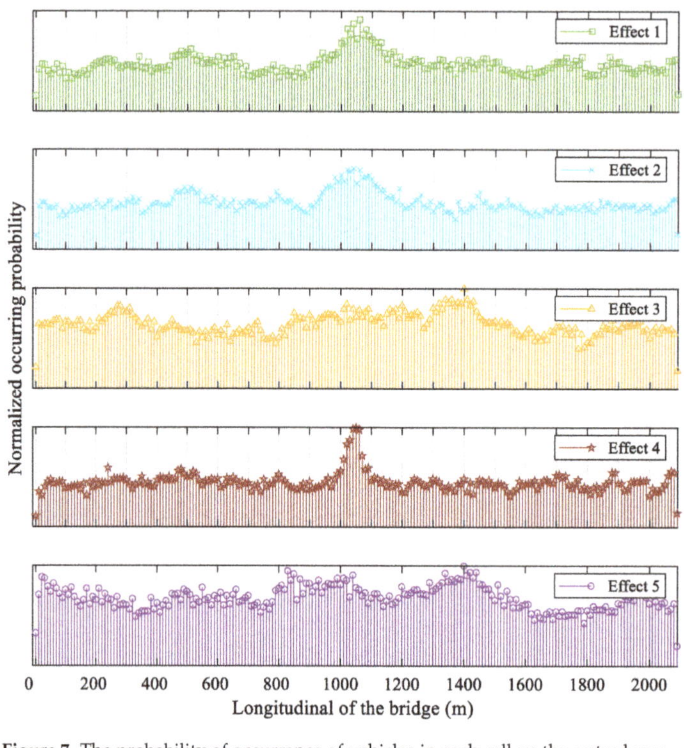

Figure 7. The probability of occurrence of vehicles in each cell on the outer lanes.

Subsequently, the probabilistic model of extreme response scenario was applied to simulate the extreme response scenarios of congestion, taking effect 1 as an example. Figure 8 depicts the probabilities of a cell being occupied by a vehicle in an extreme response scenario for the congested situation. The non-stationary characteristic of the outer lane is more pronounced than that of the inner lane. For congested traffic, the proportion of heavier vehicles is not high. Their tendency to drive in the outer lane is more pronounced.

Also, the data and the GMM fitting of GVWs for the extreme congestion response scenarios are given in Figure 9, which is the basic information for simulating GVW.

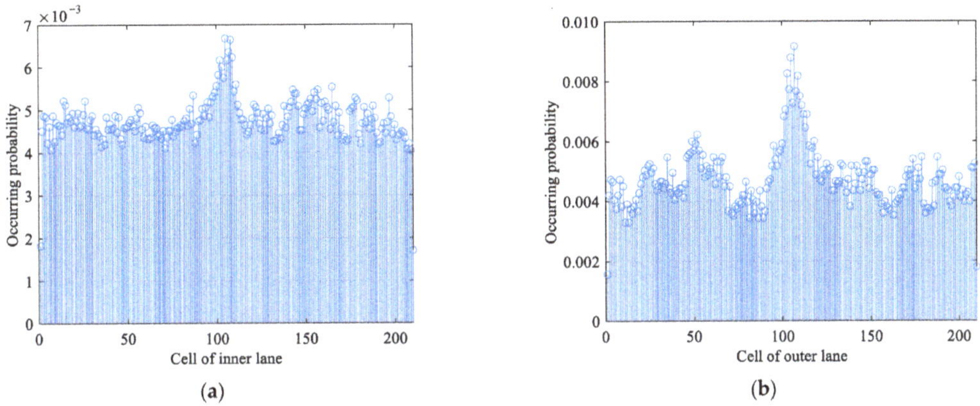

Figure 8. Probabilities of a cell to be occupied by a vehicle in an extreme response scenario and congested flow for load effect 1. (**a**) Inner lane, (**b**) outer lane.

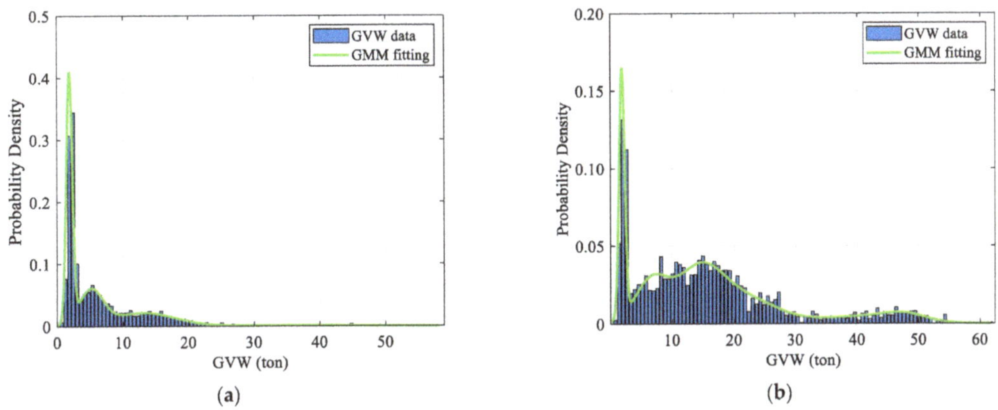

Figure 9. Data and fitting of GVW in (**a**) inner lanes and (**b**) outer lanes.

Figure 10 depicts several extreme response scenarios of effect 1 for congested traffic. As mentioned before, in the case of free-flow conditions, the dimension of the cell occupied or not by a vehicle was set to 20 m based on the normal dimensions and inter-gaps of heavy vehicles. Also, as shown in [33], the effect of light vehicles (less than 10 tonnes) could be neglected. In the simulation of the extreme response scenario under congestion, it had to be verified if these two conditions still applied. A comparison of ignoring or considering light vehicles in the congestion scenario is presented in Figure 11. If the lighter vehicles are ignored, the effect loss is in the range of 8% to 24% with a median of 15%. As shown before, during congestion, the proportion of light vehicles is higher, and the GVW of heavy vehicles is lower compared to the level of normal traffic scenarios in the free-flow situation. Therefore, the effect of light vehicles should not be ignored when performing simulations of extreme congestion response scenarios.

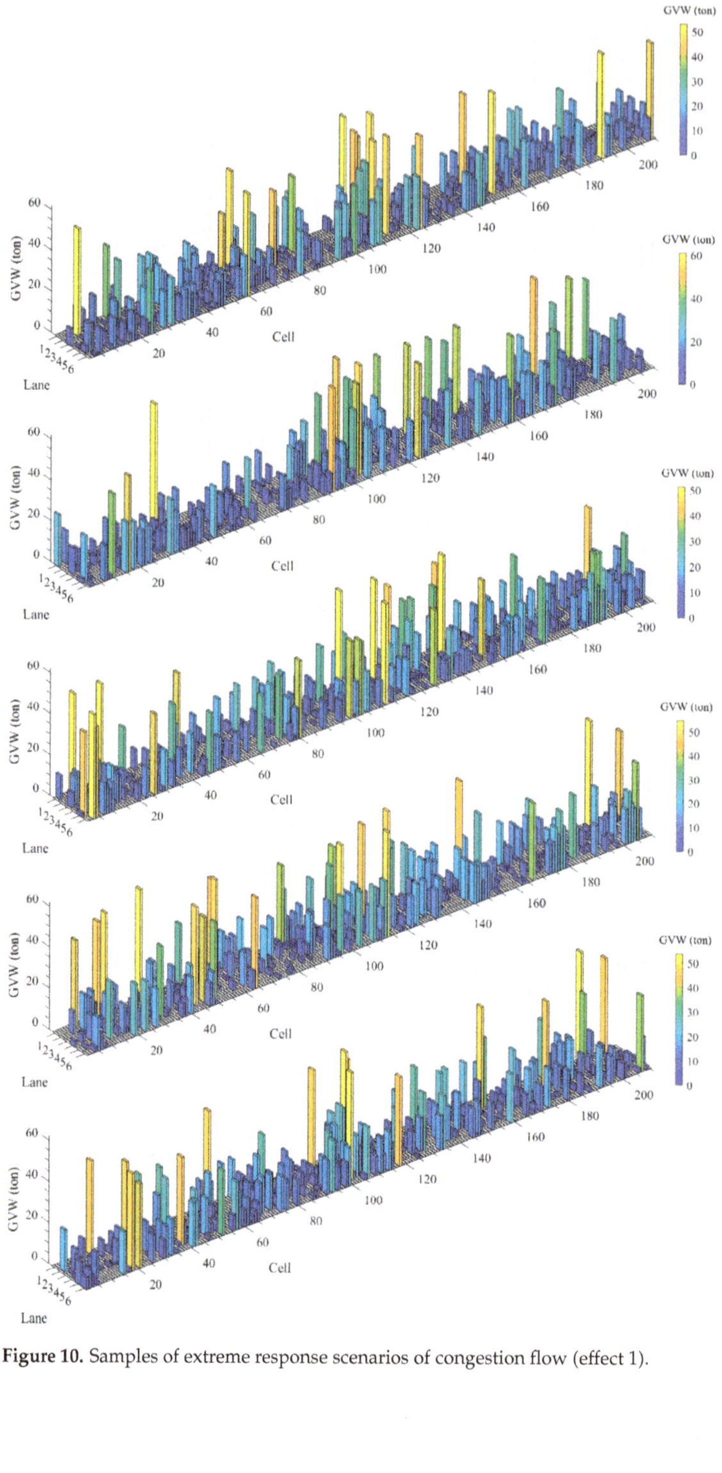

Figure 10. Samples of extreme response scenarios of congestion flow (effect 1).

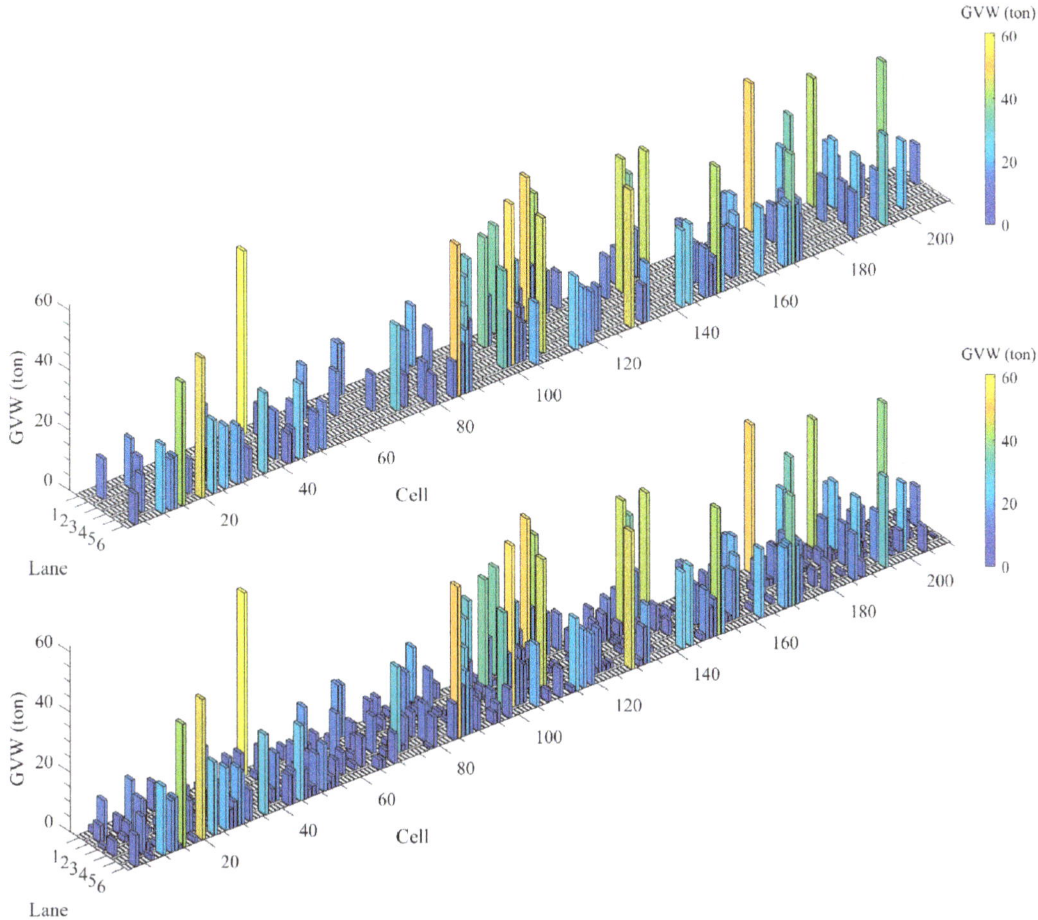

Figure 11. Comparison of ignoring light vehicles and not ignoring light vehicles in a congestion scenario.

Based on the fact that light vehicles should be considered in the simulation, then the length of the cell should be adjusted. The length of small vehicles for domestic use is generally greater than 4.5 m, and the relevant Chinese regulations recommend a safe following distance of 10 m when the speed is below 20 km/h. Therefore, the cell length is chosen as 10 m when simulating extreme congestion response scenarios. Then, the extreme response scenarios were simulated 104 times using the proposed model. Several samples of simulations for effect 1 are shown in Figure 12. The results of the proposed model show a good comparison with the extreme response scenarios obtained in the CA simulation, but at a much lower time and computation cost, therefore demonstrating the capability of the model to simulate congested-flow conditions, by just adapting some basic parameters to this new traffic situation, namely the length of the cell and the consideration of all vehicles and not only the heavier ones.

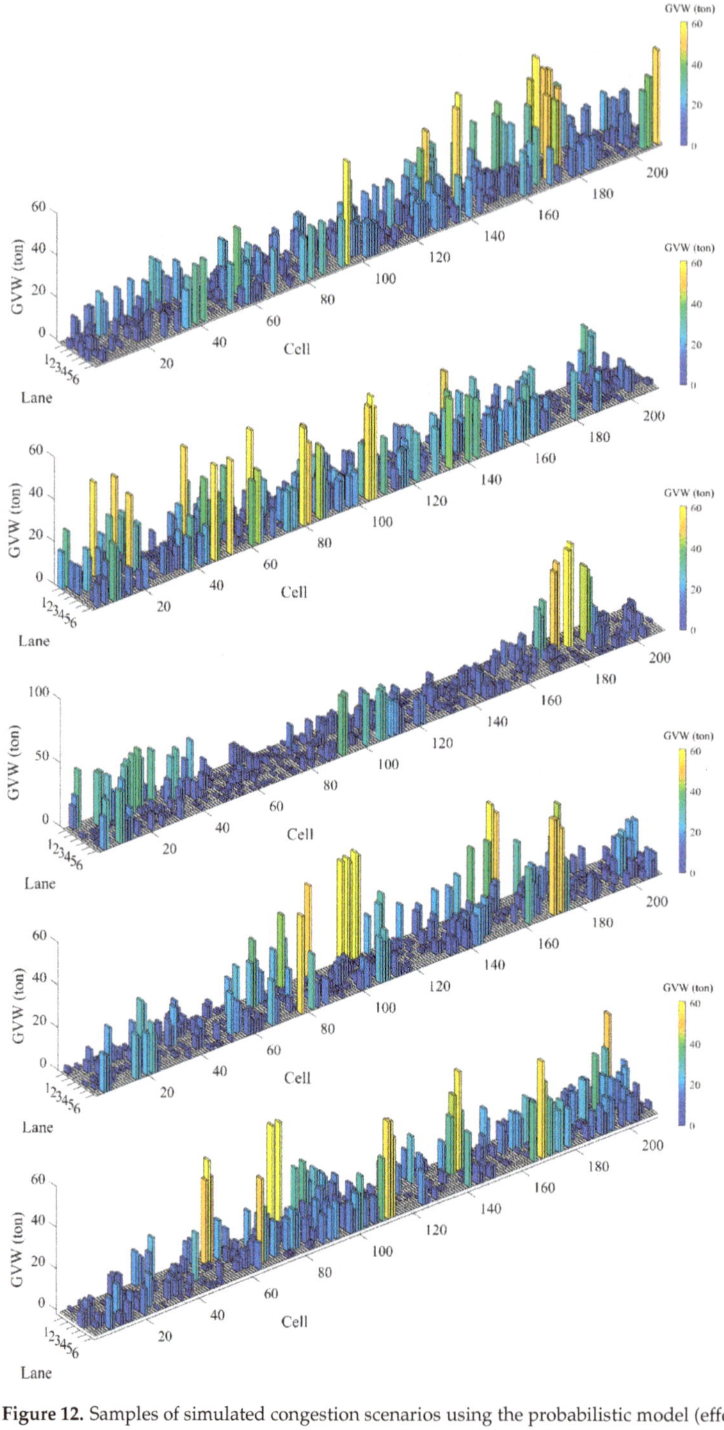

Figure 12. Samples of simulated congestion scenarios using the probabilistic model (effect 1).

In summary, this section extends the application of the extreme response scenario model previously proposed by the authors, discussing its suitability for simulating extreme response scenarios on long-span bridges in the case of congested-flow traffic. It provides an analysis of the necessary parameter selection when applying the model to this traffic condition, such as cell size and the range of vehicle weights to be simulated. In clarifying the significance of congested traffic in the structural safety assessment of long-span bridges, it demonstrates the wide applicability and flexibility of the proposed extreme response scenario model to accommodate to all traffic conditions.

5. Discussion

This study extended the probabilistic model of extreme response scenarios to account for congested traffic conditions, demonstrating its flexibility and broad applicability to various traffic situations on long-span bridges. A key finding is the necessity of adjusting key parameters when applying the model to congested traffic. For instance, the cell size used for the vehicle distribution on the bridge deck was reduced from 20 m, as used in free-flow scenarios, to 10 m to reflect the shorter inter-vehicle gaps in congested conditions. Moreover, while the contribution of light vehicles is typically negligible in free-flow conditions, they play a significant role in congested traffic. The study shows that neglecting light vehicles in congested traffic can result in a load effect loss of up to 24%, making their inclusion essential for accurate simulation and safety evaluation.

The comparison of extreme response scenarios under free-flow and congested conditions reveals that heavy vehicles are less frequent in congested traffic but tend to cluster in outer lanes, especially under extreme loading conditions. This behavior necessitates the continued use of a non-stationary Poisson distribution for vehicle location modeling. Additionally, this study confirms that the probabilistic model remains capable of efficiently simulating extreme response scenarios under congested traffic at a significantly lower computational cost compared to direct micro-simulation using the CA method.

In summary, the model demonstrates its robustness and adaptability to different traffic conditions, providing accurate predictions of extreme response scenarios with minimal computational effort. These findings underscore the model's potential for use in real-time traffic monitoring and bridge safety assessments under varying traffic conditions, especially as congested traffic becomes increasingly common on long-span bridges. Future work could focus on refining the model to account for additional factors influencing the bridge response under congested conditions, such as vehicle type and load distributions over time. Furthermore, the practical implications and contributions of the simulation results to the design and structural safety evaluation of bridges were demonstrated through a practical application of the model in a comparative analysis between observed extreme response scenarios under actual traffic conditions and model loading scenarios specified in representative international codes. The results highlight the inadequacy of the applicability of existing codes in assessing the response to highway traffic for long-span cable-stayed bridges, being too conservative in some cases and under-conservative in others [41].

6. Conclusions

The following conclusions can be drawn from the obtained results:

1. This study conducted a comprehensive investigation into the characteristics of traffic congestion on long-span bridges. Due to the lack of actual congestion traffic data for the bridge in question, a calibrated CA model was utilized to simulate congested traffic flow, creating samples for modeling extreme response scenarios. The data analysis revealed that during congestion periods, the vehicle weight composition is generally lighter than that observed under normal traffic conditions, and these events occur at a relatively low frequency. Because of this, it may happen that extreme load effects under free-flow traffic may be comparable or even higher than under congested-flow conditions. In fact, the higher presence of heavy vehicles and probability of occurrence of extreme situations (due to the shape of the influence lines of the load effects) in

the case of free-flow conditions can result in this possibility for some load effects on long-span bridges. This issue is now under investigation thanks to the simulation capabilities provided by the proposed model, valid for both free- and congested-flow traffic. It also emphasizes the need of models for free-flow traffic on long-span bridges, which almost does not exist currently because congested traffic has been considered the critical situation by most researchers in the field.

2. The application of the extreme-response-scenario probabilistic model previously proposed by the authors for free-flow traffic to the case of congested traffic demonstrated its flexible applicability and broad utility. By accurately modeling extreme response scenarios of congested flow, this approach has proven to be an effective tool for simulating extreme conditions under various traffic conditions on long-span bridges. This significantly contributes to the assessment of the safety performance of long-span bridges under traffic loads, highlighting the model's potential to enhance the understanding and management of bridge safety.
3. Through the refined simulation of extreme congestion response scenarios, this study not only corroborates the robustness of the proposed probabilistic model but also underscores its relevance and significance in facilitating accurate safety evaluations by obtaining the maximum load effects in a simpler and cheaper (in terms of computing time) way. This opens the feasibility of using the proposed model in the automatic and on-line management of existing long-span bridges in operation, giving the possibility of obtaining an expected response of the bridge in actual traffic conditions and comparing it with the experimentally obtained response measured in certain locations. The differences in both responses can warn about possible deficiencies in the normal operation of the bridge, therefore becoming a key aspect of the bridge's preventive maintenance.

Future research will aim to integrate real congestion data as advances in detection and monitoring technologies permit to further validate and refine the model's parameters, ensuring that the simulated traffic scenarios closely mirror real-world conditions and thereby enhancing the predictive accuracy essential for the structural health monitoring of long-span bridges.

Author Contributions: Conceptualization, X.W. and X.R.; methodology, X.W. and M.Z.; software, X.W.; validation, X.W. and J.R.C.; formal analysis, X.W.; investigation, X.R.; resources, X.R.; data curation, X.W. and M.Z.; writing—original draft preparation, X.W.; writing—review and editing, J.R.C.; visualization, X.W.; supervision, J.R.C.; project administration, J.R.C.; funding acquisition, J.R.C. All authors have read and agreed to the published version of the manuscript.

Funding: This research received no external funding.

Institutional Review Board Statement: Not applicable.

Informed Consent Statement: Not applicable.

Data Availability Statement: The data that support the findings of this study are available from the corresponding author upon reasonable request.

Conflicts of Interest: The authors declare no conflicts of interest.

References

1. ASCE Structurally Deficient Bridges, Bridge Infrastructure, ASCE's 2021 Infrastructure Report Card. 2021. Available online: https://infrastructurereportcard.org/cat-item/bridges/ (accessed on 17 August 2024).
2. Yang, Y.; Cai, C.S. Prediction of Extreme Traffic Load Effects of Bridges Using Bayesian Method and Application to Bridge Condition Assessment. *J. Bridge Eng.* **2019**, *24*, 04019003.
3. Farhey, D.N. Structural performances of bridge types in the U.S. national bridge inventory. *Infrastructures* **2018**, *3*, 6. [CrossRef]
4. Olivera, C.B.L.; Greco, M.; Bittencourt, T.N. Analysis of the brazilian federal bridge inventory. *Rev. IBRACON Estrut. Mater.* **2019**, *12*, 1–3. [CrossRef]

5. COST 345 (Programme). *Cost 345: Procedures Required for Assessing Highway Structures: Working Group 1 Report on the Current Stock of Highway Structures in European Countries the Cost of Their Replacement and the Annual Costs of Maintaining Repairing and Renewing Them*; Technical report; European Commission—Directorate General Research: Brussels, Belgium, 2003.
6. Mendoza-Lugo, M.A.; Nogal, M.; Morales-Nápoles, O. Estimating bridge criticality due to extreme traffic loads in highway networks. *Eng. Struct.* **2024**, *300*, 117172. [CrossRef]
7. US Dept. of Transportation. *Beyond Traffic: 2045 Final Rep.*; Office of the Secretary of Transportation: Washington, DC, USA, 2016.
8. Leahy, C.; O'Brien, E.; O'Connor, A. The effect of traffic growth on characteristic bridge load effects. *Transp. Res. Procedia.* **2016**, *14*, 3990–3999. [CrossRef]
9. Lu, N.; Beer, M.; Noori, M.; Liu, Y. Lifetime deflections of long-span bridges under dynamic and growing traffic loads. *J. Bridge Eng.* **2017**, *22*, 04017086. [CrossRef]
10. Han, W.S.; Wu, J.; Cai, C.S.; Chen, S.R. Characteristics and dynamic impact of overloaded extra heavy trucks on typical highway bridges. *J. Bridge Eng.* **2015**, *20*, 05014011. [CrossRef]
11. Ruan, X.; Zhou, J.; Shi, X.; Caprani, C.C. A site-specific traffic load model for long-span multi-pylon cable-stayed bridges. *Struct. Infrastruct. Eng.* **2017**, *13*, 494–504. [CrossRef]
12. Yuan, Y.; Han, W.; Huang, P.; Zhao, J.; Li, Y.; Zhang, J. Structure safety assessment under heavy traffic based on weigh in motion and simulation analysis. *Adv. Struct. Eng.* **2017**, *20*, 1864–1878. [CrossRef]
13. Ji, B.; Chen, D.; Ma, L.; Jiang, Z.; Shi, G.; Lv, L.; Xu, H.; Zhang, X. Research on stress spectrum of steel decks in suspension bridge considering measured traffic flow. *J. Perform. Constr. Fac.* **2011**, *26*, 65–75. [CrossRef]
14. Lu, N.; Ma, Y.; Liu, Y. Evaluating probabilistic traffic load effects on large bridges using long-term traffic monitoring data. *Sensors* **2019**, *19*, 5056. [CrossRef] [PubMed]
15. Nie, R.; Huang, Y.; Li, S. In-service condition assessment of long-span bridges based on traffic load effects using monitoring data. *Stavební Obz. Civ. Eng. J.* **2018**, *27*, 450–457. [CrossRef]
16. Deng, L.; Yan, W.; Nie, L. A simple corrosion fatigue design method for bridges considering the coupled corrosion-overloading effect. *Eng. Struct.* **2019**, *178*, 309–317. [CrossRef]
17. Deng, L.; Yan, W.C. Vehicle weight limits and overload permit checking considering the cumulative fatigue damage of bridges. *J. Bridge Eng.* **2018**, *23*, 04018045. [CrossRef]
18. Ventura, R.; Barabino, B.; Vetturi, D.; Maternini, G. Bridge safety analysis based on the function of exceptional vehicle transit speed. *Open Transp. J.* **2020**, *14*, 222–236. [CrossRef]
19. Ventura, R.; Maternini, G.; Barabino, B. Traffic hazards on main road's bridges: Real-time estimating and managing the overload risk. *IEEE T. Intell. Transp.* **2024**, *25*, 12239–12255. [CrossRef]
20. Zhang, G.; Liu, Y.; Liu, J.; Lan, S.; Yang, J. Causes and statistical characteristics of bridge failures: A review. *J. Traffic Transp. Eng.* **2022**, *9*, 388–406. [CrossRef]
21. Enright, B.; O'Brien, E.J. Monte Carlo simulation of extreme traffic loading on short and medium span bridges. *Struct. Infrastruct. Eng.* **2013**, *9*, 1267–1282. [CrossRef]
22. Nowak, A.S.; Lutomirska, M.; Sheikh Ibrahim, F.I. The development of live load for long span bridges. *Bridge Struct.* **2010**, *6*, 73–79. [CrossRef]
23. Soriano, M.; Casas, J.R.; Ghosn, M. Simplified probabilistic model for maximum traffic load from weigh-in-motion data. *Struct. Infrastruct. Eng.* **2017**, *13*, 454–467. [CrossRef]
24. Caprani, C.C. Lifetime highway bridge traffic load effect from a combination of traffic states allowing for dynamic amplification. *J. Bridge Eng.* **2013**, *18*, 901–909. [CrossRef]
25. Hwang, E.S.; Kim, D.Y. Live load model for long span steel cable bridges considering traffic congestion scenarios. *Int. J. Steel Struct.* **2019**, *19*, 1996–2009. [CrossRef]
26. Guo, D.; Caprani, C.C. Traffic load patterning on long span bridges: A rational approach. *Struct. Saf.* **2019**, *77*, 18–29. [CrossRef]
27. Zhou, J.; Caprani, C.C.; Zhang, L. On the structural safety of long-span bridges under traffic loadings caused by maintenance works. *Eng. Struct.* **2021**, *240*, 112407. [CrossRef]
28. Micu, E.A.; Malekjafarian, A.; O'Brien, E.J.; Quilligan, M.; McKinstray, R.; Angus, E.; Lydon, M.; Catbas, F.N. Evaluation of the extreme traffic load effects on the Forth Road Bridge using image analysis of traffic data. *Adv. Eng. Softw.* **2019**, *137*, 102711. [CrossRef]
29. Carey, C.; Caprani, C.C.; Enright, B. A pseudo-microsimulation approach for modelling congested traffic loading on long-span bridges. *Struct. Infrastruct. Eng.* **2018**, *14*, 163–176. [CrossRef]
30. Lipari, A.; Caprani, C.C.; O'Brien, E.J. A methodology for calculating congested traffic characteristic loading on long-span bridges using site-specific data. *Comput. Struct.* **2017**, *190*, 1–12. [CrossRef]
31. O'Brien, E.J.; Hayrapetova, A.; Walsh, C. The use of micro-simulation for congested traffic load modeling of medium- and long-span bridges. *Struct. Infrastruct. Eng.* **2012**, *8*, 269–276. [CrossRef]
32. Lipari, A. Micro-Simulation Modelling of Traffic Loading on Long-Span Bridges. Doctoral Thesis, UCD, Dublin, Germany, 2013.
33. Wang, X.; Ruan, X.; Casas, J.R.; Zhang, M. Probabilistic model of traffic scenarios for extreme load effects in long-span bridges. *Struct. Saf.* **2024**, *106*, 102382. [CrossRef]
34. Wang, F.; Xu, Y. Traffic Load Simulation for Long-Span Suspension Bridges. *J. Bridge Eng.* **2019**, *24*, 245019005. [CrossRef]

35. Chen, S.R.; Wu, J. Modeling stochastic live load for long-span bridge based on microscopic traffic flow simulation. *Comput. Struct.* **2011**, *89*, 813–824. [CrossRef]
36. Der Kiureghian, A.; Liu, P. Structural reliability under incomplete probability information. *J. Eng. Mech.* **1986**, *112*, 85–104. [CrossRef]
37. Chen, Z.; Bao, Y.; Chen, J.; Li, H. Modelling the spatial distribution of heavy vehicle loads on long-span bridges based on undirected graphical model. *Struct. Infrastruct. Eng.* **2019**, *15*, 1485–1499. [CrossRef]
38. Ruan, X.; Zhou, J.; Tu, H.; Jin, Z.; Shi, X. An improved cellular automaton with axis information for microscopic traffic simulation. *Transport. Res. C-Emer.* **2017**, *78*, 63–77. [CrossRef]
39. Enright, B.; Carey, C.; Caprani, C.C. Microsimulation evaluation of Eurocode load model for American long-span bridges. *J. Bridge Eng.* **2013**, *18*, 1252–1260. [CrossRef]
40. Tian, J.; Li, G.; Treiber, M.; Jiang, R.; Jia, N.; Ma, S. Cellular automaton model simulating spatiotemporal patterns, phase transitions and concave growth pattern of oscillations in traffic flow. *Transport. Res. B-Meth.* **2016**, *93*, 560–575. [CrossRef]
41. Wang, X.; Ruan, X.; Casas, J.R.; Zhang, M. Extreme response to traffic of long-span cable-stayed bridges: Comparison with existing traffic load models. *Struct. Infrastruct. Eng. in press.*

applied sciences

MDPI

Article

Calibration of Viscous Damping–Stiffness Control Force in Active and Semi-Active Tuned Mass Dampers for Reduction of Harmonic Vibrations

Marcin Maślanka

Department of Process Control, Faculty of Mechanical Engineering and Robotics, AGH University of Krakow, al. Mickiewicza 30, 30-059 Krakow, Poland; marcin.maslanka@agh.edu.pl

Abstract: Tuned mass dampers (TMDs) are commonly used to mitigate vibrations in civil structures. There is a growing demand for new solutions that offer similar effectiveness as TMDs but with reduced mass. In this context, this paper investigates active (ATMD) and semi-active (STMD) tuned mass dampers with relative displacement and velocity feedback. The control force of the ATMD is assumed to be the sum of viscous damping and either positive or negative stiffness forces. This control force is calibrated for a specific parameter K such that the effectiveness of the ATMD in reducing harmonic vibrations matches that of the TMD with K times larger mass. The optimal calibration is derived based on the mathematical reformulation of an existing optimal acceleration feedback control algorithm. The control approach for the ATMD is then applied to the STMD. Subsequently, the sub-optimal STMD is analyzed, with a focus on its limitations arising from the clipping of active forces. Finally, the paper presents a calibration of the STMD using a numerical optimization method. It is demonstrated that the maximum achievable performance of the numerically optimized STMD matches that of the TMD with three times larger mass.

Keywords: vibration; damping; tuned mass dampers; active; semi-active; control; negative stiffness; equivalent linearization; optimization

1. Introduction

The mitigation of vibrations of civil structures is vital for ensuring the safety and comfort of their users. Passive tuned mass dampers (TMDs) are commonly employed to reduce vibrations of buildings [1,2], bridges [3], and other structures. A notable example is the 660-ton TMD installed in the Taipei 101 Tower [4], which also serves as a popular attraction, offering visitors a chance to observe its operation.

The classical TMD [5] consists of a moving mass, a spring, and a viscous damper, all tuned to a single vibration mode of the structure. The effectiveness of the TMD depends on its mass, more specifically, the mass ratio which defines the TMD's mass in relation to the modal mass. Using a sufficiently large TMD mass ensures that vibrations are reduced to levels compliant with both technical standards and occupant comfort requirements [6]. The mass of TMDs installed in tall buildings typically amounts to about 1%, or less, of the modal mass. While this percentage may seem small, it translates to a TMD mass that is often in the order of hundreds of tons [1]. Installing large TMD masses is not only expensive but also difficult due to the limited space available for installation. One example is the 432 Park Avenue residential tower in New York, USA, where the required TMD mass of 1200 tons had to be split into two 600-ton TMDs with an innovative design to fit into two narrow spaces available on each side of the tower core [7].

As a result, there is a significant demand for innovative solutions that can provide the required vibration damping efficiency with less mass than a standard TMD. This challenge also serves as the primary motivation for the research presented in this work.

Achieving the desired level of vibration reduction using less mass than a traditional TMD is possible through the application of active and semi-active tuned mass dampers (ATMD and STMD, respectively). ATMD systems are essentially TMDs of varying designs that incorporate an active actuator. On the other hand, STMDs employ semi-active dampers. The exploration of these systems in civil engineering was initiated roughly 50 years ago by the structural control concept of Yao [8] and pioneering research activities of Kobori [9]. One of the earliest ATMD designs was presented by Chang and Soong in 1980 [10]. Pioneering implementations of ATMDs took place in Japan, with the world's first application of two ATMDs in a ten-story office building in Tokyo in 1989 [11]. Two decades later, in 2009, the list of implementations in buildings in Japan included 40 active and 13 semi-active control systems of various designs [12]. Research progress worldwide has been reviewed in, e.g., [13,14]. Most recent studies have demonstrated the effectiveness of ATMDs in mitigating the dynamic response of tall buildings subjected to strong winds [15,16]. Although numerous ATMDs have been implemented, further research into their design, control algorithms, and performance continues.

For passive TMDs, increasing their mass naturally results in a change in the force exerted on the structure. To achieve an efficiency for ATMD or STMD comparable to that of a heavier TMD, it is evident that the total force (comprising both the control force generated by actuating elements and the force from the passive elements) in the ATMD or STMD system must act on the structure in a manner similar to a larger-mass TMD. This principle was first applied in Kobori's research group, by Nishimura [17]. They introduced an optimal acceleration feedback control for ATMD using the classical H_∞ synthesis method (the so-called equal-peak design), consistent with the approach used in the optimal TMD tuning [5,18–20]. It is noticeable that their optimal ATMD exhibits a frequency response characteristic like that of a TMD with a larger mass.

Subsequent research indicated that a similar ATMD performance can be achieved using a different set of feedback signals. Chatterjee [21] presented an analytical solution for H_∞ tuning of the ATMD utilizing state feedback from the ATMD mass. Cheung et al. [22] employed the H_∞ design to an ATMD with absolute displacement feedback and, in another variant, also with relative displacement feedback. Brodersen et al. [23] used absolute displacement and relative velocity feedbacks in their optimal ATMD design. However, none of the existing studies present the optimal equal-peak design of ATMD using only internal feedback, specifically from relative displacement and relative velocity, without relying on absolute displacement or acceleration. In this work, the control approach by Nishimura et al. will be adapted to implement an ATMD using exclusively a relative displacement and velocity feedback. This concept will also be applied to the STMD.

Regarding STMDs, a significant advantage of these systems is their substantially lower power consumption compared to ATMDs. Additionally, STMDs are inherently stable. The STMDs consist of a mass, a passive spring, and a semi-active damper, for example a magneto-rheological (MR) damper. A salient feature of MR dampers is their short response time [24,25]. Various designs and operational principles of MR dampers are well-documented [26]. Rotary MR dampers have been successfully implemented in 12 STMDs, each weighing 5200 kg, to mitigate bridge vibrations [27,28]. MR damper-based STMDs have also been employed in vibration reduction systems for tall buildings [29,30]. Numerous semi-active control approaches for STMDs have been developed. In the context of harmonic vibrations, effective STMD concepts are based on the semi-active emulation of positive or negative dynamic stiffness, aiming to tune the STMD's operation to the actual frequency of vibration [31,32]. The semi-active control force of STMDs can be determined using a clipped linear quadratic regulator (LQR) [33]. Other methods include ground hook control schemes [34,35], on-off phase control algorithms [36,37], linear quadratic Gaussian control (LQG) [38], nonlinear optimal-based control [39,40], PID control [41], and more.

In previous work [42], the principles of the optimal active control by Nishimura et al. [17] are utilized to formulate a semi-active control approach for the STMD with absolute acceleration and relative motion (displacement and velocity) feedback. As a result, an optimized STMD was proposed, denoted as STMD-K, in which the parameter K defines the efficiency of STMD in such a way that for a given K value, the damping efficiency for STMD-K is the same as for TMD with a mass K times larger. In [43], a preliminary experimental validation of the STMD-K concept is presented. It was observed that the efficient implementation of the STMD-K concept requires a high-quality acceleration signal. In practice, while the acceleration signal is easily accessible, it may contain high-frequency components that make proper control implementation challenging.

The purpose of this paper is to adapt the optimal acceleration feedback control by Nishimura et al. for its alternative implementation in both ATMD and STMD, using only relative displacement and relative velocity feedback, without relying on acceleration feedback. Simultaneously, following the approach in [42], for a specified design parameter K, we aim to calibrate the control force parameters for both ATMD and STMD such that the efficiency of both systems matches that of a TMD with a K-times larger mass.

The subsequent section provides an essential introduction to Den Hartog's TMD, viewed as a reference mass damper. In Section 3, a control algorithm for ATMD is presented. The section starts with the definition of its control force, then introduces the control objective and the equations of motion. This is followed by a reformulation of Nishimura's acceleration feedback control, aiming for its alternative implementation without acceleration feedback and using only relative displacement and velocity feedback. The performance of the new ATMD is illustrated at the end of this section.

Section 4 deals with the sub-optimal STMD. This section is crucial for the understating of the limitations of the STMD and encompasses a detailed analysis of the clipped viscous damping–stiffness force. Subsequently, the equivalent linearization method of Krylov and Bogoliubov is employed to find the equivalent viscous damping and stiffness resulting from the clipped control force of the STMD. The frequency responses of the sub-optimal STMD are then discussed.

Finally, in Section 5, a calibration of the STMD using numerical optimization is introduced. To facilitate this optimization, the control force of the STMD is adjusted to incorporate correction factors aimed at minimizing the deteriorating effects of force clipping. Both the optimization procedure and the resulting outcomes are detailed. A comprehensive analysis of the STMD at its performance limits follows. Section 6 summarizes the primary findings and concludes the paper.

2. Den Hartog's TMD as a Reference Mass Damper

In this section, the Den Hartog's TMD is introduced as a reference point against which the active and semi-active mass damper solutions presented later in this work will be compared. Although the theory behind TMD is well-established and described in mechanics textbooks, the essential formulas for Den Hartog's design are provided for their subsequent use in this study.

2.1. Den Hartog's TMD Tuning

The classical Den Hartog's TMD [5] consists of a mass m_2 connected to the primary structure by a parallel configuration of a passive spring k_{2dh} and a passive viscous damper c_{2dh}. The primary structure is characterized by its mass m_1 and stiffness k_1, and represents the single-degree-of-freedom model of one targeted mode of a vibrating structure with its natural frequency $\omega_1 = \sqrt{k_1/m_1}$ and a negligible small damping.

The passive spring stiffness of the Den Hartog's TMD [5] is defined as:

$$k_{2dh} = k_1 \frac{\mu}{(\mu+1)^2} \tag{1}$$

where:

$$\mu = \frac{m_2}{m_1}, \tag{2}$$

is a mass ratio.

The viscous damping coefficient of the TMD is given by:

$$c_{2dh} = 2\zeta_{2dh}\sqrt{m_2 k_{2dh}}, \tag{3}$$

where:

$$\zeta_{2dh} = \sqrt{\frac{3\mu}{8(\mu+1)}}, \tag{4}$$

is the optimum damping ratio.

2.2. Equations of Motion and Frequency Response to Harmonic Excitation

The equations of motion for the primary structure equipped with the Den Hartog's TMD under harmonic force excitation can be written as follows:

$$\begin{aligned} m_1\ddot{x}_1 + k_1x_1 + c_{2dh}(\dot{x}_1 - \dot{x}_2) + k_{2dh}(x_1 - x_2) = f_{exc}, \\ m_2\ddot{x}_2 - c_{2dh}(\dot{x}_1 - \dot{x}_2) - k_{2dh}(x_1 - x_2) = 0, \end{aligned} \tag{5}$$

where $\ddot{x}_1$, $\dot{x}_1$, and x_1 are the absolute acceleration, velocity, and displacement of the primary structure with its mass m_1 and stiffness k_1, while $\ddot{x}_2$, $\dot{x}_2$, and x_2 are the absolute acceleration, velocity, and displacement of the TMD with mass m_2, passive spring stiffness k_{2dh} (1), and passive viscous damper c_{2dh} (3), and f_{exc} is a harmonic force excitation of amplitude F_{exc} and varying angular frequency ω:

$$f_{exc} = F_{exc} sin(\omega t). \tag{6}$$

The relative displacement of the mass m_2 with respect to the structural mass m_1 is:

$$x_d = x_1 - x_2. \tag{7}$$

The frequency response to harmonic excitation can be obtained by representing the response as harmonic components, $x_1 = X_1e^{i\omega t}$, $x_d = X_de^{i\omega t}$. Based on the equations of motion (5), the amplitude X_1 of forced response of the primary structure displacement, and the amplitude X_d of relative displacement, can be expressed as follows [44]:

$$\frac{X_1(\omega)}{F_{exc}} = \left| \frac{k_{2dh} - \omega^2 m_2 + i\omega c_{2dh}}{(k_1 - \omega^2(m_1+m_2))\,(k_{2dh} - \omega^2 m_2 + i\omega c_{2dh}) - \omega^4 m_2^2} \right|, \tag{8}$$

$$\frac{X_d(\omega)}{F_{exc}} = \left| \frac{\omega^2 m_2}{(k_1 - \omega^2(m_1+m_2))\,(k_{2dh} - \omega^2 m_2 + i\omega c_{2dh}) - \omega^4 m_2^2} \right|. \tag{9}$$

The above equations will be later used in numerical examples.

2.3. Dynamic Amplification Factor

The passive spring (1) and viscous damper (3) of Den Hartog's TMD are optimally tuned to minimize the maximum steady-state dynamic amplification factor (DAF) of primary structure displacement:

$$DAF = \frac{X_1(\omega)}{X_{1static}}, \tag{10}$$

where $X_{1static} = F_{exc}/k_1$ is a static deflection of the primary structure mass.

A crucial aspect of TMD design is the selection of its mass ratio μ (2). The maximal amplification factor, and thus the effectiveness of the TMD, is defined by μ:

$$DAF_{max} = \sqrt{\frac{\mu+2}{\mu}}. \tag{11}$$

Using the TMD of a specified mass allows for limiting the maximum amplitude of vibrations to a level defined by (11). In many cases, installing the TMD of the required mass to ensure proper efficiency is challenging due to site-specific constraints. Active and semi-active mass dampers offer a solution that can provide the required level of vibration damping efficiency while using a smaller mass.

In the subsequent section, an ATMD will be discussed that is calibrated for the given design parameter K to provide the same level of efficiency as the TMD with a mass that is K times larger.

3. Calibration of ATMD with Viscous Damping–Stiffness Control Force

The considered ATMD is illustrated in Figure 1a. It comprises a mass m_2 attached to the primary structure through a passive spring k_{2dh} of the classical Den Hartog's TMD, coupled with an ideal active actuator that applies the active control force f_a. Using the same spring k_{2dh} as in the passive TMD allows for the replacement of the active actuator in the ATMD with a viscous damper to form a TMD, without the need to tune its frequency. Conversely, the viscous damper in an existing TMD can be replaced with an active actuator to upgrade it into an ATMD of the same mass but with enhanced efficiency.

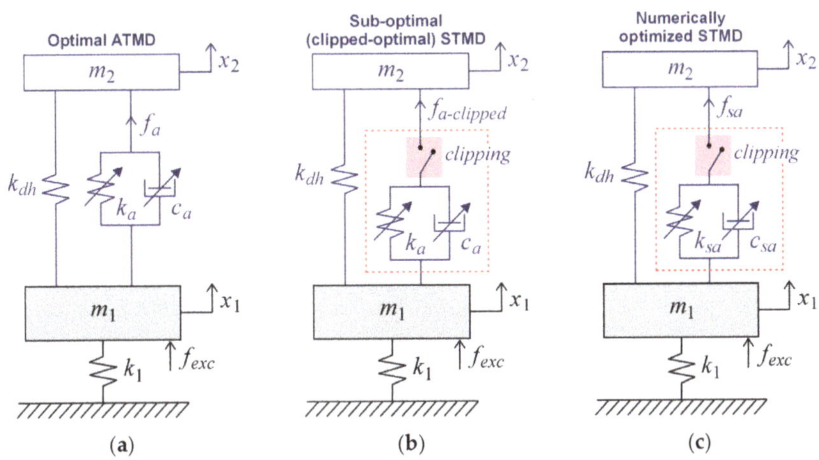

Figure 1. Primary structure with mass dampers under consideration: (**a**) Optimal ATMD with a viscous damping–stiffness force; (**b**) Sub-optimal STMD with a clipped viscous–stiffness control force; (**c**) optimized STMD with an adjusted clipped viscous–stiffness control force to compensate, as much as possible, for clipping.

The active control force f_a under consideration takes the form of a sum of two components: the viscous damping force and the stiffness force, as depicted in Figure 1 by symbols indicating a controllable spring and a controllable viscous damper.

3.1. Active Control Force

The active control force to be applied by the active actuator of the ATMD is assumed to be calculated using the relative displacement x_d and velocity $\dot{x}_d$ signals. In practice, the relative displacement can be directly measured using a single displacement sensor. The relative velocity can be then obtained from the measured relative displacement signal.

The active control force is thus expressed as:

$$f_a = c_a \dot{x}_d + k_a x_d, \tag{12}$$

where the viscous damping coefficient c_a and the positive or negative stiffness k_a represent the feedback gains. These gains require calibration to satisfy a specific control objective.

3.2. Control Objective

The control objective for the ATMD with mass m_2 and control force (12) is to achieve the same efficacy in reducing the steady-state displacement amplitude of the primary structure under harmonic excitation as can be obtained using a TMD with mass Km_2, where $K > 1$ is introduced here, like in [42], as a control algorithm parameter.

According to this control objective, the maximum value for the dynamic amplification factor (10), for the ATMD with the mass ratio $\mu = m_2/m_1$ and the given K should be:

$$DAF_{max-goal} = \sqrt{\frac{K\mu + 2}{K\mu}}. \tag{13}$$

3.3. Equations of Motion and Dynamic Amplification

The equations of motion of the two masses in Figure 1a are:

$$\begin{aligned} m_1\ddot{x}_1 + k_1x_1 + k_{2dh}(x_1 - x_2) &= f_{exc} - f_a, \\ m_2\ddot{x}_2 - k_{2dh}(x_1 - x_2) &= f_a. \end{aligned} \tag{14}$$

where $\ddot{x}_1$, $\dot{x}_1$, and x_1 are the absolute acceleration, velocity, and displacement of the primary structure with its mass m_1 and stiffness k_1, while $\ddot{x}_2$, $\dot{x}_2$, and x_2 are the absolute acceleration, velocity, and displacement of the ATMD with its mass m_2 and passive spring stiffness k_{2dh} (1), f_{exc} is the harmonic force excitation of amplitude F_{exc} and varying angular frequency ω, and f_a is the control force (12).

After accounting for the control force (12), we can express these equations as:

$$\begin{aligned} m_1\ddot{x}_1 + k_1x_1 + c_a\dot{x}_d + (k_a + k_{2dh})x_d &= f_{exc}, \\ m_2\ddot{x}_2 - c_a\dot{x}_d - (k_a + k_{2dh})x_d &= 0. \end{aligned} \tag{15}$$

Based on this system of equations, the steady-state displacement amplitude of primary structure mass and the relative displacement amplitude of the ATMD can be expressed in a similar way as previously for the case of the TMD, see (8) and (9), but after replacing k_{2dh} with $k_a + k_{2dh}$, and c_{2dh} with c_a:

$$\frac{X_1(\omega)}{F_{exc}} = \left| \frac{k_a + k_{2dh} - \omega^2 m_2 + i\omega c_a}{(k_1 - \omega^2(m_1 + m_2))\,(k_a + k_{2dh} - \omega^2 m_2 + i\omega c_a) - \omega^4 m_2^2} \right|, \tag{16}$$

$$\frac{X_d(\omega)}{F_{exc}} = \left| \frac{\omega^2 m_2}{(k_1 - \omega^2(m_1 + m_2))\,(k_a + k_{2dh} - \omega^2 m_2 + i\omega c_a) - \omega^4 m_2^2} \right|. \tag{17}$$

Later in this paper, (16) and (17) will be used to determine the frequency responses for the primary structure with the ATMD.

3.4. Calibrating Viscous Damping–Stiffness Control Force to Mimic the Optimal Acceleration Feedback Control by Nishimura et al.

To determine the optimal parameters for the control force (12) that fulfill the stated control objective, we can use the well-established theory of the so-called equal peak design of TMDs [5,18–20] that has also been adopted to ATMDs [17,21–23]. In the work by Nishimura et al. [17], this approach was applied to derive the optimal acceleration feedback control law, with a variant that also includes relative velocity feedback. The approach presented here aims to calibrate the parameters of the control force (12) so that it mimics the optimal control force outlined by Nishimura et al.

Let us consider the control force f_{Nis} discussed in [17], which takes the form of a sum of inertial and viscous damping forces:

$$f_{Nis} = -g_{Nis}m_1\ddot{x}_1 + c_{Nis}\dot{x}_d, \tag{18}$$

where $g_{Nis}m_1$ and c_{Nis} are the negative acceleration and relative velocity feedback gains.

Our goal is to find c_a and k_a such that the control force (12) is equal to the force (18). Since we assume that both forces should be equal, we can substitute (18) into the equation of motion for the ATMD mass in (14):

$$m_2\ddot{x}_2 - k_{2dh}x_d = -g_{Nis}m_1\ddot{x}_1 + c_{Nis}\dot{x}_d. \tag{19}$$

Considering that the acceleration of the ATMD mass can be expressed as $\ddot{x}_2 = \ddot{x}_1 - \ddot{x}_d$, we arrive at the following:

$$\ddot{x}_1(m_2 + g_{Nis}m_1) = m_2\ddot{x}_d + c_{Nis}\dot{x}_d + k_{2dh}x_d. \tag{20}$$

In harmonic motion, the acceleration is proportional to the displacement, as expressed by $\ddot{x}_d = -\omega^2 x_d$. Substituting this into (20) allows us to represent the absolute acceleration of the primary structure as a function of both relative displacement and relative velocity:

$$\ddot{x}_1 = \left(\frac{c_{Nis}}{m_2 + g_{Nis}m_1}\right)\dot{x}_d + \left(\frac{k_{2dh} - \omega^2 m_2}{m_2 + g_{Nis}m_1}\right)x_d. \tag{21}$$

Thus, for harmonic excitation, the control force (18) can be alternatively calculated using only relative displacement and relative velocity signals. After substituting (21) into (18), we could directly determine the parameters c_a and k_a for the control force (12), but only if the ATMD configurations in this study and in [17] are identical.

At this point, it is important to note that the ATMD configuration presented in [17] differs from the one in Figure 1a. Specifically, Nishimura et al. determine the optimal stiffness of the passive spring, denoted below as $k_{2Nis-opt}$, in contrast to the k_{2dh} selected in this work. Additionally, they introduce a passive viscous damper that is connected in parallel with both the passive spring and the active actuator. Considering these differences in ATMD configurations, the parameters we are seeking for the control force (12) can be formulated to ensure full compatibility with the optimal ATMD proposed by Nishimura et al.:

$$c_a = \frac{c_{Nis}m_2}{m_2 + g_{Nis}m_1}, \tag{22}$$

$$k_a = -g_{Nis}m_1\left(\frac{k_{2Nis-opt} - \omega^2 m_2}{m_2 + g_{Nis}m_1}\right) + \left(k_{2Nis-opt} - k_{2dh}\right), \tag{23}$$

where $\left(k_{2Nis-opt} - k_{2dh}\right)$ is added to eliminate a small discrepancy between the spring stiffness k_{2dh} and the optimal value $k_{2Nis-opt}$, and the gain g_{Nis} determined in [17] is:

$$g_{Nis} = \frac{2 + \mu - \mu \cdot DAF_{max}^2}{DAF_{max}^2 + 1}, \tag{24}$$

with DAF_{max} which expresses the resulting maximum dynamic amplification factor that can be achieved with the ATMD.

Considering the control objective and the desired maximal dynamic amplification factor given by (13), the expression for the gain (24) can be simplified to a concise form, dependent on the parameter K:

$$g_{Nis} = \frac{K\mu - \mu}{K\mu + 1}. \tag{25}$$

The spring stiffness $k_{2Nis-opt}$ in (23) is:

$$k_{2Nis-opt} = k_1\frac{\mu}{(1+\mu)^2}(1 - g_{Nis}), \tag{26}$$

and can be expressed as a function of K as:

$$k_{2Nis-opt} = k_{2dh}\frac{1+\mu}{1+K\mu} = k_1\frac{\mu}{(1+K\mu)(\mu+1)}. \tag{27}$$

The viscous damping coefficient in (22) equals:

$$c_{Nis} = 2\zeta_{Nis} m_2 \omega_2, \tag{28}$$

where:

$$\omega_2 = \omega_1 \frac{\sqrt{1 - g_{Nis}}}{1 + \mu} = \frac{\omega_1}{\sqrt{(1+\mu)(K\mu+1)}}, \tag{29}$$

and

$$\zeta_{Nis} = \sqrt{\frac{3(\mu + g_{Nis})}{8(1+\mu)}} = \sqrt{\frac{3}{8}\frac{K\mu}{K\mu+1}}. \tag{30}$$

After substituting (25), (27) and (28) into (22) and (23), we obtain the sought feedback gains conveniently expressed as functions of the primary structure parameters m_1, ω_1, the mass ratio μ, the control parameter K, and, in the case of k_a, the actual frequency of vibration ω:

$$c_a = m_1 \omega_1 \sqrt{\frac{3\mu^3}{2K(1+\mu)^3}}, \tag{31}$$

$$k_a = m_1 \frac{K\mu - \mu}{K\mu + K}\left(\omega^2 - \frac{\omega_1^2}{\mu+1}\right). \tag{32}$$

The ATMD with the viscous damping–stiffness control force defined by (12) and the parameters (31) and (32), represents a new ATMD that is calibrated to replicate the performance of the ATMD by Nishimura et al. under harmonic excitation. Despite their different feedback signals, both ATMD designs, the one presented in [17] and the one introduced here, are fully equivalent. Depending on the actual frequency, the active control force (12) consists of a viscous damping force combined with either a negative, for $\omega^2 < \omega_1^2/(\mu+1)$, or positive, for $\omega^2 > \omega_1^2/(\mu+1)$, stiffness force. The case of $\omega^2 = \omega_1^2/(\mu+1)$ results in a pure viscous damping force.

3.5. Numerical Demonstration

Figure 2 presents the frequency responses of the primary structure equipped with the ATMD with mass ratio $\mu = 1\%$, which was designed using the feedback gains (31) and (32). These results are presented for two distinct values of the control parameter K, namely $K = 2$ and $K = 4$. To provide a more comprehensive evaluation of the ATMD's performance, its characteristics are compared with those for the TMD with mass ratios $\mu = 1\%$, 2%, and 4%. All the characteristics are normalized by $X_{1static} = F_{exc}/k_1$.

Figure 2a confirms that the maximum displacement amplitude of the primary structure for the ATMD with a mass ratio of μ and design parameter K matches that of the TMD with a larger mass ratio of $K\mu$. The peak value of the structure's displacement amplitude is effectively governed by the ATMD's design parameter K.

In the case of Den Hartog's TMD, the peak response of the primary structure is solely dependent on the mass ratio μ, as described by (11). A larger mass ratio results in a smaller maximum relative displacement, as shown in Figure 2b. For the ATMD, its maximum relative displacement does not exceed that of the TMD with the same mass ratio, and this maximum value remains the same for any K, as depicted in Figure 2b.

Since the frequency responses in Figure 2 are normalized, the specific parameters used in this numerical example are not crucial. However, for the sake of completeness, note that the sample parameters used in this work (also in subsequent sections) are as follows: $f_1 = 0.5$ Hz, $m_1 = 500{,}000$ kg, $F_{exc} = 8$ kN.

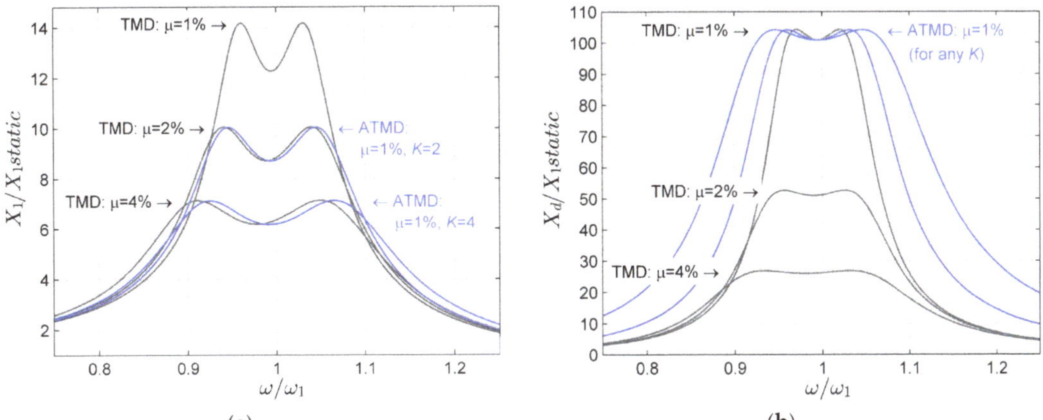

Figure 2. Frequency responses of the primary structure with the ATMD with mass ratio μ = 1%, for K = 2 and K = 4, in comparison to the frequency responses for the TMD with μ = 1%, 2%, and 4%: (**a**) Dynamic amplification of primary structure displacement amplitude; (**b**) dynamic amplification of relative displacement amplitude.

Appendix A contains the MATLAB code that allows for reproducing the calculations of the frequency responses.

4. Sub-Optimal STMD with Clipped Viscous Damping–Stiffness Control Force

In accordance with the terminology used in the literature on semi-active control, see, for example, [45], the term "sub-optimal" or "clipped-optimal" STMD refers to a type of STMD where a semi-active damper is used to realize, to the extent that it is possible, an optimal active control strategy.

It is evident that implementing the active control force (12) in an STMD using a semi-active damper requires the clipping of the non-dissipative forces. This limitation inherently results in inferior performance compared to an ATMD equipped with an active actuator. Despite this, no modifications to the control force are introduced in the sub-optimal STMD system to compensate for the adverse effect of clipping.

The sub-optimal STMD is separately depicted in Figure 1b to distinguish it from the optimized STMD examined later in this paper. Both STMDs, shown in Figure 1b,c, have the same structure. They consist of mass m_2, the same passive spring k_{2dh}, as in Den Hartog's TMD, and a semi-active damper to apply a semi-active force. However, the semi-active force is calibrated differently in each of the two STMD concepts.

In this section, an analysis of the sub-optimal STMD is provided, with a focus on its clipped control force. Examples of force characteristics are described, and a clipping coefficient is introduced to analytically quantify the degree of clipping. The increase in energy dissipation due to the clipping is also calculated. Most notably, the Krylov–Bogoliubov method of equivalent linearization is employed to derive an equivalent model for a clipped viscous-damping control force. This model is useful for demonstrating the influence of force clipping on both the resulting damping and stiffness. These analyses are essential for understanding the limitations of the sub-optimal STMD and for its subsequent optimization.

4.1. Semi-Active Control Force of Sub-Optimal STMD

The semi-active force for the sub-optimal STMD is considered in the same form as it was for the ATMD (12), but with the non-dissipative forces clipped to zero. The clipping operation is crucial as it ensures that the resulting semi-active control force is fully dissipative, thereby making it possible to realize this force using a semi-active damper.

Accordingly, the control force of the sub-optimal STMD can be formulated as:

$$f_{a-clipped} = \begin{cases} f_a: & f_a \dot{x}_d > 0 \\ 0: & otherwise' \end{cases} \quad (33)$$

where f_a is the active control force defined by (12) with the optimal control gains (31) and (32).

From (33), it is evident that the clipping deactivates the semi-active damper in some parts of a vibration cycle, as graphically represented by the on-off switch symbol in Figure 1b,c.

4.2. Examples of Force Characteristics

Figures 3 and 4 show example characteristics of the control force f_a and its clipped form $f_{a-clipped}$ (33) as functions of time, relative displacement, and relative velocity.

The force characteristics in Figure 3 are presented for the case of positive stiffness k_a, under conditions of moderate clipping. Figure 4 presents these characteristics for the case of negative stiffness k_a, for a relatively small degree of clipping. Capital letters are used to label characteristic points in the figures to facilitate tracing the effect of force clipping simultaneously in terms of time, displacement, and velocity.

In Figures 3a and 4a, the time intervals are marked where the force f_a satisfies the condition $f_a \dot{x}_d > 0$, indicating that it has a dissipative character and can therefore be realized by a semi-active damper. The width of these intervals is not constant and depends on the frequency of vibrations ω as well as the parameters k_a and c_a.

For $k_a > 0$, the clipping necessitates an abrupt reduction of the semi-active damper force to zero whenever the sign of the relative velocity changes (points C and F in Figure 3). The clipping is then maintained until the control force f_a, whose absolute value is decreasing, crosses zero (points D and G).

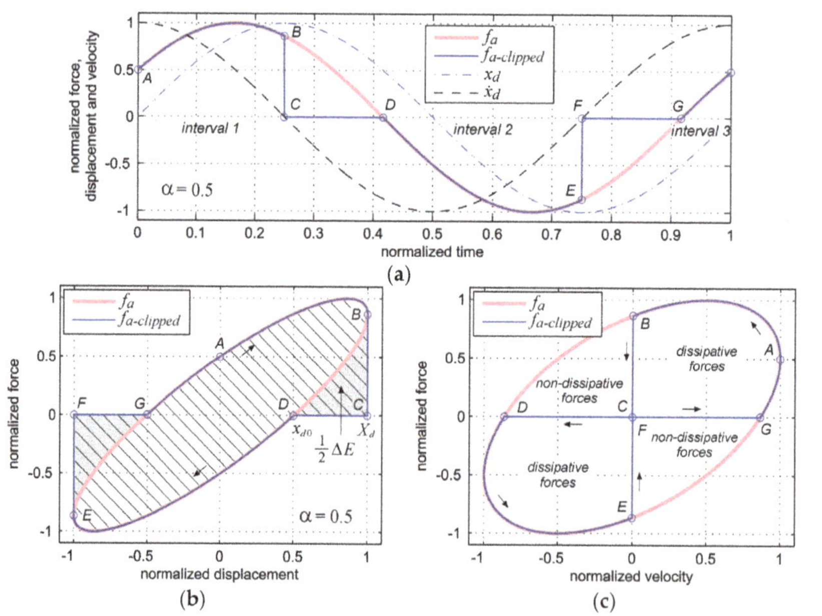

Figure 3. Example of viscous damping—positive stiffness force characteristics before and after clipping: (**a**) Force, displacement and velocity versus time; (**b**) force versus displacement; (**c**) force versus velocity. Clipping coefficient: $\alpha = 0.5$.

In the case of $k_a < 0$, the clipping occurs from the moment the control force f_a crosses zero (points J and M, Figure 4) and initially does not require abrupt change of the force.

However, resuming operation of the semi-active damper requires a sudden change in force from zero to the instantaneous value of f_a (points K and N). More about the clipped viscous damping with negative stiffness can be found in [46,47].

The clipping operation represents a significant constraint and, if the active control force f_a in (33) is optimal, then the clipped force (33) is no longer optimal. However, it is not obvious how exactly the clipping affects the resulting stiffness and damping, depending on the degree of clipping. This issue will be addressed below.

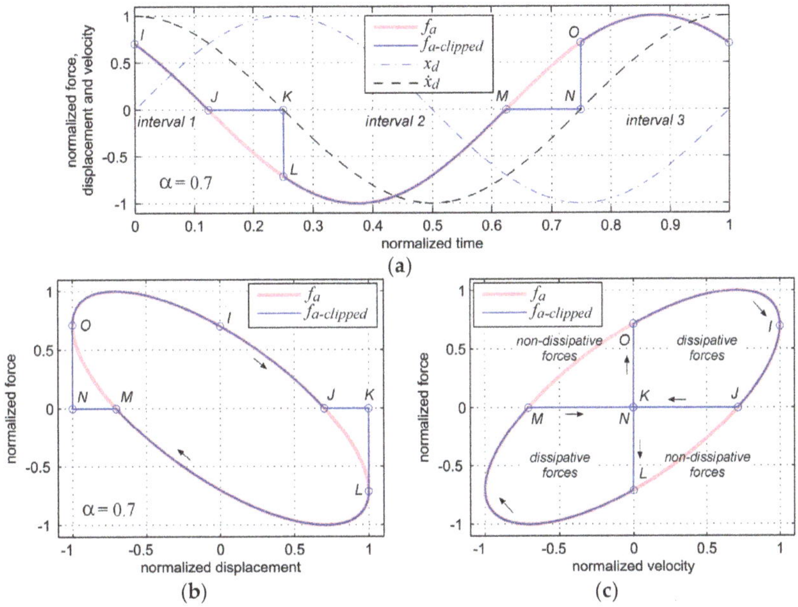

Figure 4. Example of viscous damping—negative stiffness force characteristics before and after clipping: (**a**) Force, displacement and velocity versus time; (**b**) force versus displacement; (**c**) force versus velocity. Clipping coefficient: $\alpha = 0.7$.

4.3. Description of Force Clipping Degree Using the Clipping Coefficient

A description of the clipping should begin with the introduction of a clipping coefficient, which will allow for an assessment of how much the clipped force differs from the unclipped force. For the purposes of this work, the clipping coefficient will be introduced based on the analysis of the force as a function of displacement.

Figures 3b and 4b show an ellipse representing the force f_a as a function of the relative displacement x_d. For $k_a > 0$, this ellipse can be described by:

$$f_{e1} = c_a\omega\sqrt{X_d^2 - x_d^2} + k_a x_d, \tag{34}$$

$$f_{e2} = -c_a\omega\sqrt{X_d^2 - x_d^2} + k_a x_d, \tag{35}$$

where f_{e1} and f_{e2} denote the parametric representation of the higher and lower branches of the ellipse, respectively.

The positive root of f_{e2}, that is, the positive displacement at which the force f_{e2} becomes zero (see point D in Figure 3b) is:

$$x_{d0} = X_d \frac{c_a\omega}{\sqrt{k_a^2 + c_a^2\omega^2}}. \tag{36}$$

Based on (36), the clipping coefficient α is introduced as follows:

$$\alpha = \frac{x_{d0}}{X_d} = \frac{c_a\omega}{\sqrt{k_a^2 + c_a^2\omega^2}}. \tag{37}$$

The clipping coefficient ranges from 1 (no clipping, the case of pure viscous damping, $k_a = 0$), through a middle value of 0.5 (moderate clipping, for $k_a = \sqrt{3}c_a\omega$), to 0 (maximal clipping, for $c_a = 0$).

Figure 5 presents α as a function of normalized frequency for the sub-optimal STMD with mass ratio μ of 1% and 2%, for three selected values of K. Due to their presentation in terms of normalized frequency, the plots are independent of the structural parameters (m_1, ω_1). They depend solely on μ and K, which determine the values of k_a and c_a. The degree of clipping, described by α, is also independent of the forcing amplitude. Figure 5 indicates small and moderate degree of clipping for the sub-optimal STMD for typical values of μ and K in the considered frequency range. In Figure 5a, the two values of α are marked, corresponding to the cases of force characteristics in Figures 3 and 4.

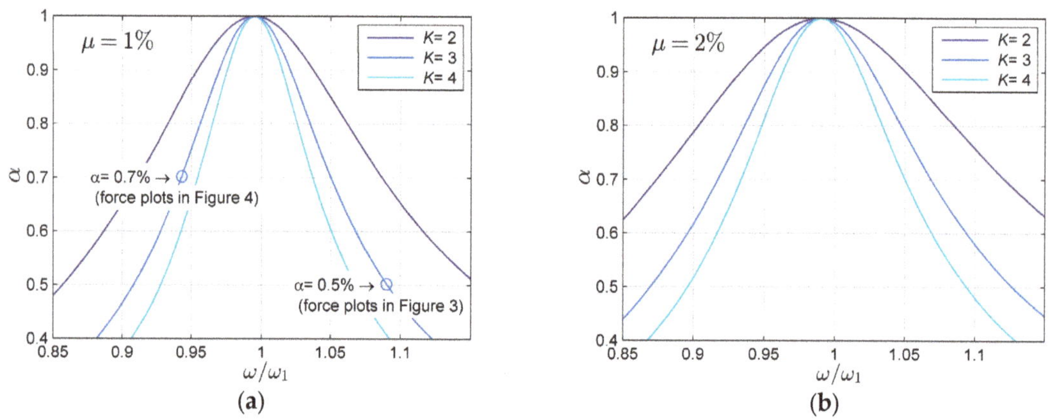

Figure 5. Clipping coefficient versus normalized frequency of vibration calculated for the sub-optimal STMD for different K: (**a**) For $\mu = 1\%$; (**b**) for $\mu = 2\%$.

4.4. Energy Dissipation Resulting from Clipping

Another aspect in the analysis of the clipping operation concerns its impact on the energy dissipation during a vibration cycle. As illustrated by the force-displacement loops in Figures 3b and 4b, the clipping augments cycle energy dissipation. This is evident from the enlargement of the area enclosed by the force-displacement plot. This area is quantitatively equal to the cycle energy.

The total cycle energy resulting from the clipped force (33) is represented by the dashed area in Figure 3b and can be expressed as:

$$E_{clipped} = E_{unclipped} + \Delta E, \tag{38}$$

where $E_{unclipped} = \pi\omega c_a X_d^2$ is the cycle energy due to the active control force f_a, and ΔE denotes the increase in cycle energy due to the clipping. This additional energy ΔE is represented by the shaded areas in Figure 3b and, for $k_a > 0$, can be calculated as:

$$\Delta E = 2\int_{x_{d0}}^{X_d} f_{e2} dx_d. \tag{39}$$

where f_{e2} and x_{d0} are given by (35) and (36), respectively.

Evaluating the integral in (39) yields an analytical expression for the increase in cycle energy due to the clipping that is valid for both positive and negative stiffness k_a:

$$\Delta E = X_d^2\left(c_a\omega\left(\alpha\sqrt{1-\alpha^2} + arcsin(\alpha) - \frac{\pi}{2}\right) - |k_a|\left(\alpha^2 - 1\right)\right). \tag{40}$$

After substituting (40) into (38), the total cycle energy resulting from the clipped force (33) is expressed as ((40) with the sign before the $\pi/2$ term changed):

$$E_{clipped} = X_d^2\left(c_a\omega\left(\alpha\sqrt{1-\alpha^2} + arcsin(\alpha) + \frac{\pi}{2}\right) - |k_a|\left(\alpha^2 - 1\right)\right). \tag{41}$$

The above expression can be employed to derive the energy-equivalent viscous damping coefficient, c_{eq}, by equalizing $E_{clipped} = \pi\omega c_{eq} X_d^2$. However, to establish a full equivalent model for the clipped force (33), an adequate equivalent linearization method must be employed to also derive an equivalent stiffness.

4.5. Calculation of Equivalent Viscous Damping and Equivalent Stiffness for the Clipped Viscous Damping–Stiffness Force Using Harmonic Balance Method

The method of equivalent linearization by Krylov and Bogoliubov [48] is a well-established analytical technique based on the harmonic balance method for linearizing non-linear systems. It can be used to replace a non-linear function of strongly non-linear systems, see e.g., [49], with an equivalent linear function. In the context of semi-active vibration control systems, a good example can be found in [50].

Using this method, the equivalent damping and stiffness are determined by multiplying the force (33) by $cos(\omega t)$ and $sin(\omega t)$, respectively, and then integrating the resulting expressions over a full vibration period:

$$c_{eq} = \frac{1}{X_d\pi}\int_0^{2\pi/\omega} f_{a-clipped}cos(\omega t)dt, \tag{42}$$

$$k_{eq} = \frac{\omega}{X_d\pi}\int_0^{2\pi/\omega} f_{a-clipped}sin(\omega t)dt. \tag{43}$$

Detailed analytical calculations for integrals (42) and (43) are provided in Appendix B. As an outcome of these calculations, the following expressions are derived:

$$c_{eq} = \frac{c_a}{2} + \frac{c_a arcsin(\alpha)}{\pi} + \frac{c_a sin(2arcsin(\alpha))}{2\pi} - \frac{|k_a|(\alpha^2-1)}{\pi\omega}, \tag{44}$$

$$k_{eq} = \frac{k_a}{2} + \frac{k_a arcsin(\alpha)}{\pi} - \frac{k_a sin(2arcsin(\alpha))}{2\pi} - \frac{sgn(k_a)c_a\omega(\alpha^2-1)}{\pi}. \tag{45}$$

It should be noted that although the amplitude of relative motion X_d appears in (42) and (43), the resulting equivalent model is independent of the amplitude X_d. Consequently, the equivalent model is independent of the amplitude of force excitation. This feature holds true only for harmonic motion. The equivalent model, represented by equations (44) and (45), will be used later to analyze the frequency responses of the STMD.

In Figure 6, sample calculations of the equivalent viscous damping and the equivalent stiffness are presented for the sub-optimal STMD with $\mu = 1\%$ and $K = 3$. The calculations were conducted using the sample parameters listed in Section 3.5.

The greater the difference between ω and ω_1, the more the degree of clipping increases. Consequently, both the amount of additional energy dissipated and the resulting equivalent viscous damping coefficient c_{eq} rise. As an effect, c_{eq} significantly exceeds c_a (Figure 6a). In the case of stiffness (Figure 6b), the resulting equivalent stiffness, k_{eq}, is noticeably smaller in magnitude than k_a. Furthermore, the difference between k_a and k_{eq} becomes larger as the difference between ω and ω_1 increases.

The cumulative impact of clipping on both damping and stiffness adversely affects the performance of the sub-optimal STMD, as will be demonstrated in the following.

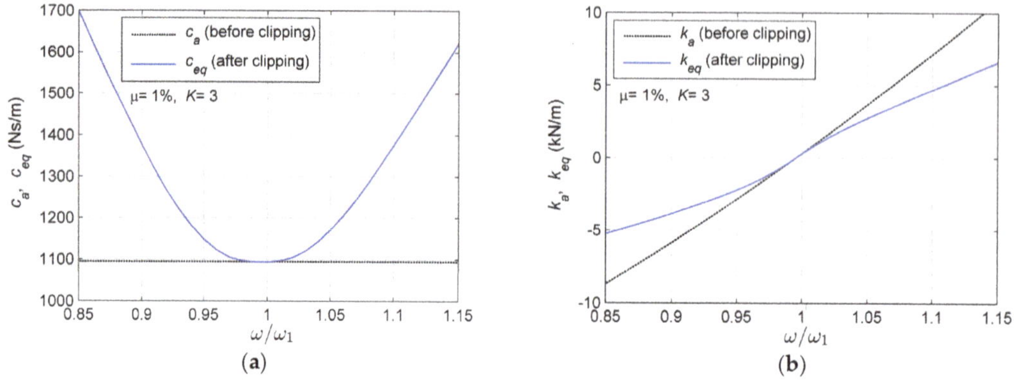

Figure 6. Force parameters before clipping and their equivalent counterparts after clipping versus normalized frequency of vibration for the sub-optimal STMD with μ = 1% and K = 3: (**a**) Viscous damping coefficient c_a and equivalent viscous damping coefficient c_{eq}; (**b**) stiffness k_a and equivalent stiffness k_{eq}.

4.6. Numerical Analysis of Sub-Optimal STMD

Figure 7 presents the frequency responses calculated for the sub-optimal STMD with μ = 1%, for six different values of K from 2 to 20, in comparison to the frequency responses obtained for the TMD. These results clearly indicate that the performance of the sub-optimal STMD is noticeably worse than that of the ATMD (Figure 2). This is attributed to the described increase in damping and reduction in stiffness as the result of clipping.

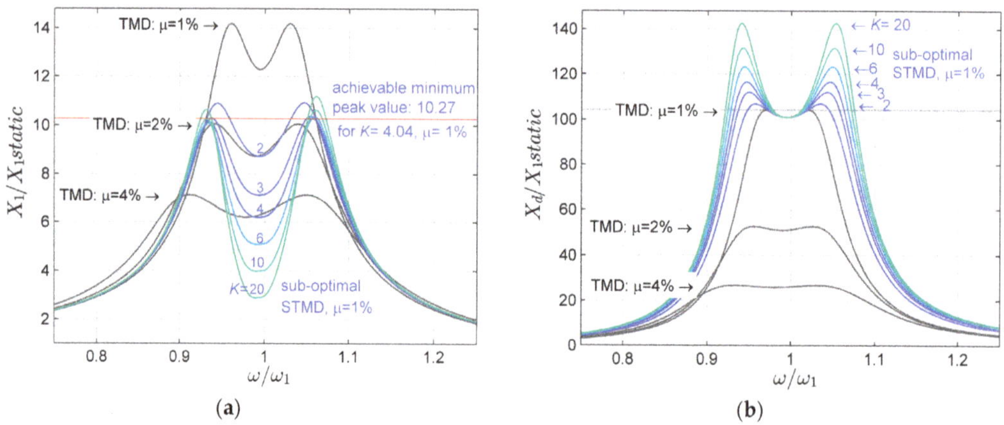

Figure 7. Frequency responses of the primary structure with the sub-optimal STMD with μ = 1%, for six different K ranges from 2 to 20, in comparison to the frequency responses due to the TMD with μ = 1%, 2% and 4%: (**a**) Dynamic amplification of primary structure displacement; (**b**) dynamic amplification of relative displacement.

For normalized frequencies ω/ω_1 close to 1, the control force (33) is minimally affected by clipping. As a result, in this frequency range, the sub-optimal STMD operates as intended, and the resulting vibration amplitude X_1 is consistent with that for the TMD with an enlarged mass ratio $K\mu$ (Figure 7a). Specifically, when $\omega = \omega_1/\sqrt{\mu+1}$, the stiffness k_a becomes zero, making the control force of the STMD purely dissipative. At this central frequency, the performance of the STMD is equivalent to that of the ATMD.

The numerical analysis revealed that the sub-optimal STMD has a clear limit in effectiveness, determined as the smallest achievable value of maximum dynamic amplification factor (DAF_{max}). For $\mu = 1\%$, the calculations show that the minimum DAF_{max} of 10.27 is achieved for $K = 4.04$, as indicated by the red line in Figure 7a. Increasing K beyond 4.04 can only result in an increase in DAF_{max}, with a slightly higher right peak.

Additional significant limitations of the sub-optimal STMD are evident in Figure 7b. The maximum relative displacement for the sub-optimal STMD far exceeds that of a TMD with the same mass. Such performance from STMD is considered unacceptable. In practice, constraining the relative motion is crucial due to spatial limitations at the installation site of the mass damper. Active and semi-active solutions are expected to fit within the same space as a TMD with the same mass. This consideration is also taken into account in the subsequent numerical optimization of the STMD.

Note that the frequency responses shown in Figure 7 were calculated using (16) and (17), as previously in Section 3.5. However, in these calculations, c_a and k_a were substituted with the equivalent model c_{eq} (44) and k_{eq} (45), respectively. To assess the accuracy of the equivalent model, analogous steady-state frequency responses were also determined through simulations in MATLAB/Simulink. Based on this comparison, it was determined that regardless of the amplitude of force excitation, the maximum discrepancy between the calculated and simulated amplitude X_1 in the steady state does not exceed 0.58% for $K = 3$, 0.93% for $K = 4$, and 2.7% for $K = 20$. Therefore, the equivalent model (44) and (45) is assumed to be sufficiently accurate and useful in predicting the frequency responses of the primary structure equipped with the STMD.

Appendix A provides the MATLAB code for calculating the frequency responses for the sub-optimal STMD.

5. Calibration of STMD Using Numerical Optimization

The control force of the sub-optimal STMD, as previously discussed, lacks adjustments to mitigate the effects of force clipping. In this section, an optimized STMD featuring necessary modifications to its control force is presented. These modifications are aimed at compensating for the effects of clipping to the extent possible, thereby improving the calibration and overall performance of the STMD.

A modified semi-active control force with two incorporated correction factors is subsequently introduced, designed to scale the control force before clipping occurs. The numerical optimization procedure used to calibrate these correction factors are then outlined. Finally, the optimization results as well as an analysis of the optimized STMD at its performance limit are presented.

5.1. Semi-Active Control Force

The semi-active control force f_{sa} for the STMD depicted in Figure 1c is defined as:

$$f_{sa} = \begin{cases} c_{sa}\dot{x}_d + k_{sa}x_d : & (c_{sa}\dot{x}_d + k_{sa}x_d)\dot{x}_d > 0 \\ 0 : & otherwise \end{cases}, \tag{46}$$

where c_{sa} and k_{sa} are the modified force parameters, formulated similarly as before, see (31) and (32), but incorporating correction factors K_{cor} and c_{cor}:

$$c_{sa} = c_{cor} m_1 \omega_1 \sqrt{\frac{3\mu^3}{2K_{cor}K(1+\mu)^3}}, \tag{47}$$

$$k_{sa} = m_1 \frac{K_{cor}K\mu - \mu}{K_{cor}K\mu + K_{cor}K} \left(\omega^2 - \frac{\omega_1^2}{\mu + 1} \right). \tag{48}$$

The adoption of the modified control force in the form (46) is based on an analysis of the operation of the sub-optimal STMD. The introduction of K_{cor} is necessary to achieve the proper effectiveness, which is defined for the design parameter K in the same manner as for the ATMD (see (13)). The parameter c_{cor} allows for adjustment of damping, which will enable control of the frequency response of the STMD around the central frequency between the two peaks.

It should be noted that the introduction of the correction factors in the notations of relative velocity and displacement feedback gains for the control force of STMD was previously presented in [42]. The method used in [42] is very similar; however, it was applied to a differently defined control force with an additional feedback loop from acceleration.

5.2. Objective and Procedure of Optimization

In the case of the sub-optimal STMD, a clear performance limit was observed and manifested as the smallest achievable value of DAF_{max} (Figure 7a). Therefore, the first natural objective of optimizing the STMD was to determine the best calibration of the STMD force parameters for the highest possible efficiency, exceeding that of the sub-optimal STMD.

The main objective of optimization was to determine the STMD force parameters for various K and μ, in such a way as to minimize the differences in efficiency between the STMD and the ATMD for a given K. It was also assumed that the optimized STMD must satisfy the condition concerning the amplitude of relative displacement X_d, which cannot exceed that for a TMD with the same mass and for the same excitation.

The numerical optimization was carried out in MATLAB using the unconstrained minimization method implemented as the *fminsearch* function. The objective function $J(v)$ to be minimized using the *fminsearch* function was defined based on the dynamic amplification of the primary structure displacement, as introduced in [42]. This objective function consists of the sum of squared differences between the local peaks, and the local minimum between those peaks, for both the STMD with specific μ and K, and the Den Hartog's TMD with mass ratio $K\mu$:

$$J(v) = (P_{sa}(v) - P)^2 + (Q_{sa}(v) - Q)^2 + (S_{sa}(v) - S)^2, \tag{49}$$

where $v = [K_{cor}\ c_{cor}]$ is a vector of the optimization variables; $P_{sa}(v)$, $Q_{sa}(v)$ and $S_{sa}(v)$ are, respectively, the left and right local peaks and the local minimum in between those peaks of the frequency response $X_1/X_{1static}$ of the primary structure with the STMD of mass ratio μ and control parameter K; while P, Q and S are the corresponding quantities for the TMD with mass ratio $K\mu$.

Specifically, the objective function (49) assumes a zero value when the two peaks of the frequency response for the STMD are equal and match the value given by (13), and when the local minimum is as for the TMD with mass ratio of $K\mu$.

The *fminsearch* function was run with an initial estimate of the optimization variables v = [1 1]. The maximum number of iterations allowed was set to 400. In each iteration of the optimization, the actual feedback gains c_{sa} (47), k_{sa} (48) and the corresponding equivalent model c_{eq} (44), k_{eq} (45) are calculated for the current estimate of optimization variables K_{cor}, c_{cor}. Then, the frequency response of the primary structure with the STMD is determined, as well as the current value of the objective function (49). The frequency responses are calculated using (16), but with c_{eq} and k_{eq} instead of c_a and k_a. This approach allows for rapid frequency response calculations in each iteration, leading to quickly obtained optimization results even for large number of iterations.

The optimization was conducted for the sample model parameters listed in Section 3.5, for various K, and for μ ranging from 0.4% to 3%.

5.3. Optimization Results

An important observation from the optimization is that the optimized STMD reaches a clear efficiency limit around $K = 3$. For $K > 3$, the optimization yields disproportionately large values of K_{cor} and c_{cor} without achieving further improvement in the STMD efficiency.

The optimized correction factors K_{cor} and c_{cor} are plotted in Figure 8a,b versus K from 1 to 3 and for various μ from 0.4% to 3%. Within the range of values presented in these figures, both K_{cor} and c_{cor} increase with K in a similar manner for each mass ratio μ.

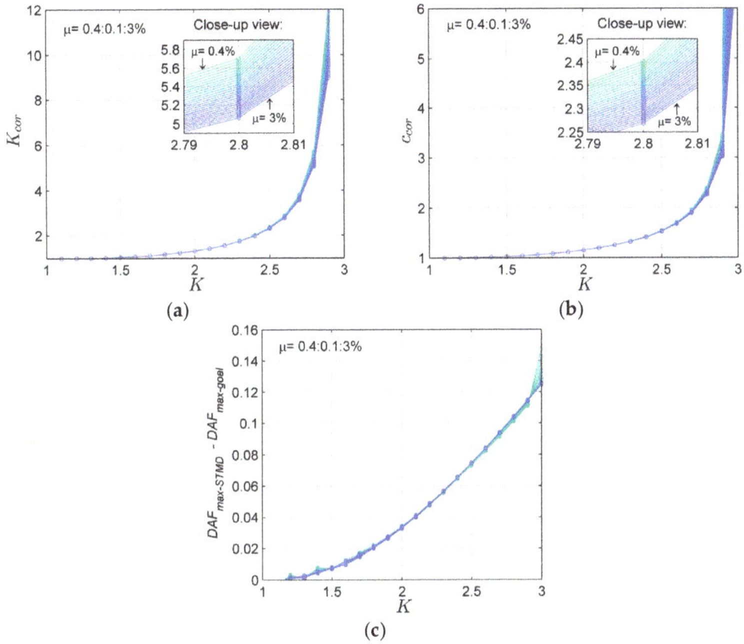

Figure 8. Results of the numerical optimization for different mass ratios μ ranging from 0.4% to 3% versus design parameter K: (**a**) Correction factor K_{cor}; (**b**) correction factor c_{cor}; (**c**) resulting difference between maximum dynamic amplification factors for the optimized STMD with mass ratio μ and the TMD with mass ratio $K\mu$.

Figure 8c shows the difference between the maximum DAF (10) values calculated for the optimized STMD with mass ratio μ and the reference TMD with mass ratio $K\mu$. This difference serves as a straightforward measure for assessing the results of optimization. The maximum value in Figure 8c is 0.148, obtained for $K = 3$ and $\mu = 0.4\%$. This value implies that the peak response of the STMD exceeds the target $DAF_{max} = 12.95$ by 1.14%. Similarly, when $K = 3$ and $\mu = 3\%$, the maximum difference is 0.125, indicating that the STMD's peak response is 2.6% above the targeted value of 4.82. Given that the maximum percentage error is under 3%, the optimization outcomes are considered acceptable.

Table 1 lists the optimized values of K_{cor} and c_{cor} for selected K ranging from 1.5 to 3, and for three typical μ of 0.5%, 1%, and 2%. It is noticeable that the values for $K = 3$ are exceptionally high compared to the rest of the results. When K is slightly reduced to 2.98, these values become significantly smaller. This is because, for $K = 3$, the optimized STMD is already at its performance limit.

Table 1. Optimized parameters K_{cor} and c_{cor} for $\mu = 0.5\%$, 1%, and 2% and for different K.

K	Correction Factor K_{cor}			Correction Factor c_{cor}		
	μ=0.5%	μ=1%	μ=2%	μ=0.5%	μ=1%	μ=2%
1.5	1.0585	1.0582	1.0582	1.0285	1.0283	1.0283
2	1.3336	1.3331	1.3321	1.1546	1.1539	1.1530
2.4	2.0131	2.0058	1.9934	1.4201	1.4177	1.4129
2.6	2.9001	2.8751	2.8307	1.7085	1.7019	1.6891
2.8	5.6752	5.5289	5.2808	2.3970	2.3679	2.3161
2.9	11.715	10.991	9.8577	3.4490	3.3446	3.1709
2.98	106.41	62.555	35.590	10.407	7.9897	6.0348
3	1239.4	890.90	106.90	35.534	30.165	10.463

5.4. Analysis of the Optimized STMD at Its Performance Limit

5.4.1. Frequency Responses

Figure 9 presents the frequency responses for the optimized STMD with mass ratio $\mu = 1\%$ and the highest achievable efficiency at $K = 3$. The results for the optimized STMD are compared with those for the sub-optimal STMD and the ATMD (both with the same μ and K). For additional context, this figure also includes results for the TMD with μ = 1% and 3%.

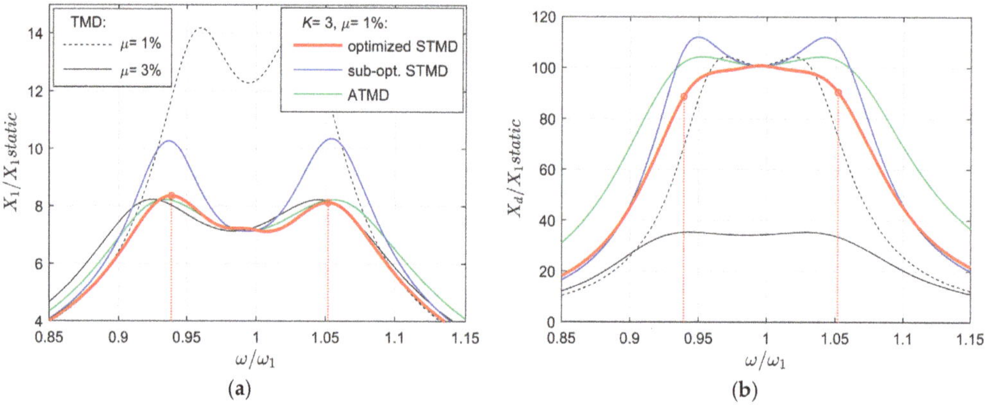

Figure 9. Frequency responses for the optimized STMD in comparison to those for the sub-optimal STMD and ATMD, all with $\mu = 1\%$ and $K = 3$ and TMD with $\mu = 1\%$ and 3%: (**a**) Dynamic amplification of primary structure displacement; (**b**) dynamic amplification of relative displacement.

Figure 9a confirms that the optimized STMD performs significantly better than the sub-optimal STMD and nearly as well as the ATMD with the same K. More precisely, while the shapes of the responses for the STMD and ATMD are slightly different, the left peak of the STMD response is 2.4% higher and the right peak is slightly smaller than that of the ATMD.

Significantly, the maximum amplitude of relative displacement for the optimized STMD is not greater than that for the TMD and ATMD with the same mass ratio. In fact, the relative displacement for the optimized STMD is lower than that for the ATMD across the entire frequency range, as shown in Figure 9b.

5.4.2. Characterization of the Control Force

It is underlined that the enhanced efficiency of the optimized STMD is not associated with extensive control force demands. In Figure 10a, the force demands for the semi-active damper in the STMD are compared with those for the active actuator in the ATMD and the viscous damper in the TMDs. The figure displays the maximum force requirements at steady-state as a function of frequency, assuming a force excitation of F_{exc} = 8 kN.

Figure 10a reveals that the maximum control force of the optimized STMD is only 6% greater than that of the ATMD and 9% greater than that of the sub-optimal STMD. This occurs even though the optimized STMD at $K = 3$ is operating at its performance limit and the correction factors are significantly high. The only slight increase observed in the control force of the optimized STMD can be attributed to its dependency on relative motion, which is considerably reduced compared to both the sub-optimal STMD and ATMD, as previously explained. Note that the results in Figure 10a are not normalized. The actual force demands will depend on the assumed worst-case excitation force and the specific model parameters.

Figure 10b provides further insights into the control force characteristics of the optimized STMD by comparing the clipping coefficients α for both the optimized and sub-optimal STMDs. Throughout the entire frequency range, the degree of force clipping for the optimized STMD is noticeably higher (α is smaller) than for the sub-optimal STMD. This difference arises because the optimized stiffness k_{sa} used in the calculations of the control force (46) is significantly greater than its corresponding value k_a in the sub-optimal STMD. The active force component, $k_{sa}x_d$, consequently intensifies, leading to a higher degree of clipping. In the following, we discuss how this increased stiffness force enhances the efficiency of the optimized STMD, despite the higher degree of force clipping.

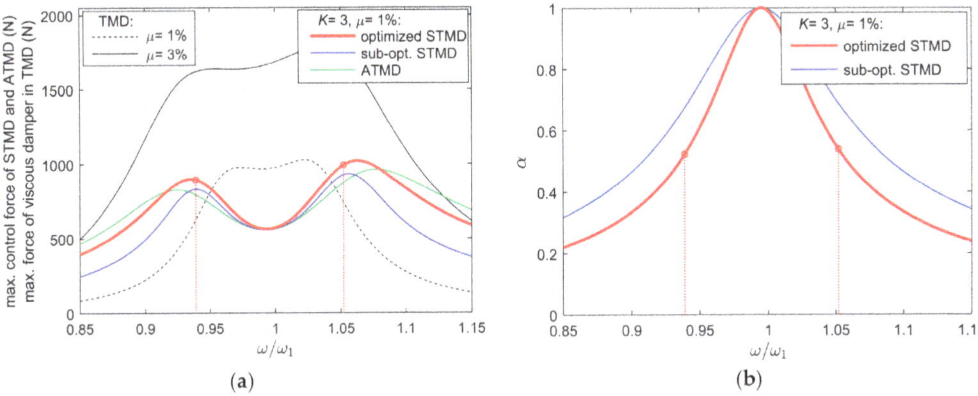

Figure 10. Characterization of the control force of the optimized STMD: (**a**) Maximum control force versus normalized frequency for the optimized and sub-optimal STMDs and ATMD, all with $\mu = 1\%$ and $K = 3$, and for TMD with $\mu = 1\%$ and 3%; (**b**) force clipping coefficient for the optimized and sub-optimal STMDs.

5.4.3. Analysis of the Force–Displacement Characteristics at Crucial Frequencies

Figure 11a,b provide a comparison of the force–displacement characteristics between the optimized STMD and the sub-optimal STMD at two specific frequencies. These frequencies correspond to the left and right peaks of the STMD response, as indicated by the red dotted lines in Figures 9 and 10.

The force plots in Figure 11 are supplemented with values for relevant parameters: k_a (the optimal controlled stiffness of the ATMD, also used in force calculation for the sub-optimal STMD), k_{eq} (the resultant equivalent stiffness after force clipping achieved for the sub-optimal STMD), k_{sa} (the stiffness used in force calculation for the optimized STMD), and k_{eq-opt} (introduced here to denote the equivalent stiffness resulting from the semi-active force of the optimized STMD, calculated using (45) but for k_{sa} and c_{sa}). These parameters clearly demonstrate that the resultant equivalent stiffness k_{eq-opt} for the optimized STMD is approximately equal to the optimal stiffness k_a of the ATMD, for both positive and negative stiffness k_a.

For a better illustration of this effect, Figure 11a,b include plots of the linear stiffness force $k_a x_d$ of the optimal ATMD, as well as the equivalent linear stiffness forces $k_{eq}x_d$ and $k_{eq-opt}x_d$ for both the sub-optimal and the optimized STMD, respectively. Additionally, a comparison of the optimal stiffness k_a, with the resulting equivalent stiffnesses k_{eq} and k_{eq-opt}, is presented as a function of frequency in Figure 11c. Based on these results, it can be concluded that the presented calibration method for the STMD enables precise control of both positive and negative stiffness. Around crucial frequencies of both peaks, the resulting equivalent stiffness for the optimized STMD is closely aligned with the optimal stiffness of the reference ATMD with the same mass ratio and K parameter. However, it should be emphasized that as the value of K increases, the absolute value of the optimal stiffness k_a also increases (at a given frequency), while the demanded viscous damping c_a decreases. According to the previous analysis, increasing k_a leads to a higher degree of clipping. This, in turn, results in greater cycle energy dissipation, as well as increased the resultant equivalent viscous damping. As a result, it is not possible to satisfy both the demands for increased stiffness and decreased damping simultaneously. This leads to an inherent performance limit of the STMD.

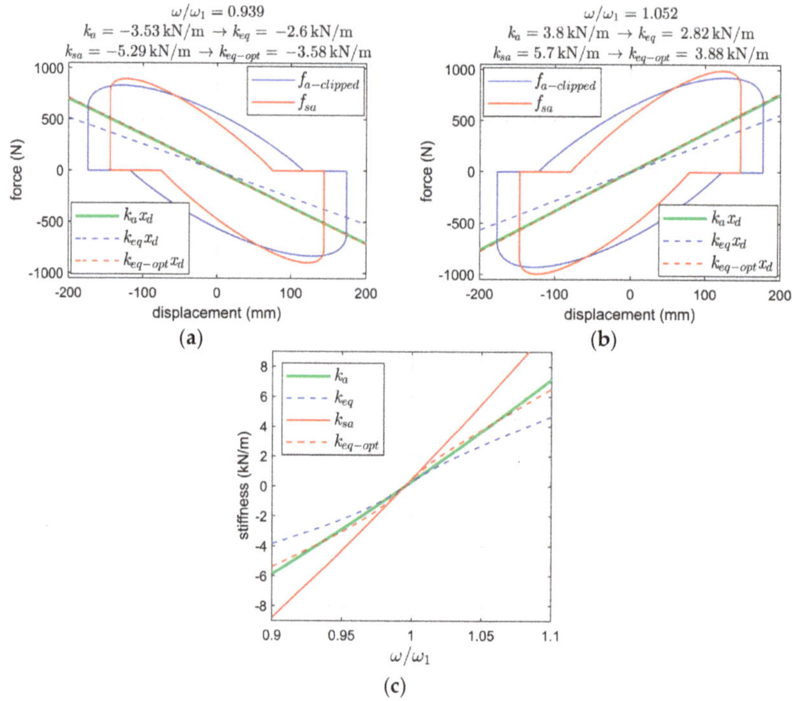

Figure 11. Further characterization of the control force of the optimized STMD in comparison to that of the sub-optimal STMD: (**a**) Force–displacement loops at the frequency corresponding to the left peak of the frequency response; (**b**) force–displacement loops at the frequency corresponding to the right peak; (**c**) comparison of the optimal controlled stiffness of the ATMD with the equivalent stiffnesses resulting from control forces of both STMDs. All results for $\mu = 1\%$ and $K = 3$.

6. Conclusions

This paper investigated both active and semi-active tuned mass dampers with relative displacement and velocity feedback. Following the approach in [42], both mass dampers were calibrated for the specific parameter K, ensuring that their effectiveness in reducing harmonic vibrations is comparable to that of the Den Hartog's TMD with K times larger mass. The key conclusions drawn from this work are as follows:

1. The optimal ATMD with acceleration feedback proposed by Nishimura et al. [17] can alternatively be realized based on a relative displacement and velocity feedback, without relying on acceleration feedback.
2. Both the sub-optimal STMD and the optimized STMD with the clipped viscous damping–stiffness control force have a clearly defined efficiency limit.
3. The application of the numerically optimized correction factors in the semi-active control force (46)–(48) allowed us to achieve an enhanced performance of the optimized STMD compared to the sub-optimal STMD.
4. The highest effectiveness of the optimized STMD in reducing harmonic vibrations corresponds to the TMD with roughly three times greater mass.

Although the results presented in this work are promising, they should be treated as primarily theoretical due to the assumptions made regarding the use of an ideal semi-active damper in the STMD and an ideal active actuator in the ATMD. In practice, the effectiveness of both systems will be limited, in part, due to the force tracking errors, which are discrepancies between the ideal control force and its actual value. These aspects require further investigation in future research.

Funding: This research was funded by the AGH University of Krakow within the scope of the Research Program No. 16.16.130.942.

Institutional Review Board Statement: Not applicable.

Informed Consent Statement: Not applicable.

Data Availability Statement: Data are contained within the article.

Conflicts of Interest: The author declares no conflict of interest.

Appendix A. MATLAB Code

For the reader's convenience, the MATLAB code comparing the frequency responses for the ATMD, sub-optimal STMD, optimized STMD, and TMD is provided. The main script, responsible for calculating and plotting the responses, can be found in Table A1. This script utilizes the following functions:

- 'mm_FRF.m' (Table A2): Calculates the frequency response;
- 'mm_fa_gains.m' (Table A3) and 'mm_fsa_gains.m' (Table A4): Compute the parameters for active and semi-active control forces, respectively;
- 'mm_DH_tuning.m' (Table A5): Implements the Den Hartog tuning;
- 'mm_equiv_linear.m' (Table A6): Performs the equivalent linearization of the clipped viscous damping–stiffness force.

For each formula implemented in the code, the corresponding equation number from the paper is provided.

Table A1. MATLAB script comparing the frequency responses for the ATMD, STMD, and TMD.

```
% Paper_demo.m
% Calculations and plots of the frequency responses for the ATMD, STMD, and TMD
close all, clear all
%% Design parameters (adjust, see Table 1 for Kcor and ccor)
K = 3; mu = 1/100; Kcor = 890.90; ccor = 30.165;
% K = 2.8; mu =1/100; Kcor = 5.5289; ccor = 2.3679;
% K = 2.6; mu = 1/100; Kcor = 2.8751; ccor = 1.7019;
% K = 2; mu = 1/100; Kcor = 1.3331; ccor = 1.1539;
%% Model parameters
f1 = 0.5; % natural frequency, Hz
om1 = 2*pi*f1; % natural frequency, rad/s
m1 = 500e3; % structure mass, kg
Fexc = 8e3; % amplitude of force excitation, N
k1 = om1^2*m1; % structure stiffness, N/m
m2 = m1*mu; % mass, kg
om = om1*[0.7:2e-3:1.3]; % frequency of excitation vector, rad/s
%% TMDs design and frequency responses for mass ratio mu and K*mu
[k2dh, c2dh] = mm_DH_tuning(m1, om1, mu);
[k2dhK,c2dhK] = mm_DH_tuning(m1, om1, K*mu);
[TMD1_DAFX1,TMD1_DAFXd] = mm_FRF(m1, om1, m2, k2dh, c2dh, om, Fexc);
[TMD2_DAFX1,TMD2_DAFXd] = mm_FRF(m1, om1, K*m2, k2dhK, c2dhK, om, Fexc);
%% ATMD design and frequency responses for mass ratio mu
[ca, ka] = mm_fa_gains(K, m1, om1, mu, om);
[ATMD_DAFX1, ATMD_DAFXd] = mm_FRF(m1, om1, m2, (k2dh+ka), ca, om, Fexc);
%% Sub-optimal STMD - frequency responses for mass ratio mu
[keq, ceq, alpha] = mm_equiv_linear(ka, ca, om);
[STMD_DAFX1sub,STMD_DAFXdsub]=mm_FRF(m1,om1,m2,(k2dh+keq),ceq,om,Fexc);
%% Optimized STMD design and frequency responses for mass ratio mu
```

Table A1. *Cont.*

```
[csa, ksa] = mm_fsa_gains(K, m1, om1, mu, om, Kcor, ccor);
[keqopt, ceqopt, alphaopt] = mm_equiv_linear(ksa, csa, om);
[STMD_DAFX1,STMD_DAFXd]=mm_FRF(m1,om1,m2,(k2dh+keqopt),ceqopt,om,Fexc);
%% Plots
figure
plot(om/om1, TMD1_DAFXd,'–k'), hold on
plot(om/om1, TMD2_DAFXd,'k')
plot(om/om1, ATMD_DAFXd,'g')
plot(om/om1, STMD_DAFXdsub,'b')
plot(om/om1, STMD_DAFXd,'r')
xlabel('omega/omega1'), ylabel('Xd/X1static'), xlim([0.7 1.3])
figure
plot(om/om1, TMD1_DAFX1,'--k'), hold on
plot(om/om1, TMD2_DAFX1,'k')
plot(om/om1, ATMD_DAFX1,'g')
plot(om/om1, STMD_DAFX1sub,'b')
plot(om/om1, STMD_DAFX1,'r')
xlabel('omega/omega1'), ylabel('X1/X1static'), xlim([0.7 1.3])
legend(['TMD,\mu=',num2str(100*mu),'%'],['TMD,\mu=',num2str(100*K*mu),'%'],...
['ATMD,\mu=',num2str(100*mu),'%,K=',num2str(K)],...
['STMD-sub,\mu=',num2str(100*mu),'%,K=',num2str(K)],...
['STMD-opt,\mu=',num2str(100*mu),'%,K=',num2str(K)])
```

Table A2. MATLAB function for the frequency response calculation.

```
function [DAFX1, DAFXd] = mm_FRF(m1, om1, m2, k2, c2, om, Fexc)
% Frequency response
% Outputs: DAFX1, DAFXd - dynamic amplifications of disp. amplitudes X1 and Xd
k1 = om1^2*m1; % structure' stiffness, N/m
% Complex amplitudes of harmonic response:
num = k2-om.^2*m2 + 1i*om.*c2;
den = (k1-om.^2*(m1 + m2)).*(k2-om.^2*m2 + 1i*om.*c2) - om.^4*m2^2;
x1 = Fexc * num./den; % structural mass complex amplitude, Equation (8)
xd = Fexc * om.^2*m2 ./den; % relative complex amplitude, Equation (9)
% Displacement amplitudes:
X1 = abs(x1); % absolute amplitude X1 of mass m1, m
Xd = abs(xd); % relative amplitude Xd, m
% Dynamic amplification factors:
X1static = Fexc/k1; % static deflection due to the excitation force, m
DAFX1 = X1/X1static; % dynamic amplification of X1, Equation (10)
DAFXd = Xd/X1static; % dynamic amplification of Xd
```

Table A3. MATLAB function for calculation of the control force parameters of ATMD.

```
function [ca, ka] = mm_fa_gains(K, m1, om1, mu, om)
% Viscous damping–stiffness force parameters
% Outputs: ca, ka - parameters of the control force given by Equation (12)
ca = m1*om1*sqrt(3*mu^3/(2*K*(1 + mu)^3)); % Equation (31)
ka = m1* (K*mu-mu)/(K*mu + K) * (om.^2 - om1^2/(mu + 1)); % Equation (32)
```

Table A4. MATLAB function for calculation of the control force parameters of STMD.

```
function [csa, ksa] = mm_fsa_gains(K, m1, om1, mu, om, Kcor, ccor)
% Clipped viscous damping–stiffness force parameters
% Outputs: csa, ksa - parameters of the control force given by Equation (46)
csa = ccor*m1*om1*sqrt(3*mu^3/(2*K*Kcor*(1 + mu)^3));% Equation (47)
ksa = m1*(K*Kcor*mu-mu)/(K*Kcor*mu + K*Kcor)*(om.^2-om1^2/(mu + 1));% Equation (48)
```

Table A5. MATLAB function for calculation of the TMD parameters.

```
function [k2dh, c2dh] = mm_DH_tuning(m1, om1, mu)
% Den Hartog tuning, Equations (1) and (3)
% Outputs: k2dh - stiffness, N/m; c2dh - viscous damping, Ns/m
m2 = mu*m1; % mass, kg
om2 = om1/(1 + mu); % frequency, rad/s
k2dh = om2^2*m2; % spring stiffness, N/m
dzeta2 = sqrt(3/8*mu/(1 + mu)); % damping ratio
c2dh = 2*dzeta2*m2*om2; % viscous damping coefficient, Ns/m
```

Table A6. MATLAB function for the equivalent linearization of the clipped control force.

```
function [keq, ceq, alpha] = mm_equiv_linear(ka, ca, om)
% Equivalent linearization of the clipped viscous damping–stiffness control force
% Outputs: keq, ceq - equivalent stiffness and equivalent viscous damping,
% alpha - clipping coefficient
alpha = ca.*om ./sqrt(ka.^2 + om.^2.*ca.^2);% Equation (37)
ceq = ca/2 + ca.*asin(alpha)/pi + ca.*sin(2*asin(alpha))/(2*pi) -...
sign(ka).*ka.*(alpha.^2 -1)./(pi*om); % Equation (44)
keq = ka/2 - sign(ka).*ca.*om.*(alpha.^2 -1)/pi + ka.*asin(alpha)/pi -...
ka.*sin(2*asin(alpha) )/(2*pi); % Equation (45)
```

Appendix B. Analytical Calculations of Integrals in the Equivalent Linearization for a Clipped Viscous Damping–Stiffness Force

We calculate the integrals in (42) and (43) as the sum of the definite integrals in those time intervals where the force is not clipped to zero. For harmonic displacement $x_d = X_d sin(\omega t)$, and velocity $\dot{x}_d = \omega X_d cos(\omega t)$, (42) and (43) can be expressed as:

$$c_{eq} = \frac{1}{X_d \pi} \int_0^{2\pi/\omega} (c_a \omega X_d cos(\omega t) + k_a X_d sin(\omega t)) cos(\omega t) dt, \tag{A1}$$

$$k_{eq} = \frac{\omega}{X_d \pi} \int_0^{2\pi/\omega} (c_a \omega X_d cos(\omega t) + k_a X_d sin(\omega t)) sin(\omega t) dt. \tag{A2}$$

To facilitate the description, let us assume:

$$k_{eq} = A + B, \tag{A3}$$

$$c_{eq} = C + D, \tag{A4}$$

where for the harmonic motion, the integrals A, B, C, D become independent of the amplitude X_d:

$$A = \frac{c_a \omega^2}{\pi} \int_0^{2\pi/\omega} sin(\omega t) cos(\omega t) dt, \tag{A5}$$

$$B = \frac{k_a \omega}{\pi} \int_0^{2\pi/\omega} sin^2(\omega t) dt, \tag{A6}$$

$$C = \frac{c_a \omega}{\pi} \int_0^{2\pi/\omega} cos^2(\omega t) dt, \tag{A7}$$

$$D = \frac{k_a}{\pi} \int_0^{2\pi/\omega} sin(\omega t) cos(\omega t) dt. \tag{A8}$$

Integrals (A5)–(A8) will be calculated in the three intervals marked in Figure 4a, for the case of negative stiffness k_a. The interval 1 is defined from 0 to the time instant corresponding to point J in Figure 4a and can be given by $(0, arcsin(\alpha)/\omega)$, where α is the clipping coefficient (37). Integration over the time interval 1 yields:

$$A_1 = \frac{c_a \omega^2}{\pi} \int_0^{arcsin(\alpha)/\omega} sin(\omega t) cos(\omega t) dt = \frac{\alpha^2 c_a \omega}{2\pi}, \tag{A9}$$

$$B_1 = \frac{k_a\omega}{\pi}\int_0^{\frac{arcsin(\alpha)}{\omega}} sin^2(\omega t)dt = \frac{k_a arcsin(\alpha)}{2\pi} - \frac{k_a sin(2arcsin(\alpha))}{4\pi}, \tag{A10}$$

$$C_1 = \frac{c_a\omega}{\pi}\int_0^{\frac{arcsin(\alpha)}{\omega}} cos^2(\omega t)dt = \frac{c_a arcsin(\alpha)}{2\pi} + \frac{c_a sin(2arcsin(\alpha))}{4\pi}, \tag{A11}$$

$$D_1 = \frac{k_a}{\pi}\int_0^{\frac{arcsin(\alpha)}{\omega}} sin(\omega t)cos(\omega t)dt = \frac{\alpha^2 k_a}{2\pi\omega}. \tag{A12}$$

The interval 2, from point K to M in Figure 4a, is given by $(\frac{\pi}{2\omega}, \frac{\pi+arcsin(\alpha)}{\omega})$. By calculating the integrals within this interval, we obtain:

$$A_2 = \frac{c_a\omega^2}{\pi}\int_{\frac{\pi}{2\omega}}^{\frac{\pi+arcsin(\alpha)}{\omega}} sin(\omega t)cos(\omega t)dt = \frac{c_a\omega(\alpha^2-1)}{2\pi}, \tag{A13}$$

$$B_2 = \frac{k_a\omega}{\pi}\int_{\frac{\pi}{2\omega}}^{\frac{\pi+arcsin(\alpha)}{\omega}} sin^2(\omega t)dt = \frac{k_a arcsin(\alpha)}{2\pi} - \frac{k_a sin(2arcsin(\alpha))}{4\pi} + \frac{k_a}{4}, \tag{A14}$$

$$C_2 = \frac{c_a\omega}{\pi}\int_{\frac{\pi}{2\omega}}^{\frac{\pi+arcsin(\alpha)}{\omega}} cos^2(\omega t)dt = \frac{c_a arcsin(\alpha)}{2\pi} + \frac{c_a sin(2arcsin(\alpha))}{4\pi} + \frac{c_a}{4}, \tag{A15}$$

$$D_2 = \frac{k_a}{\pi}\int_{\frac{\pi}{2\omega}}^{\frac{\pi+arcsin(\alpha)}{\omega}} sin(\omega t)cos(\omega t)dt = \frac{(\alpha^2-1)k_a}{2\pi\omega}. \tag{A16}$$

Similarly, we compute the integrals in interval 3, which is defined in Figure 4a from point N to the end of the period. This interval is given by $(\frac{3\pi}{2\omega}, \frac{2\pi}{\omega})$. The integrals are:

$$A_3 = \frac{c_a\omega^2}{\pi}\int_{\frac{3\pi}{2\omega}}^{\frac{2\pi}{\omega}} sin(\omega t)cos(\omega t)dt = -\frac{c_a\omega}{2\pi}, \tag{A17}$$

$$B_3 = \frac{k_a\omega}{\pi}\int_{\frac{3\pi}{2\omega}}^{\frac{2\pi}{\omega}} sin^2(\omega t)dt = \frac{k_a}{4}, \tag{A18}$$

$$C_3 = \frac{c_a\omega}{\pi}\int_{\frac{3\pi}{2\omega}}^{\frac{2\pi}{\omega}} cos^2(\omega t)dt = \frac{c_a}{4}, \tag{A19}$$

$$D_3 = \frac{k_a}{\pi}\int_{\frac{3\pi}{2\omega}}^{\frac{2\pi}{\omega}} sin(\omega t)cos(\omega t)dt = -\frac{k_a}{2\pi\omega}. \tag{A20}$$

The sums of the integrals in the respective intervals are:

$$A = A_1 + A_2 + A_3 = c_a\omega\frac{(\alpha^2-1)}{\pi}, \tag{A21}$$

$$B = B_1 + B_2 + B_3 = \frac{k_a arcsin(\alpha)}{\pi} - \frac{k_a sin(2arcsin(\alpha))}{2\pi} + \frac{k_a}{2}, \tag{A22}$$

$$C = C_1 + C_2 + C_3 = \frac{c_a arcsin(\alpha)}{\pi} + \frac{c_a sin(2arcsin(\alpha))}{2\pi} + \frac{c_a}{2}, \tag{A23}$$

$$D = D_1 + D_2 + D_3 = \frac{\alpha^2 k_a}{2\pi\omega} + \frac{(\alpha^2-1)k_a}{2\pi\omega} - \frac{k_a}{2\pi\omega}. \tag{A24}$$

Ultimately, for $k_a < 0$, we obtain:

$$c_{eq} = C + D = \frac{c_a}{2} + \frac{c_a arcsin(\alpha)}{\pi} + \frac{c_a sin(2arcsin(\alpha))}{2\pi} + \frac{k_a(\alpha^2-1)}{\pi\omega}, \tag{A25}$$

$$k_{eq} = A + B = \frac{k_a}{2} + \frac{k_a arcsin(\alpha)}{\pi} - \frac{k_a sin(2arcsin(\alpha))}{2\pi} + c_a\omega\frac{(\alpha^2-1)}{\pi}. \tag{A26}$$

After conducting the same calculations for $k_a > 0$, for the three intervals marked in Figure 3a, we can derive c_{eq} and k_{eq} in the same form but with an opposite sign for the last term in both (A25) and (A26). Considering this change of the sign allows us to present the final expressions, for both positive and negative k_a, as follows:

$$c_{eq} = \frac{c_a}{2} + \frac{c_a arcsin(\alpha)}{\pi} + \frac{c_a sin(2arcsin(\alpha))}{2\pi} - \frac{|k_a|(\alpha^2 - 1)}{\pi\omega}, \quad \text{(A27)}$$

$$k_{eq} = \frac{k_a}{2} + \frac{k_a arcsin(\alpha)}{\pi} - \frac{k_a sin(2arcsin(\alpha))}{2\pi} - \frac{sgn(k_a)c_a\omega(\alpha^2 - 1)}{\pi} \quad \text{(A28)}$$

References

1. Elias, S.; Matsagar, V. Research developments in vibration control of structures using passive tuned mass dampers. *Annu. Rev. Control* **2017**, *44*, 129–156. [CrossRef]
2. Koutsoloukas, L.; Nikitas, N.; Aristidou, P. Passive, semi-active, active and hybrid mass dampers: A literature review with associated applications on building-like structures. *Dev. Built Environ.* **2022**, *12*, 100094. [CrossRef]
3. Fujino, Y.; Siringoringo, D. Vibration Mechanisms and Controls of Long-Span Bridges: A Review. *Struct. Eng. Int.* **2013**, *23*, 248–268. [CrossRef]
4. Li, Q.S.; Zhi, L.H.; Tuan, A.Y.; Kao, C.S.; Su, S.C.; Wu, C.F. Dynamic behavior of Taipei 101 tower: Field measurement and numerical analysis. *J. Struct. Eng.* **2010**, *137*, 143–155. [CrossRef]
5. Den Hartog, J.P. *Mechanical Vibrations*; McGraw-Hill: New York, NY, USA, 1956.
6. Burton, M.D.; Kwok, K.C.S.; Abdelrazaq, A. Wind-induced motion of tall buildings: Designing for occupant comfort. *Int. J. High-Rise Build.* **2015**, *4*, 1–8.
7. Hayes, J. Tall storeys. *Eng. Technol.* **2018**, *13*, 70–73. [CrossRef]
8. Yao, J.T. Concept of structural control. *J. Struct. Div.* **1972**, *98*, 1567–1574. [CrossRef]
9. Nishitani, A. Obituary for Professor Takuji Kobori: Great pioneer in structural dynamics and control. *Struct. Control Health Monit.* **2008**, *15*, 117–119. [CrossRef]
10. Chang, J.C.H.; Soong, T.T. Structural control using active tuned mass dampers. *J. Eng. Mech. Div.* **1980**, *106*, 1091–1098. [CrossRef]
11. Kobori, T.; Koshika, N.; Yamada, K.; Ikeda, Y. Seismic-response-controlled structure with active mass driver system. Part 1: Design. *Earthq. Eng. Struct. Dyn.* **1991**, *20*, 133–149. [CrossRef]
12. Ikeda, Y. Active and semi-active vibration control of buildings in Japan—Practical applications and verification. *Struct. Control Health Monit.* **2009**, *16*, 703–723. [CrossRef]
13. Spencer, B.F.; Nagarajaiah, S. State of the art of structural control. *J. Struct. Eng.* **2003**, *129*, 845–856. [CrossRef]
14. Casciati, F.; Rodellar, J.; Yildirim, U. Active and semi-active control of structures—Theory and applications: A review of recent advances. *J. Intell. Mater. Syst. Struct.* **2012**, *23*, 1181–1195. [CrossRef]
15. Zhou, K.; Zhang, J.W.; Li, Q.S. Control performance of active tuned mass damper for mitigating wind-induced vibrations of a 600-m-tall skyscraper. *J. Build. Eng.* **2021**, *45*, 103646. [CrossRef]
16. Zhou, K.; Li, Q.S. Vibration mitigation performance of active tuned mass damper in a super high-rise building during multiple tropical storms. *Eng. Struct.* **2022**, *269*, 114840. [CrossRef]
17. Nishimura, I.; Kobori, T.; Sakamoto, M.; Koshika, N.; Sasaki, K.; Ohrui, S. Active tuned mass damper. *Smart Mater. Struct.* **1992**, *1*, 306–311. [CrossRef]
18. Asami, T.; Nishihara, O.; Baz, A.M. Analytical solutions to H_∞ and H_2 optimization of dynamic vibration absorbers attached to damped linear systems. *J. Vib. Acoust.* **2002**, *124*, 284–295. [CrossRef]
19. Deraemaeker, A.; Soltani, P. A short note on equal peak design for the pendulum tuned mass dampers. *Proc. Inst. Mech. Eng. Part K J. Multi-Body Dyn.* **2016**, *231*, 285–291. [CrossRef]
20. Habib, G.; Detroux, T.; Viguié, R.; Kerschen, G. Nonlinear generalization of Den Hartog's equal-peak method. *Mech. Syst. Signal Process.* **2015**, *52*, 17–28. [CrossRef]
21. Chatterjee, S. Optimal active absorber with internal state feedback for controlling resonant and transient vibration. *J. Sound Vib.* **2010**, *329*, 5397–5414. [CrossRef]
22. Cheung, Y.L.; Wong, W.O.; Cheng, L. Design optimization of a damped hybrid vibration absorber. *J. Sound Vib.* **2012**, *331*, 750–766. [CrossRef]
23. Brodersen, M.L.; Bjørke, A.-S.; Høgsberg, J. Active tuned mass damper for damping of offshore wind turbine vibrations. *Wind Energy* **2017**, *20*, 783–796. [CrossRef]
24. Koo, J.H.; Goncalves, F.D.; Ahmadian, M.A. Comprehensive analysis of the response time of MR dampers. *Smart Mater. Struct.* **2006**, *15*, 351–358. [CrossRef]
25. Strecker, Z.; Roupec, J.; Mazurek, I.; Machacek, O.; Kubik, M.; Klapka, M. Design of magnetorheological damper with short time response. *J. Intell. Mater. Syst. Struct.* **2015**, *26*, 1951–1958. [CrossRef]
26. Zhu, X.; Jing, X.; Cheng, L. Magnetorheological fluid dampers: A review on structure design and analysis. *J. Intell. Mater. Syst. Struct.* **2012**, *23*, 839–873. [CrossRef]

27. Weber, F.; Maślanka, M. Frequency and damping adaptation of a TMD with controlled MR damper. *Smart Mater. Struct.* **2012**, *21*, 055011. [CrossRef]
28. Weber, F.; Distl, H.; Maślanka, M. Semi-active TMD concept for Volgograd Bridge. *Top. Dyn. Civ. Struct.* **2013**, *4*, 79–88.
29. Weber, F.; Distl, H.; Fischer, S.; Braun, C. MR damper controlled vibration absorber for enhanced mitigation of harmonic vibrations. *Actuators* **2016**, *5*, 27. [CrossRef]
30. Zemp, R.; de la Llera, J.C.; Saldias, H.; Weber, F. Development of a long-stroke MR damper for a building with tuned masses. *Smart Mater. Struct.* **2016**, *25*, 105006. [CrossRef]
31. Weber, F.; Maślanka, M. Precise stiffness and damping emulation with MR dampers and its application to semi-active tuned mass dampers of Wolgograd Bridge. *Smart Mater. Struct.* **2014**, *23*, 015019. [CrossRef]
32. Weber, F. Semi-active vibration absorber based on real-time controlled MR damper. *Mech. Syst. Signal Process.* **2014**, *46*, 272–288. [CrossRef]
33. Pinkaew, T.; Fujino, Y. Effectiveness of semi-active tuned mass dampers under harmonic excitation. *Eng. Struct.* **2001**, *23*, 850–856. [CrossRef]
34. Kang, J.W.; Kim, H.S.; Lee, D.G. Mitigation of wind response of a tall building using semi-active tuned mass dampers. *Struct. Des. Tall Spec.* **2011**, *20*, 552–565. [CrossRef]
35. Viet, L.D.; Nghi, N.B.; Hieu, N.N.; Hung, D.T.; Linh, N.N.; Hung, L.X. On a combination of ground-hook controllers for semi-active tuned mass dampers. *J. Mech. Sci. Technol.* **2014**, *28*, 2059–2064. [CrossRef]
36. Moutinho, C. Testing a simple control law to reduce broadband frequency harmonic vibrations using semiactive tuned mass dampers. *Smart Mater. Struct.* **2015**, *24*, 055007. [CrossRef]
37. Ferreira, F.; Moutinho, C.; Cunha, Á.; Caetano, E. Proposal of optimum tuning of semiactive TMDs used to reduce harmonic vibrations based on phase control strategy. *Struct. Control Health Monit.* **2018**, *25*, e2131. [CrossRef]
38. Rosół, M.; Martynowicz, P. Implementation of the LQG controller for a wind turbine tower-nacelle model with an MR tuned vibration absorber. *J. Theor. Appl. Mech.* **2016**, *54*, 1109–1123. [CrossRef]
39. Martynowicz, P. Nonlinear optimal-based vibration control for systems with MR tuned vibration absorbers. *J. Low Freq. Noise Vib. Act. Control* **2019**, *38*, 1607–1628. [CrossRef]
40. Martynowicz, P. Real-time implementation of nonlinear optimal-based vibration control for a wind turbine model. *J. Low Freq. Noise Vib. Act. Control* **2019**, *38*, 1635–1650. [CrossRef]
41. Demetriou, D.; Nikitas, N.; Tsavdaridis, K.D. Performance of fixed-parameter control algorithms on high-rise structures equipped with semi-active tuned mass dampers. *Struct. Des. Tall Spec. Build.* **2016**, *25*, 340–354. [CrossRef]
42. Maślanka, M. Optimised semi-active tuned mass damper with acceleration and relative motion feedbacks. *Mech. Syst. Signal Process.* **2019**, *130*, 707–731. [CrossRef]
43. Maślanka, M. Measured performance of a semi-active tuned mass damper with acceleration feedback. In Proceedings of the Active and Passive Smart Structures and Integrated Systems, Portland, OR, USA, 25–29 March 2017; Volume 10164, p. 1016423.
44. Krenk, S. Frequency analysis of the tuned mass damper. *J. Appl. Mech.* **2005**, *72*, 936–942. [CrossRef]
45. Tseng, H.E.; Hedrick, J.K. Semi-active control laws-optimal and sub-optimal. *Veh. Syst. Dyn.* **1994**, *23*, 545–569. [CrossRef]
46. Weber, F.; Boston, C. Clipped viscous damping with negative stiffness for semi-active cable damping. *Smart Mater. Struct.* **2011**, *20*, 045007. [CrossRef]
47. Weber, F.; Boston, C.; Maślanka, M. An adaptive tuned mass damper based on the emulation of positive and negative stiffness with an MR damper. *Smart Mater. Struct.* **2011**, *20*, 015012. [CrossRef]
48. Krylov, N.M.; Bogoliubov, N.N. *Introduction to Non-Linear Mechanics*; Princeton University Press: Princeton, NJ, USA, 1955.
49. Tunç, C.; Tunç, O. A note on the stability and boundedness of solutions to non-linear differential systems of second order. *J. Assoc. Arab. Univ. Basic Appl. Sci.* **2017**, *24*, 169–175. [CrossRef]
50. Høgsberg, J. The role of negative stiffness in semi-active control of magneto-rheological dampers. *Struct. Control Health Monit.* **2011**, *18*, 289–304. [CrossRef]

applied sciences

MDPI

Review

State-of-the-Art Review of Structural Vibration Control: Overview and Research Gaps

Neethu B. Dharmajan [1,2] and Mohammad AlHamaydeh [1,*]

1 Department of Civil Engineering, College of Engineering, American University of Sharjah, Sharjah P.O. Box 26666, United Arab Emirates; bneethu2022@gmail.com
2 Department of Civil and Environmental Engineering, Khalifa University, Abu Dhabi P.O. Box 127788, United Arab Emirates
* Correspondence: malhamaydeh@aus.edu

Abstract

This paper comprehensively reviews structural vibration control systems for earthquake mitigation in civil engineering structures. Structural vibration control is vital for enhancing the resilience and safety of infrastructure subjected to seismic activity. This study examines various control strategies, including passive, active, and hybrid methods, with a focus on the advantages of semi-active systems, which offer a balance of energy efficiency and adaptive capabilities. Semi-active devices, such as magnetorheological dampers, are highlighted for their ability to offer adaptive control without the high energy demands of fully active systems. The review discusses challenges like time delays, sensor placement, and model uncertainties that can impact the practical implementation of these systems. Experimental studies and real-world applications demonstrate the effectiveness of semi-active systems in reducing seismic responses. This paper emphasizes the need for further research into optimizing control algorithms and addressing practical challenges to enhance the reliability and robustness of these systems. It concludes that semi-active control systems are a promising solution for enhancing structural resilience in earthquake-prone areas, offering a practical alternative that strikes a balance between performance and energy requirements.

Keywords: structural control; semi-active control; seismic control; earthquakes; dampers; control algorithms

1. Introduction

1.1. General

Structural vibrations in civil engineering have a significant impact on the performance, safety, and serviceability of various infrastructures, including buildings, bridges, and towers. Seismic vibrations pose a significant threat to the safety and performance of civil structures, particularly in regions prone to earthquakes. Structural vibration control has recently emerged as an effective technology for earthquake mitigation. Over the years, extensive research has been conducted to develop effective approaches for mitigating seismic vibrations and minimizing the adverse effects of earthquakes on structures. Researchers have studied different passive, active, semi-active, and hybrid control methods for the seismic response control of structures. Comprehensive, state-of-the-art reviews on structural control in general [1–4] and specific controls [5–7] are presented, providing an overview of the various control strategies and their applications. A list of all state-of-the-art review papers is presented in Table 1. Housner et al. [8] presented an in-depth review covering diverse control systems, sensors, and research requirements from 1990 to 1996. Notably, this study pioneers the Physical Design Problem (PDP) concept, subsequently becoming a fundamental principle in various research endeavors. The authors advocate for prioritizing specific topics, including energy-efficient strategies for control devices and algorithms, integrating intelligent sensors for distributed sensing and control in dispersed systems, and addressing concerns related to near-field strong earthquake ground motion that influences structural control applications. Their recommendations set the stage for subsequent research, shaping the trajectory of advancements in the field during the specified period. Regarding active structural control, recent influential papers by Soong [4,9,10] offer a comprehensive overview of the various active structural control systems developed up to 1990. Additionally, Datta [11] provides a thorough review of the active control of structures subject to earthquake excitations. This review discusses different types of active control and their theoretical foundations, highlights key findings from parametric studies on various control strategies, addresses the limitations and practical challenges of active control systems, briefly examines the more promising semi-active control methods, and concludes with an outline of several active control strategies implemented in practice. Korkmaz [12] provides a comprehensive review of active control, delving into the challenges posed by engineering informatics. The review emphasizes the imperative for multi-objective control strategies, highlighting their dual focus on ensuring the structure's safety and serviceability while enhancing the robustness of the control system. Reviews on semi-active control are also reported [13,14]. Other notable review papers include state-of-the-art reviews on the base isolation of structures [15–17].

Table 1. List of publications on the state of the art of structural control.

List (40)	Publications
General (13)	[4,8,10,12,14,18–23]
Passive (4)	[15,16,24,25]
Active (3)	[11,12]
Active and Semi-active (3)	[2,26–28]
Semi-active (3)	[13,28–30]
MR Damper (7)	[6,31–36]
Hybrid (2)	[3,5]
Control Strategies (5)	[27,29,32,37,38]

The development of active, hybrid, and semi-active control systems has reached the stage of full-scale applications in actual structures. Nevertheless, every structural control system and its applications encounter challenges and limitations across various aspects. The challenges in implementing control strategies have hindered their expected applications.

To maximize effectiveness, the controller must be designed with consideration for the issues and should be robust, stable, reliable, and simple to design and implement. This paper presents a comprehensive examination of these challenges and limitations, providing insights into future research requirements and directions.

Structural control constitutes a vast and interdisciplinary research domain within civil engineering systems, making it impractical to discuss or reference all relevant publications and applications comprehensively. Therefore, the initial section of this paper provides an overview of various structural control systems, outlining their general limitations and challenges. Upon reviewing the literature, it becomes evident that semi-active control systems have emerged as promising techniques, offering the versatility and adaptability of active control systems without the need for a substantial energy supply while retaining the reliability characteristic of passive control systems. The following section provides a comprehensive examination of various semi-active control devices and their associated advantages. The practical implementation of a semi-active control system crucially depends on the chosen control strategy, presenting a foundational challenge. The following section provides a systematic review of the control theories and algorithms for semi-active control systems.

This section summarizes the advantages and disadvantages of each control algorithm, highlighting efforts to address the distinct challenges associated with each. In discussions related to control theory, emphasis is placed on issues directly tied to the physical behavior of civil structures, rather than delving into intricate developments in control system theory. The subsequent section explores the ongoing efforts of various researchers to refine control strategies tailored to specific issues and requirements, outlining different approaches to modify algorithms for enhanced system performance. The final section consolidates the research findings, offering a conclusion and outlining potential avenues for future research. This structured approach thoroughly explores the diverse aspects of structural control, ranging from devices to theories and strategies, and provides valuable insights for researchers and practitioners.

1.2. Research Significance

Numerous surveys have explored vibration control strategies for civil structures, with a predominant focus on distinct control system categories. Most of these review papers emphasize the limitations of control systems, impeding their anticipated implementation. Unfortunately, none comprehensively address the experimental challenges of implementing semi-active control for structures. This study presents a state-of-the-art review paper to address this research gap, offering a broader perspective on research efforts to overcome practical implementation challenges in semi-active control systems.

The contributions of this review include studying and summarizing the various types of semi-active control devices and control algorithms in civil structures, identifying key challenges and limitations in the application of control systems, and reviewing emerging trends in semi-active control strategies to address these challenges and ensure robust performance. This paper provides a realistic evaluation of semi-active control systems, discussing practical issues such as delays, uncertainties, algorithm selection, and system integration. It emphasizes methods to mitigate these undesirable effects and highlights the advantages of semi-active systems in seismic applications. Additionally, it highlights the increasing importance of multi-objective adaptive control algorithms, which address multiple challenges, such as optimizing control parameters, mitigating uncertainties, managing sensor and actuator configurations, facilitating decentralized control, and minimizing time delays.

2. Overview of Structural Control Systems

It is crucial to recognize that the control system and the structure do not operate as isolated dynamic entities; instead, they significantly interact and mutually influence each other. The control system, integrated with the structure, manages seismic forces by providing counteracting forces or dissipating seismic energy, thereby preventing a catastrophic failure of the structural system. Structural control class selection depends on various factors, including budget constraints, desired performance, specific structural characteristics, and project requirements. Each class has its advantages and limitations, and informed choices are to be made based on the type of structure, control goals, and constraints. This section presents the classification of the control systems and their critical features.

Structural control can be broadly classified based on the amount of external energy required; the different control systems can be categorized as passive, active, semi-active, hybrid, and passive control systems (Figure 1) [13]. Based on the mathematical modeling of the control systems, active and semi-active systems are further classified according to their mathematical modeling [11], as shown in Figure 2.

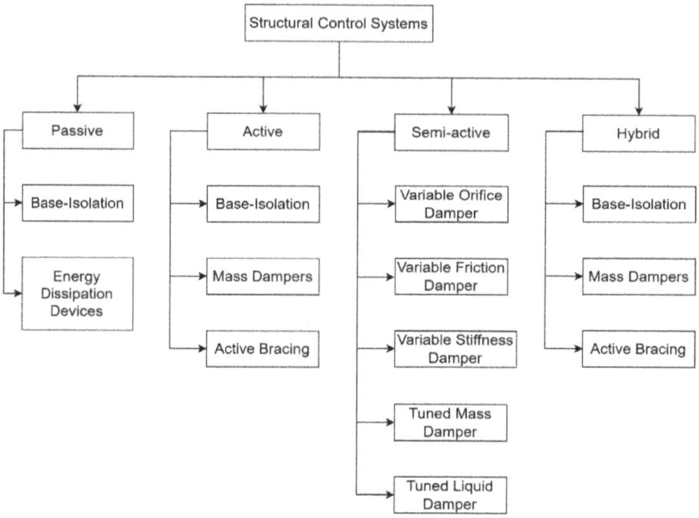

Figure 1. Classification of control systems.

2.1. Passive Control System

Passive control systems are the basic structural control systems that do not need any external power to operate. The passive systems do not generate energy within the structure, thereby maintaining the stability of the control system. Passive energy-dissipating systems are external add-on damping devices commonly used to dissipate energy from structural vibrations. Passive devices generate control forces in response to the structure's motion. Passive control devices are broadly classified into two categories: passive isolation devices and energy-dissipating devices [22]. Some of the most widely used passive devices are viscoelastic dampers, viscous fluid dampers, friction dampers, metallic dampers, tuned mass dampers, liquid column dampers, and base isolators. Researchers have conducted state-of-the-art reviews on the application of passive isolation [15–17]. Rahimi et al. [7] presented a critical review of the application of tuned mass dampers for the structural vibration control of structures subjected to wind and earthquakes, providing a comparison of their efficiency and the comparative advantages and disadvantages. The potential of TMDs for improving the wind and seismic behaviors of prototype civil structures was also reviewed by Elias et al. [24]. The review highlights the dynamic characteristics and unique features of different systems, namely single-tuned mass dampers (STMDs), multiple-tuned mass dampers (MTMDs), and spatially distributed MTMDs (d-MTMDs), which have been explored both theoretically and through experimental investigations.

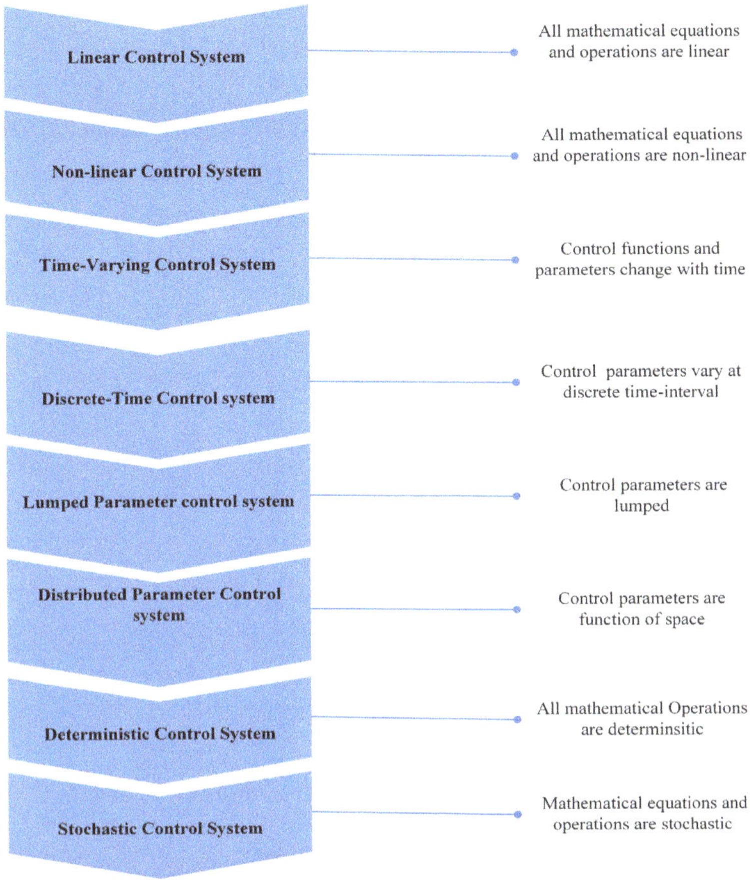

Figure 2. Classification of control systems based on mathematical modeling.

2.2. Active Control System

Active control systems are more sophisticated and require a significant power source for operation. These control systems use sensors to measure structural responses. Based on feedback from the sensors, the control forces are computed according to a prescribed controller and are then applied to the structure through actuators. These systems have the advantage over passive systems in adapting to varying load conditions [8]. However, it can also destabilize the structure if the control forces are not applied in the proper position and at the appropriate time. Some control devices are active mass dampers, active tuned mass dampers, and active tendon systems. Fisco et al. [2] presented a review of active and semi-active devices for structures since 1997 and identified that research in recent years has moved toward semi-active and hybrid control systems. Ikeda [28] provided a report on the application of active and semi-active devices in Japan, which discusses the practical application of active control devices in structures.

2.3. Semi-Active Control System

These systems require low power supplies for their activation compared to typical active control systems. These systems generate counteracting forces by reactive devices with variable damping and/or stiffness characteristics. These systems are also referred to as controllable passive systems. These systems can operate during large earthquakes, as they require a minimal magnitude of external energy, which can be supplied with the help of battery power. Semi-active systems do not impart energy into the structure and therefore do not cause structural instability. The semi-active control devices include hydraulic dampers, electrorheological and magnetorheological dampers, semi-active stiffness control devices, friction control devices, semi-active tuned mass dampers, and tuned liquid dampers [13]. It is noted that these control systems are adaptable to active systems and possess fail-safe features similar to those of passive systems. A detailed review of dynamic models of MR dampers and control algorithms is presented by Jung et al. [6]. The MR-fluid-based dampers are shown to be highly effective for full-scale civil engineering structures. Gkatzogias et al. [30] presented a current state-of-the-art review of semi-active control in bridges, focusing on some full-scale applications of semi-active control devices and relevant benchmark studies. Among the semi-active control devices, the MR damper is suggested to be a promising alternative due to its reduced adaptability and reliability compared to passive and active devices.

2.4. Hybrid Control System

Hybrid control systems are developed by combining a passive system with an active or semi-active one. Hybrid systems employ multiple control devices, meaning both active control devices and passive devices are installed within the same structure. Hence, combining the systems can eliminate some of the limitations and disadvantages of each system that arise when they are used separately, thereby achieving better performance. Some of the combinations studied include base isolation and active actuators, tuned mass dampers and base isolation, and viscoelastic dampers and base isolation, among others. Fisco et al. [3] provided a review of the hybrid control systems and the control algorithms for the structures. The authors also noted that most research on smart structures is concentrated in the US, Japan, Taiwan, China, and Korea, with lesser attention given to Europe.

Passive and semi-active controls have better preference than active control due to their reliability and inherent stability. An active control system may develop asynchronous control forces due to time delays, which can render the structure unstable. Out of semi-active, passive, and hybrid control, semi-active control has garnered more attention because it can offer a more effective reduction in seismic responses. Therefore, this paper reviews the advancement in semi-active control applications for earthquake mitigation. The following section presents a review of semi-active control systems for structures.

3. General Limitations of Structural Control Systems

Although the structural control of civil engineering structures has emerged as a notably efficient method to mitigate the potentially devastating effects of earthquakes, most papers report that the limitations of the control system have hindered the expected implementation of the systems. Jiang and Wang [5] state that while structural control has advanced significantly, three major concerns still impede the application of active or hybrid control techniques to full-scale structures: system complexity, reliable measurements, and the capacity of the actuators. One of the most challenging aspects of active control research in civil engineering is its interdisciplinary nature [4]. It requires integrating knowledge from various fields beyond traditional civil engineering, including computer science, data processing, control theory, material science, sensing technology, stochastic processes, structural dynamics, and wind and earthquake engineering.

The successful implementation of structural control systems is not without its intricate challenges that demand careful consideration and resolution. For the practical implementation of structural control systems, aside from the availability of ample power sources for implementing control schemes, several intricate real-time application challenges require careful consideration. These challenges encompass a broad spectrum of issues that engineers and researchers face when designing and deploying effective strategies to mitigate structural vibrations and enhance the resilience of buildings and infrastructure. This section discusses the challenges associated with implementing structural control.

3.1. Modeling Errors

The accuracy of structural control depends on the ability to model the actual structure effectively. However, limitations in modeling, often arising from a reduced number of degrees of freedom, can introduce errors. Consequently, control algorithms based on idealized models may struggle to effectively manage the actual structure's dynamic behavior. The modeling errors are classified into four types: parameter errors, model order errors, neglected disturbances, and neglected nonlinearities. No controller can treat all four at once [8]. Disregarding various errors and concentrating solely on one category may result in inaccurate conclusions. Most hardware and design suggestions and results are based on satisfactory computer simulations. It is challenging to draw clear conclusions about hardware and design recommendations because the research papers heavily rely on numerous referenced hardware tests. The available results predominantly stem from computer simulations, which are effective in conservative systems where errors in design parameters cannot destabilize the system. However, active feedback control can introduce instability or suboptimal performance if significant modeling errors and control energy are substantial. Relying solely on computer simulations may not offer sufficient evidence for a robust design. Simulations may or may not encompass enough modeling errors to assess the stability of the actual system. The lack of addressing the interdependency between modeling and control design is one of the severe issues in control systems. The study by Housner et al. [8] stated that robust control theories, such as adaptive control and H_∞ can treat the errors and stabilize the system if the unbounded error is known. However, such systems usually trade performance for robustness.

3.2. Time Delay

A standard structural control system for civil engineering structures comprises multiple components, including sensors, filters, controllers, and control devices. Since each element requires a finite amount of time to function, there is always a delay between measuring response quantities and applying control forces. In seismic control systems, these delays stem from the time required to measure state vector responses during earthquakes and to process this data using digital computer units. Furthermore, time delays are introduced when calculating active forces through control algorithms, activating control devices, and applying them to the plant or system.

Experimental studies have shown that time delay is inevitable in operating control systems, which can deteriorate control performance. The existence of a time lag between sensing structural responses and applying control forces is a critical issue. From previous studies [39–41], it is observed that time delay can cause degradation in control performance. These time delays can occur in control system inputs, outputs, or states, and their characteristics can vary from being discrete to distributed, constant, time-varying, known, or unknown, and from deterministic to stochastic, depending on the specific features of the system [42].

Study [39] demonstrated that the decentralized output feedback polynomial controller, the Lyapunov controller, and the simple passive controller are highly robust to time delay issues, even without implementing any compensation to mitigate the effects of the delays. Also, various compensation methods have been proposed to address the instability and performance degradation caused by these delays. Extensive real-scale semi-active tests and simulations were conducted [43] to estimate the time delay effects in the control electronics and the mechanical and electrical parts of the MR damper within a closed control loop using a Bingham model of 50 kN MR dampers. They identified three distinct time delays in the real-scale 50 kN MR damper: Control Electronics Time Delay, which included signal acquisition, processing of the acquired signal, and power supply operations; Electrical Time Delay, which pertains to the electrical part of the damper or its electromagnetic circuit; and Mechanical Time Delay, which is the interval between the moment the current to the device starts changing and the moment the device begins to adjust its mechanical behavior.

3.3. Sensor and Controller Placement

The practical constraints of installing sensors at all critical points for comprehensive feedback collection are formidable. Balancing the optimization of sensors and controller locations with limited resources can be a complex task, and using observers to construct a state vector from sparse measurements may introduce additional challenges. The strategic positioning of actuators and sensors not only enhances energy efficiency and cost-effectiveness but also plays a crucial role in influencing the stability and reliability of a control system. The methods for designing optimal damper placement can be categorized into three primary groups: evolutionary, analytical, and heuristic. A review of these methods has been outlined by Kookalani et al. [44]. The authors recommend that more studies be conducted on damper placement for algorithms such as the artificial bee colony algorithm (ABCA) and the firefly algorithm (FA). Soto and Adeli [45] presented a review of placement control devices for structural control techniques, stating that most of the available literature on semi-active control focuses on MR dampers. The study also states that the research on the optimal placement of control devices in semi-active and hybrid control systems remains largely unexplored. While structural control systems show promise in effectively mitigating external dynamic forces, such as those induced by winds and earthquakes, the precise optimization of device placement has not received extensive investigation.

3.4. Parametric Uncertainties

Structural parameters, such as material properties and nonlinear behaviors, often exhibit uncertainties. Additionally, online identification complicates the management of time-dependent degradation and further complicates maintaining the effectiveness of control schemes in the face of evolving structural conditions. The mathematical model of a system is typically an approximation of its actual dynamic behavior, with differences arising from unmodeled and uncertain parameters. These uncertainties can significantly affect the performance and stability of the control system. These uncertainties can be classified into structured and unstructured categories.

Navigating these multifaceted challenges is crucial for successfully implementing structural control systems, ultimately enhancing the safety and resilience of civil infrastructure. In summary, the structural control of civil engineering structures presents a promising solution for earthquake mitigation. However, it encompasses intricate challenges related to control strategy, mathematical modeling, sensor networks, input variables, and parameter uncertainties. Successfully addressing these challenges is crucial for designing effective control systems that safeguard structures and enhance their resilience against seismic events. These efforts are crucial in enhancing the overall safety and longevity of critical infrastructure. Experimental studies form the basis for studying these challenges. This paper emphasizes the significance of experimentally assessing semi-active control systems. It illustrates how numerous issues that may arise in real-world applications are also evident during the experimental testing of these systems. The following section presents experimental studies on the semi-active control of structures and discusses how the issues related to these studies are addressed.

4. Semi-Active Control Systems

The introduction of structural control in civil engineering is widely acknowledged on a global scale, attributed to the groundbreaking article authored by Yao [46] and the subsequent promotion of its implementations by Kobori [47]. Despite the successful application of numerous structural control strategies, their widespread adoption has been hindered by challenges related to cost, dependence on external power sources, and the complexity of mechanical aspects throughout the structure's lifespan. Although active control effectively reduces structural responses, its requirement for an extensive external power source and its potential to destabilize the structure are significant concerns regarding its applicability. Semi-active control systems offer the advantages of versatility and adaptability of active control systems, without requiring an ample energy supply, while maintaining the reliability of passive control systems [41]. Comprehensive state-of-the-art reviews on active and semi-active control [2,26–28] and MR are notable [6,13,29,31,32]. Symans and Constantinou [13] conducted an extensive literature review that examined dynamic characteristics and distinctive attributes of various semi-active systems. The review presents the control systems experimentally tested at the component level and within scaled-down structural models, and it considers semi-active systems, including stiffness control devices, electrorheological dampers, magnetorheological (MR) dampers, friction control devices, fluid viscous dampers, tuned mass dampers (TMDs), and tuned liquid dampers. Table 2 lists the different semi-active control devices and their salient features.

Table 2. Critical features of semi-active control devices.

Control Systems	Key Features
Stiffness control devices	• Used to alter the stiffness and, consequently, the natural vibration properties of the structure to which they are connected. • The system mainly regulates the rigidity of a structure to achieve a non-resonant state during earthquakes. • The semi-active stiffness devices are activated or deactivated to incorporate or remove, respectively, the stiffness of the structure's bracing system.
Electrorheological dampers	• Comprise a hydraulic cylinder that holds micron-sized dielectric particles dispersed in a fluid (typically oil). • In a strong electric field, the particles align and polarize, thereby providing greater resistance to flow. • Altering the electric field can adjust the dynamic performance of an ER damper.
Magnetorheological (MR) dampers	• MR dampers generally comprise a hydraulic cylinder filled with micron-sized, magnetically polarizable particles dispersed in a fluid (commonly oil). • The behavior of MR fluid is managed by applying a magnetic field to the fluid. • Without a magnetic field, the MR fluid moves freely; however, when subjected to a magnetic field, it behaves like a semi-solid. • MR fluids can withstand a maximum yield stress in the range of 50–100 kPa [35,43]

Table 2. *Cont.*

Control Systems	Key Features
Friction control devices	• Employed either as energy dispersers in the lateral bracing of a structure or as elements in sliding isolation systems.
Friction-controllable sliding bearing	• The pressure between the two sliding surfaces can be adjusted to manage the friction between the bearing and the ground.
Tuned mass dampers	• Tuned mass dampers essentially comprise a single-degree-of-freedom mass–spring–damper arrangement usually positioned on the uppermost floor of a multi-story building. • The dynamic traits of the system are adjusted to manage the movement of the structure to which it is connected.
Tuned liquid dampers	• An effective control approach for managing a variety of dynamic loading scenarios.
Semi-active slip bracing system	• Commonly used when the lateral force-resisting consists of braces. • It slips at the interface when the brace's axial forces reach a predetermined threshold (the friction coefficient multiplied by the clamping force). • Installing this type of device allows for the adjustment of a brace's strength without changing its stiffness. • Another benefit of this device over a passive one is its ability to operate effectively under minor forces (such as small earthquakes or winds) while still delivering effective performance under greater demands.
Electro-inductive device	• Key factors in utilizing electro-inductive devices rather than fluid dampers include the following: • The behavior is essentially unaffected by outside temperature, as the device's operating temperature consistently exceeds the air temperature and is achieved in just a few seconds. • Maintenance of the device is minimized due to the absence of aging or leakage issues, which are present in fluid dampers.
Air-jet actuators	• They were introduced as an effective solution to the traditional concern that power would not be available when needed unless it was stored in advance. • The flow of compressed air supplies the actuator force, to be transmitted to the structure through the principle of momentum preservation.
SMA actuators	• They are commonly adopted to realize passive devices, while the different (and long) reaction times for cooling and heating prevent their adoption in semi-active devices.

Semi-active systems offer a strong alternative to both passive and active systems, as they outperform the former in terms of structural performance and the latter in terms of power consumption. This has been demonstrated in several scaled-down experimental investigations involving bridges and framed structures, e.g., [44–49]. Table 3 lists publications on semi-active control for buildings and bridges. It has been observed that most of these studies focus on the controller's efficiency in mitigating seismic vibrations. However, little attention is given to addressing the practical issues in implementing the systems.

Table 3. List of publications on the semi-active control of structures.

List	Publications
Buildings	[29,35,36,39,45,48,50–120]
Bridges	[20,41,49,55,96,121–157]

Experimental studies with controllable devices demand focused efforts to address distinct aspects within the three stages of the semi-active control loop: acquisition, processing, and command. Additionally, the control system's efficiency is dramatically dependent on the control algorithm. Dyke et al. [158] examined the effectiveness of different control strategies for MR dampers, illustrating that the performance of the control system is significantly influenced by the algorithm selected for implementation. The subsequent section offers a comprehensive overview of the control algorithms employed in semi-active vibration control.

5. Control Algorithms Used for Different Semi-Active Control Systems

Semi-active control systems employ various control algorithms to achieve their intended purposes. These control algorithms are designed to adaptively regulate the damping or stiffness properties of structural components, thereby mitigating vibrations and enhancing the overall performance of structures. Creating an accurate mathematical model of the structure–controller system for real-world structures, while accounting for the various uncertainties associated with process parameters, is a challenging endeavor. Nonetheless, these issues must be addressed to create and develop controllers that operate effectively in these complex systems. Various control strategies have been documented in the literature, including fuzzy logic, neural networks, instantaneous optimal control, the sliding mode technique, linear quadratic Gaussian (LQG) controllers, and linear quadratic regulators (LQRs). Jansen and Dyke [29] examined and assessed the effectiveness of different semi-active control techniques for MR dampers, including the Lyapunov controller, decentralized bang-bang controller, modulated homogeneous friction algorithm, and a clipped optimal controller. Numerous recognized algorithms in control engineering have been applied to manage systems, including optimal control, LQR or LQG, pole placement, sliding mode control, the H_2 and H_1 methods, fuzzy control, and various additional techniques. Soong [4] and Casciati et al. [159] detail the most appropriate algorithms for structural applications and the practical factors that need consideration. The selection of a control algorithm relies on the unique features of the semi-active control system, the type of structural vibrations, and the goals of control. Engineers and researchers select the most suitable algorithm based on the application and desired performance outcomes.

Classification of Control Algorithms

The key factor in determining the overall performance of an innovative adaptive structure is the implementation of a control algorithm that can effectively counteract both known and unknown external excitations by providing additional force input through control devices. This control algorithm should be robust, versatile, and easy to design and implement. Furthermore, it should allow flexibility in selecting performance objectives to ensure a comprehensive and effective response reduction. The issue of modeling for controllers is a complex and ongoing challenge because no established theory offers a model ideally suited for control design, primarily due to the complexity of modeling the system and designing the controller. After all, no established theory offers a model ideally suited for control design, primarily because modeling the system and designing the controller are inherently interconnected problems. The optimal choice of a control algorithm may hinge on factors such as the specific nonlinearity exhibited by the semi-active device, the accessibility of the feedback measurements, or the number of devices intended for incorporation into the structure. Furthermore, the specific structural feedback data, such as acceleration, velocity, and displacement, are of great importance in determining the control strategy for a feedback control system. Control algorithms are classified based on various factors or criteria in different literature sources. Control algorithms can be classified as linear and nonlinear based on the measured response and the control signal. Additionally, control algorithms can be classified as open-loop, closed-loop, or open–closed-loop, depending on the type of information used to determine their output control signals (Figure 3). Another common practice among researchers in classifying control algorithms is to categorize them as optimal, stochastic, adaptive, intelligent, and hybrid. Figure 4 shows this classification and the sub-classification for this type.

Some of the commonly used algorithms and their features are discussed below. The linear quadratic regulator (LQR) and linear quadratic Gaussian (LQG) algorithms are the most widely used optimal control algorithms, appreciated for their simplicity [29]. Neither of these control approaches require prior knowledge of the external excitation, as they are based solely on the system's state variables. However, they are primarily suitable for linear systems [160]. Additionally, the LQG control requires accurate feedback or structural models, and its primary flaw is that the uncertainty is modeled as white noise, which is not ideal [161].

Sliding mode control (SMC) is a form of nonlinear control that generates intermittent control signals to modify a system's dynamic behavior. Sliding mode control (SMC) operates on the concept of switching between various independent structural configurations. The control approach drives the system to follow a trajectory along predefined behavioral surfaces. The design ensures that system trajectories are continually directed toward a nearby region of this surface, with varying control intensities, rather than remaining confined to a single control regime [162,163]. As a result, the system navigates the boundaries of different control structures while minimizing errors. SMC stands out for its robustness and ability to handle both linear and nonlinear systems without requiring precise models [164], although it suffers from issues like the chattering effect and limited optimization capabilities.

A robust control algorithm is a frequency-domain strategy designed to effectively handle uncertainties. These uncertainties—arising from external excitations or inaccuracies in structural response measurements—can degrade control system performance. Robust control addresses these challenges during the design phase to maintain reliable operation. Among the prominent methods are the H_∞ and H_2 controllers. The H_∞ controller uses a combination of time- and frequency-domain approaches to tackle a wide range of control problems. It excels in disturbance rejection and guarantees system stability under diverse operating conditions [165]. Its popularity stems from its flexibility—performance goals can be incorporated as weighting functions during the design process. The H_∞ framework also includes process and measurement noise, making it robust to model uncertainties. It is suitable for both single-input single-output (SISO) and multi-input multi-output (MIMO) systems. The H_2 controller is a specific subset of the H_∞ approach. It minimizes the H_2 norm, aiming for optimal performance while stabilizing the system. Like H_∞, the H_2 controller typically follows a two-input, two-output configuration and is particularly effective for systems where the emphasis is on minimizing overall energy or variance in the output. Robust control is favored for its ability to function without a structural model and handle system uncertainties, but its complex design poses practical challenges. One of the limitations of a robust control system is that time-domain constraints cannot be integrated easily.

Adaptive control algorithms are designed to respond to changes, uncertainties, and the random nature of control systems, whether in the controller itself or the feedback and feedforward signals. In recent years, adaptive control strategies have been widely explored both numerically and experimentally in the field of structural control for various civil engineering applications [166–168]. Adaptive control demonstrates strong real-time adaptability and robustness under both linear and nonlinear conditions by dynamically responding to uncertainties, albeit at the cost of increased design complexity. This comparison underscores the importance of selecting control algorithms based on specific structural needs, performance criteria, and implementation feasibility [169].

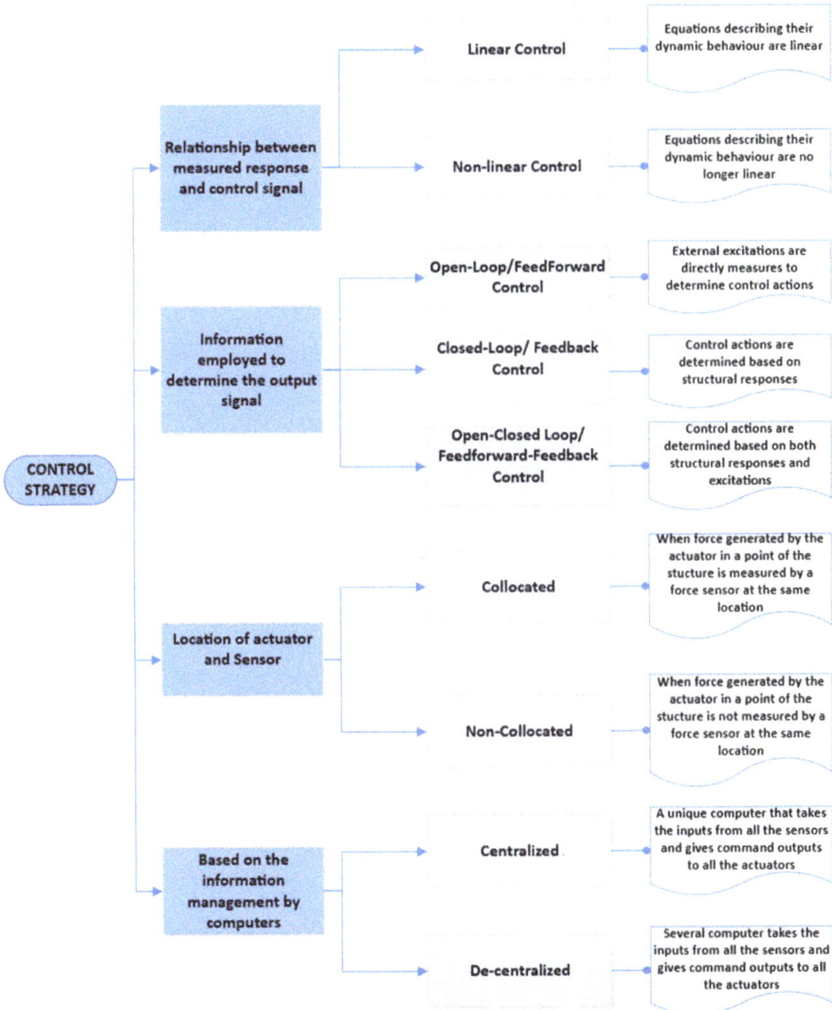

Figure 3. Classification of control strategies based on the type of input.

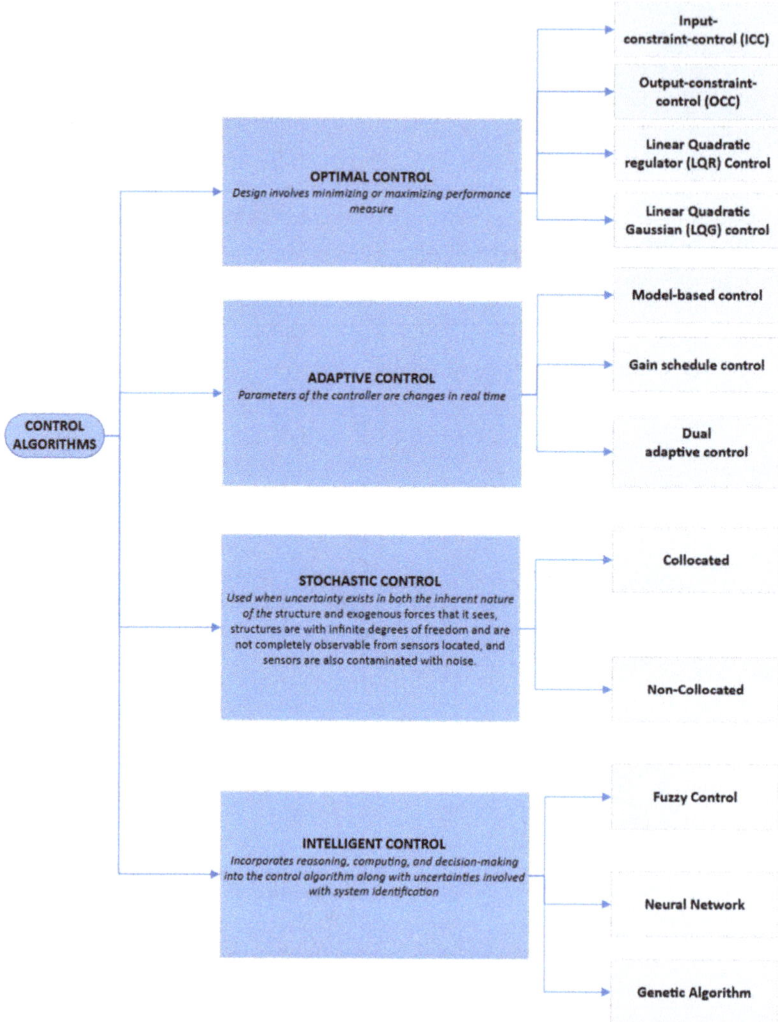

Figure 4. Classification of control algorithms based on different factors.

The accuracy of nonlinear system modeling has a significant impact on the effectiveness of traditional control algorithms. While conventional control strategies account for uncertainties in measuring both input excitations and system responses, their success still largely depends on the specific control law used. In contrast, intelligent control systems enhance these algorithms by incorporating elements such as reasoning, computation, and decision-making, while also addressing uncertainties in system identification and modeling. This makes intelligent controls a more realistic approach, offering a wider range of solutions. Intelligent systems are influenced by human input, which introduces the potential for error due to reliance on the designer's experience. Intelligent control methods are typically classified into three main types: (a) fuzzy control, (b) neural network control, and (c) genetic algorithms. Neural network control and fuzzy logic control both eliminate the need for predefined structural models, relying instead on data-driven or rule-based approaches [122,170–173]. While they offer flexibility for nonlinear systems, they face challenges related to training complexity, stability, and design intricacy. These algorithms do not consider feedback from the actuator and rely solely on the structural responses, which can be difficult to obtain during seismic events.

The effectiveness of a specific controller can vary for different structures, indicating that the performance of control algorithms relies on the characteristics of the structural system and the earthquake data it experiences. It is crucial to compare structural responses using ground motions with the same hazard level. For example, the performance of a controller on the first floor may decrease as the number of building stories increases, making it more suitable for lower floors. Therefore, it is crucial to investigate the optimal number and placement of dampers based on the specific structures. To achieve a more robust performance evaluation for semi-actively controlled structures using MR dampers, it is necessary to expand the dataset by including more seismic ground motions, building more comprehensive models, and developing more advanced control algorithms. The most effective way to demonstrate the robustness of the control algorithms is through experimental testing. The review shows that several novel control algorithms have been developed to enhance the efficiency of the control systems [117,134,158–166]. However, most of them are not tested experimentally. Experimental testing of semi-active control systems in structures requires addressing specific aspects to achieve a more realistic evaluation of their effectiveness in controlling earthquake-induced vibrations [174]. The practicality of a semi-active control system in mitigating the seismic impact on structures is studied through experimental testing at both the component level and with small-scale model structures affixed to a shake table [165]. It has been observed from the literature that control systems are developed to address specific challenges, such as uncertainty, time delay, and sensor optimization. The following section reviews the controllers developed to address specific issues.

6. Optimized Control System Designed to Tackle Practical Implementation Challenges

Upon reviewing the literature, it becomes evident that each control algorithm exhibits strengths and weaknesses. Researchers have consistently focused on enhancing control system performance by taking various factors into account [28,39,42,56,61,122,169–171,175–183]. The overall effectiveness of the control system can be improved by considering a range of factors, which are elaborated in this section.

To successfully apply control systems to structures, a balance must be found between reliability and robustness. Reliability, as a crucial concept, depends on the control system's sustained efficiency over time, encompassing all aspects of frequent system use, regardless of whether it is used for controlling minor, major, near-field, or far-field earthquake ground motions. However, increasing the reliability of a system does not automatically ensure optimal performance under varying earthquake conditions.

In contrast, system robustness is crucial for preventing failure, malfunction, or counterproductive effects in structural response during exposure to a wide range of ground motions, including extreme earthquakes. While enhancing reliability ensures consistent system operation, enhancing robustness guarantees effective performance under more substantial seismic events.

The subsequent section discusses research-driven strategies to elevate seismic control performance. These strategies are tailored to enhance the robustness of semi-active control systems. Notable approaches include minimizing the impact of uncertainty, time delay, and the optimal placement of sensors and actuators. This paper aims to shed light on how advancements in robustness can strengthen the effectiveness of active and semi-active control systems under seismic conditions.

6.1. Control Systems to Ensure Robustness to Uncertainties

This review observes that most control algorithms developed are investigated for their robustness and stability in response to system uncertainties, such as stiffness and mass. Conventional control algorithms rely on nominal models and may not work well in uncertain models [42]. Various methods, including H_∞, H_2, H_∞/H_2, μ-synthesis, and linear matrix inequalities, are employed for robust controller design. Table 4 presents a comprehensive overview of advanced control algorithms specifically designed to address various types of uncertainties commonly encountered in structural systems. These uncertainties may include variations in structural damping, changes in frequency content due to ground motion characteristics, the nonlinear behavior of materials and components, and inherent time delays in sensing, computation, and actuation.

Table 4. Robust control algorithms for uncertainties.

Publication	Method Used	Issues Addressed
(Amini and Ghaderi) [184]	Discrete Wavelet Transform and particle swarm optimization	Uncertainty in structural damping and earthquake
(Bagheri and Amini) [185]	Wavelet analysis and pattern search method	Uncertainty in the frequency content of earthquakes
(Amiri et al.) [186], (Bitaraf and Hurlebaus) [187]	Fuzzy logic-based control algorithm	Nonlinearity
(Oliveira et al.) [188]	Predictive control	Uncertainty and time delay
(Nguyen et al.) [189]	New fuzzy sliding mode controller Adaptive neuro-fuzzy inference system	Uncertainty and time delay
(Ghaderi and Amini) [190]	Adaptive block backstepping	Unknown parameters
(Ramezani et al.) [104]	Fuzzy type-1 and type-2	Uncertainties
(Shan et al.) [191]	Model reference adaptive backstepping control algorithm	Structural nonlinearity, structural uncertainty, and the influence of actuator saturation
(Zabihi-Samani and Ghanooni-Bagha) [101]	Adjustable cuckoo search wavelet-based fuzzy logic controller (ACSWBFLC)	Nonlinear modeling of MR damper Optimization of placement and number of MR dampers
(Darbanian et al.) [170]	Fuzzy-LQR Algorithm	Uncertainties of Structural Parameters

To manage these complex and dynamic challenges, the following table highlights several sophisticated control strategies that have shown promising results in both theoretical and experimental studies. Among these, the Discrete Wavelet Transform (DWT) stands out for its ability to decompose non-stationary signals and extract relevant features in both time and frequency domains, making it suitable for real-time monitoring and adaptive control during seismic events. Fuzzy logic control is another method, known for its effectiveness in handling imprecise or ambiguous input data. This approach does not require a precise mathematical model of the system; instead, it relies on rule-based reasoning, making it particularly suitable for systems that are uncertain or nonlinear in nature.

Additionally, Adaptive Neuro-Fuzzy Inference Systems (ANFISs) are employed as a hybrid method, combining the learning capabilities of neural networks with the interpretability of fuzzy logic. This approach enables the controller to adapt to changing system dynamics by updating its rule base and membership functions over time, resulting in enhanced control accuracy and robustness.

It is observed that the use of frequency domain control methods involves analyzing and designing control systems in the frequency domain, where the behavior of a system is examined in terms of its response to different frequencies. H_∞ control is one such method that is often used for robust control system design, particularly in the presence of uncertainties and disturbances. Robust H_∞ control seeks to minimize the impact of disturbances on the controlled system while maintaining stability and optimal performance. It provides a framework for handling uncertainties and disturbances to optimize the system's response in the presence of these factors. Studies have also shown that the use of frequency-shaping filters incorporated into the control system influence or shape the system's frequency response. These filters allow us to tailor the system's behavior at specific frequencies, making it more responsive or less sensitive to specific input frequencies. Narasimhan and Nagarajaiah [165] developed a novel control algorithm based on H_∞ for a variable friction-based semi-active control system. The controller was designed to determine the optimal control force for reducing responses to near-fault earthquakes. The study's results, considering stiffness uncertainty, indicate that both active and semi-active H_∞ controllers demonstrate robustness and effectiveness in reducing responses for intelligent base-isolated structures in the face of near-fault earthquakes.

Loh et al. [192] conducted an experimental validation of wireless communications for real-time structural control, assessing the control performance of the wireless system in comparison to a conventional tethered control system. The efficacy of the wireless control system is evaluated through a shaking table test performed on a three-story steel frame equipped with an MR damper on each floor. The study demonstrates that wireless sensor networks represent a promising technology capable of operating in real-time environments. Additionally, the study highlights the advantages of decentralized control approaches due to their robustness against failure, ensuring that the control system can still operate even if one damper fails to function correctly.

Wang et al. [107] studied the mechanical performance of a piezoelectric ceramic friction damper installed on a three-story structure. An adaptive fuzzy neural network controller (FNNC) is proposed, and its performance is compared with the LQR optimal control. To study the robustness and stability of the proposed controller in the presence of model uncertainties, a stiffness adjustment of $\pm 10\%$ is considered. The results indicate that FNCC exhibits good robustness and stability, particularly in the face of uncertainty, and can effectively reduce the responses.

6.2. Control Systems to Handle Time-Delay Issues

The existence of a time lag between sensing structural responses and applying control forces is a critical issue. Research focused on mitigating time delays in control systems has been reviewed, and it is seen that control algorithms such as the time-delay compensation method based on Newmark's method decentralized output feedback polynomial control (DOFPC), Lyapunov control, the Taylor series expansion of the control force, the decentralized H_1 controller, adaptive control, and the Proportional–Integral–Derivative (PID) controller demonstrate robustness to time delay. Wang [40] conducted a study employing decentralized dynamic output feedback controllers to reduce the H_∞ norm of the closed-loop system. The formulation of the problem considers the impact of feedback time delay, which is consequently addressed in the design of the controllers. The controller's performance is found to be more efficient when compared to a time-delayed LQG controller. Di Paolo and Pirrotta [193] studied the effect of time delay under random excitation. Comparing three approaches, the study suggests that the Taylor expansion method can serve the dual purpose of assessing both the critical time delay at which control effectiveness diminishes and of evaluating the response with variance. Cha et al. [39] studied the effects of time delay on large-scale semi-active control strategies. This study explored the impact of time delay on the performance of a structure equipped with a large-scale MR damper. The study utilized numerical simulations for earthquakes occurring in both near-field and far-field scenarios. The MR damper operates via four unique semi-active control methods: simple passive control (SPC), decentralized output feedback polynomial control (DOFPC), Lyapunov control, and clipped-optimal control (COC). The findings reveal that all controllers, except for COC, exhibit significant robustness in the face of time delay. In contrast, the clipped-optimal controller requires integration with a compensator to enhance performance when dealing with time delays. An experimental verification of the adaptive neuro-fuzzy inference system (ANFIS) and a novel fuzzy sliding mode controller (FSMC) was demonstrated in the study by Nguyen et al. [189] to mitigate the effects of time delay and uncertainties.

6.3. *Control Systems for Optimal Placement and Number of Control Devices*

Strategically positioning dampers optimally enhances control performance and reduces the system's reliance on external power sources, thus lowering the overall cost of the control system and creating additional available space. Furthermore, an optimally placed, compact set of dampers can deliver the same level of performance as a complete set, typically requiring less energy to generate control forces. Ideal locations for these dampers may include the lower floors and other vulnerable areas of civil structures, where seismic excitations often induce significant nonlinear deformations [51,194–196].

An overview of the optimal damper distribution as a passive energy dissipation system for retrofitting structures against earthquakes is presented in [44]. Ribakov and Agranovich [197] introduced an algorithm for determining damper locations based on their maximum contribution to total seismic energy dissipation. Chat et al. [4,198] proposed a multi-objective genetic algorithm to optimize the placement of control devices and sensors, considering cost and seismic control performance when designing structural control systems. To study the optimal placement of devices, the linear quadratic Gaussian (LQG) algorithm emerges as the most employed in the studies reviewed, as seen in the works of [4,5,197–199] and [44].

7. Research Gaps and Future Directions

Research in the field of structural control systems has made significant progress; however, challenges to practical implementation require further investigation. In theory, structural control offers significant potential for enhancing the performance and safety of structures, particularly under dynamic loading such as earthquakes. However, its practical implementation often presents challenges that are not fully addressed in academic research. Therefore, greater emphasis should be placed on translating theoretical advancements into practical solutions. Future research should prioritize the development of control systems that are not only technically effective but also feasible and straightforward to implement from the perspective of builders and engineers in the field. Bridging this gap between theory and practice is essential to facilitate the broader adoption of structural control technologies in real-world construction projects.

Several significant gaps remain in the existing literature on adaptive control systems for structural engineering applications. One key area lacking sufficient research is the use of adaptive and intelligent controllers for mitigating seismic responses in structures employing control systems. This topic warrants further study to develop robust strategies suited for time-varying systems with inherent uncertainties. Another underexplored area is the use of indirect adaptive control, which is especially useful when the system model is either unknown or subject to change. Despite its potential to enhance performance and seismic resilience, its application remains limited. Addressing these shortcomings is crucial for advancing adaptive control technologies that aim to mitigate seismic impacts on structures.

To address these challenges and close the existing research gaps, several directions for future work can be considered. First, there is a need for the development of robust, multi-functional control systems that combine intelligent and adaptive control strategies. Such systems could offer more flexible and responsive solutions to structural vibrations. Second, the formulation of multi-objective control schemes that optimize both vibration mitigation performance and cost-efficiency would be highly beneficial, particularly for practical implementation. Third, since ensuring a stable energy supply during earthquakes is a significant challenge, future research should explore energy harvesting techniques and their seamless integration with vibration control systems to enable autonomous or semi-autonomous operation. Lastly, combining vibration control systems with structural health monitoring technologies can create a more holistic and responsive approach to managing structural performance, especially in post-earthquake scenarios.

8. Concluding Remarks

An overview of seismic vibration mitigation systems for structures has been provided in this review. Based on the current research landscape, there is a preference for semi-active control systems in structural vibration control over other control systems. Both numerical and experimental studies consistently demonstrate the superior performance of semi-active systems compared to passive systems, which require significantly less external energy than active control systems. Amongst the semi-active control devices, MR dampers have emerged as the most extensively studied.

Despite advancements in research on structural control strategies, their widespread adoption has been hindered by practical implementation challenges. This paper outlines the limitations and challenges of implementing control systems, identifies critical issues in experimental execution, and proposes improved control systems tailored to address each challenge. The careful selection of a control algorithm is crucial in ensuring the effective and robust performance of these control systems. Various control algorithms have been investigated in the pursuit of effective response control, revealing that no single algorithm can be deemed optimal.

This paper makes a clear and valuable contribution by critically evaluating the current state of seismic vibration mitigation systems, with a strong emphasis on the growing prominence of semi-active control strategies. This paper highlights the advantages of semi-active systems and addresses the significant gap between theoretical development and practical implementation. Key challenges in real-world applications, including experimental limitations, algorithmic selection, and system integration, are identified. This paper offers a comprehensive assessment of existing limitations. This paper also reviews improved control strategies to address these challenges, emphasizing the importance of selecting the appropriate algorithm. Notably, multi-objective adaptive control algorithms are gaining increased attention in seismic control research. These algorithms target challenges such as optimizing various parameters, minimizing uncertainties, optimizing the quantity and distribution of sensors and actuators, establishing decentralized control networks, and reducing time delays. This paper provides an overview of existing structural vibration control systems and enhances their practicality and robustness for mitigating seismic vibrations.

Author Contributions: Conceptualization, N.B.D.; methodology, N.B.D.; formal analysis, N.B.D.; investigation, N.B.D.; resources, N.B.D. and M.A.; data curation, N.B.D.; writing—original draft preparation, N.B.D.; writing—review and editing, M.A. and N.B.D.; visualization, N.B.D.; supervision, M.A.; project administration, M.A.; funding acquisition, M.A. All authors have read and agreed to the published version of the manuscript.

Funding: This research was financially supported by the Postdoctoral Fellowship Award (PDFA) program [Grant Number: PDF-RA-23-E7] and the Open Access Program (OAP) at the American University of Sharjah (AUS).

Institutional Review Board Statement: Not applicable.

Informed Consent Statement: Not applicable.

Data Availability Statement: The original contributions presented in the study are included in the article, further inquiries can be directed to the corresponding author.

Acknowledgments: The authors greatly appreciate financial support from AUS. This paper represents the opinions of the authors and does not mean to represent the position or opinions of AUS.

Conflicts of Interest: The authors declare no conflicts of interest.

References

1. Saeed, M.U.; Sun, Z.; Elias, S. Research developments in adaptive intelligent vibration control of smart civil structures. *J. Low Freq. Noise Vib. Act. Control* **2022**, *41*, 292–329. [CrossRef]
2. Fisco, N.R.; Adeli, H. Smart structures: Part I—Active and semi-active control. *Sci. Iran.* **2011**, *18*, 275–284. [CrossRef]
3. Fisco, N.R.; Adeli, H. Smart structures: Part II—Hybrid control systems and control strategies. *Sci. Iran.* **2011**, *18*, 285–295. [CrossRef]
4. Soong, T.T.; Spencer, B.F. Supplemental energy dissipation: State-of-the-art and state-of-the- practice. *Eng. Struct.* **2002**, *24*, 243–259. [CrossRef]
5. Jiang, H.; Cheng, F.Y.; Wang, Z. State-of-the-Art in Hybrid Control and Issues on Optimum Seismic Observer and Device Placement. In Proceedings of the 12th World Conference on Earthquake Engineering, Auckland, New Zealand, 30 January–4 February 2000; pp. 1–8.
6. Jung, H. State-of-the-art of semiactive control systems using MR fluid dampers in civil engineering applications. *Struct. Eng. Mech.* **2004**, *17*, 493–526. [CrossRef]
7. Rahimi, F.; Aghayari, R.; Samali, B. Application of tuned mass dampers for structural vibration control: A state-of-the-art review. *Civ. Eng. J.* **2020**, *6*, 1622–1651. [CrossRef]
8. Housner, B.G.W.; Bergman, L.A.; Caughey, T.K.; Chassiakos, A.G.; Claus, R.O.; Masri, S.F.; Skelton, R.E.; Soong, T.T.; Spencer, B.F.; Yao, J.T.P. Structural control: Past, present, and future. *J. Eng. Mech.* **1997**, *123*, 897–971. [CrossRef]
9. Spencer, B.F.; Soong, T.T. New applications and development of active, semi-active, and hybrid control techniques for seismic and non-seismic vibration in the USA. In Proceedings of the International Post-SMiRT Conference Seminar on Seismic Isolation, Passive Energy Dissipation and Active Control of Vibration of Structures, Cheju, Republic of Korea, 23–25 August 1999.
10. Soong, T.T.; Spencer, B.F. Active, semi-active and hybrid control of structures. In Proceedings of the 12th World Conference on Earthquake Engineering, Auckland, New Zealand, 30 January–4 February 2000.
11. Datta, T.K. On Active Control of StructuresA State-of-the-Art Review. *ISET J. Earthq. Technol.* **2003**, *40*, 1–17. [CrossRef]
12. Korkmaz, S. A review of active structural control: Challenges for engineering informatics. *Comput. Struct.* **2011**, *89*, 2113–2132. [CrossRef]
13. Symans, M.D.; Constantinou, M.C. Semi-active control systems for seismic protection of structures: A state-of-the-art review. *Eng. Struct.* **1999**, *21*, 469–487. [CrossRef]
14. Spencer, B.F., Jr.; Nagarajaiah, S. State of the Art of Structural Control. *J. Struct. Eng.* **2003**, *129*, 845–856. [CrossRef]
15. Patil, S.; Reddy, G. State of Art Review—Base isolation systems for structures. *Int. J. Emerg. Technol. Adv. Eng.* **2012**, *2*, 438–453.
16. Girish, M.; Pranesh, M. Sliding Isolation Systems: State-of-the-Art Review. *IOSR J. Mech. Civ. Eng.* **2015**, *12*, 30–35.

17. Kunde, M.C.; Jangid, R.S. Seismic behavior of isolated bridges: A-state-of-the-art review. *Electron. J. Struct. Eng.* **2003**, *3*, 140–170. [CrossRef]
18. Kumar, G.; Kumar, R.; Kumar, A. A Review of the Controllers for Structural Control. *Arch. Comput. Methods Eng.* **2023**, *30*, 3977–4000. [CrossRef]
19. Koutsoloukas, L.; Nikitas, N.; Aristidou, P. Passive, semi-active, active and hybrid mass dampers: A literature review with associated applications on building-like structures. *Dev. Built Environ.* **2022**, *12*, 100094. [CrossRef]
20. Javanmardi, A.; Ghaedi, K.; Huang, F.; Hanif, M.U.; Tabrizikahou, A. Application of Structural Control Systems for the Cables of Cable-Stayed Bridges: State-of-the-Art and State-of-the-Practice. *Arch. Comput. Methods Eng.* **2022**, *29*, 1611–1641. [CrossRef]
21. Kavyashree, B.; Patil, S.; Rao, V.S. Review on vibration control in tall buildings: From the perspective of devices and applications. *Int. J. Dyn. Control* **2021**, *9*, 1316–1331. [CrossRef]
22. Saaed, T.E.; Nikolakopoulos, G.; Jonasson, J.E.; Hedlund, H. A state-of-the-art review of structural control systems. *JVC/J. Vib. Control* **2015**, *21*, 919–937. [CrossRef]
23. Banerjee, T.; Ghosh, D.; Das, D. A state-of-the-art review and future challenges on application of devices for vibration control of bridges. *Mech. Adv. Mater. Struct.* **2025**, *0*, 1–27. [CrossRef]
24. Elias, S.; Matsagar, V. Research developments in vibration control of structures using passive tuned mass dampers. *Annu. Rev. Control* **2017**, *44*, 129–156. [CrossRef]
25. Liu, C.; Jing, X.; Daley, S.; Li, F. Recent advances in micro-vibration isolation. *Mech. Syst. Signal Process* **2015**, *56*, 55–80. [CrossRef]
26. Casciati, F.; Rodellar, J.; Yildirim, U. Active and semi-active control of structures-theory and applications: A review of recent advances. *J. Intell. Mater. Syst. Struct.* **2012**, *23*, 1181–1195. [CrossRef]
27. Morales-Beltran, M.; Paul, J. Technical Note: Active and Semi-Active Strategies to Control Building Structures under Large Earthquake Motion. *J. Earthq. Eng.* **2015**, *19*, 1086–1111. [CrossRef]
28. Ikeda, Y. Active and semi-active vibration control of buildings in Japan-practical applications and verification. *Struct. Control Health Monit.* **2009**, *16*, 703–723. [CrossRef]
29. Jansen, L.M.; Dyke, S.J. Semiactive Control Strategies for MR Dampers: Comparative Study. *J. Eng. Mech.* **2000**, *126*, 795–803. [CrossRef]
30. Gkatzogias, K.I.; Kappos, A.J. Semi-active control systems in bridge engineering: A review of the current state of practice. *Struct. Eng. Int.* **2016**, *26*, 290–300. [CrossRef]
31. Ratna, K.S.; Daniel, C.; Ram, A.; Yadav, B.S.K.; Hemalatha, G. Analytical Investigation of MR Damper for Vibration Control: A Review. *J. Appl. Eng. Sci.* **2021**, *11*, 49–52. [CrossRef]
32. Masa'id, A.; Lenggana, B.W.; Ubaidillah, U.; Susilo, D.D.; Choi, S.B. A Review on Vibration Control Strategies Using Magnetorheological Materials Actuators: Application Perspective. *Actuators* **2023**, *12*, 113. [CrossRef]
33. Kori, J.G.; Jangid, R.S. Semi-active mr dampers for seismic control of structures. *Bull. N. Z. Soc. Earthq. Eng.* **2009**, *42*, 157–166. [CrossRef]
34. Cha, Y.-J.; Zhang, J.; Agrawal, A.K.; Dong, B.; Friedman, A.; Shirley; Dyke, J.; Ricles, J. Comparative Studies of Semi-active Control Strategies for MR dampers. *J. Chem. Inf. Model.* **2013**, *53*, 1689–1699. [CrossRef]
35. Dyke, S.J.; Spencer, B.F., Jr. A Comparison of Semi-Active Control Strategies for the MR Damper. In Proceedings of the Intelligent Information Systems IIS'97, Grand Bahama Island, Bahamas, 8–10 December 1997; pp. 580–584. [CrossRef]
36. Spencer, B.F., Jr.; Dyke, S.J.; Sain, M.K.; Carlson, J.D. Phenomenological Model for Magnetorheological Dampers. *J. Eng. Mech.* **1997**, *123*, 230–238. [CrossRef]
37. Wani, Z.R.; Tantray, M.; Noroozinejad Farsangi, E.; Nikitas, N.; Noori, M.; Samali, B.; Yang, T. A Critical Review on Control Strategies for Structural Vibration Control. *Annu. Rev. Control* **2022**, *54*, 103–124. [CrossRef]
38. Liu, J.; Xia, K.; Zhu, C. The state-of-the-art review of structural control strategy. In Proceedings of the 2009 International Conference on E-learning, E-Business, Enterprise Information Systems, and E-Government, EEEE, Hong Kong, China, 5–6 December 2009; pp. 211–213. [CrossRef]
39. Cha, Y.J.; Agrawal, A.K.; Dyke, S.J. Time delay effects on large-scale MR damper based semi-active control strategies. *Smart Mater. Struct.* **2013**, *22*, 015011. [CrossRef]
40. Wang, Y. Time-delayed dynamic output feedback H∞controller design for civil structures: A decentralized approach through homotopic transformation. *Struct. Control Health Monit.* **2011**, *18*, 121–139. [CrossRef]
41. Lee, T.; Kawashima, K. Semiactive Control of Nonlinear Isolated Bridges with Time Delay. *J. Struct. Eng.* **2007**, *133*, 235–241. [CrossRef]
42. Palizvan Zand, J.; Sabouri, J.; Katebi, J.; Nouri, M. A new time-domain robust anti-windup PID control scheme for vibration suppression of building structure. *Eng. Struct.* **2021**, *244*, 112819. [CrossRef]
43. Occhiuzzi, A.; Spizzuoco, M.; Serino, G. Experimental analysis of magnetorheological dampers for structural control. *Smart Mater. Struct.* **2003**, *12*, 703–711. [CrossRef]

44. Kookalani, S.; Shen, D.; Zhu, L.L.; Lindsey, M. An Overview of Optimal Damper Placement Methods in Structures. *Iran. J. Sci. Technol.—Trans. Civ. Eng.* **2022**, *46*, 1785–1804. [CrossRef]
45. Soto, M.G.; Adeli, H. Placement of control devices for passive, semi-active, and active vibration control of structures. *Sci. Iran.* **2013**, *20*, 1567–1578.
46. Yao, J.T.P. Concept of Structural Control. *J. Struct. Div-ASCE.* **1972**, *98*, 7. [CrossRef]
47. Kurata, N.; Kobori, T.; Takahashi, M.; Niwa, N.; Midorikawa, H. Actual seismic response controlled building with semi-active damper system. *Earthq. Eng. Struct. Dyn.* **1999**, *28*, 1427–1447. [CrossRef]
48. Yang, G.; Spencer, B.F.; Carlson, D.; Sain, K. Large-scale MR uid dampers: Modeling and dynamic performance considerations. *Eng. Struct.* **2002**, *24*, 309–323. [CrossRef]
49. Iemura, H.; Pradono, M.H. Passive and semi-active seismic response control of a cable-stayed bridge. *J. Struct. Control* **2002**, *9*, 189–204. [CrossRef]
50. Das, D.; Datta, T.K.; Madan, A. ANN-cum-Fuzzy Control of Seismic Response using MR Dampers. In Proceedings of the 15th World Conference on Earthquake Engineering, Lisbon, Portugal, 24–28 September 2012.
51. Das, D.; Datta, T.K.; Madan, A. Semiactive fuzzy control of the seismic response of building frames with MR dampers. *Earthq. Eng. Struct. Dyn.* **2012**, *41*, 99–118. [CrossRef]
52. Wongprasert, N.; Symans, M.D. Experimental Evaluation of Adaptive Elastomeric Base-Isolated Structures Using Variable-Orifice Fluid Dampers. *J. Struct. Eng.* **2005**, *131*, 867–877. [CrossRef]
53. Villarreal, K. Effects of MR damper placement on structure vibration parameter. *Proc. IEEE Conf. Control.* **2005**.
54. Dyke, S.J.; Spencer, B.F.; Sain, M.K.; Carlson, J.D. Modeling and control of magnetorheological dampers for seismic response reduction. *Smart Mater. Struct.* **1999**, *5*, 565–575. [CrossRef]
55. Nagarajaiah, B.S.; Riley, M.A.; Member, S.; Reinhorn, A. Control of Sliding—Isolated bridge with absolute acceleration feedback. *J. Eng. Mech.* **1994**, *119*, 2317–2332. [CrossRef]
56. Kwok, N.M.; Nguyen, T.H.; Ha, Q.P.; Li, J.; Samali, B. MR Damper Structural Control Using a Multi-Level Sliding Mode Controller. In *Australian Earthquake Engineering Conference*; Australian Earthquake Engineering Society: Melbourne, Australia, 2007; pp. 1–7.
57. Yoshioka, H.; Ramallo, J.C.; Spencer, B.F., Jr. "Smart" Base Isolation Strategies Employing Magnetorheological Dampers. *J. Eng. Mech.* **2002**, *128*, 540–551. [CrossRef]
58. Zahrai, S.M.; Salehi, H. Semi-active seismic control of mid-rise structures using magneto-rheological dampers and two proposed improving mechanisms. *Iran. J. Sci. Technol. Trans. Civ. Eng.* **2014**, *38*, 21–36.
59. Bagherkhani, A.; Baghlani, A. Probability of Failure of High-Rise Buildings Equipped with Magnetorheological Dampers by Considering Uncertainty in Stationary and Nonstationary Earthquakes. *ASCE-ASME J. Risk Uncertain. Eng. Syst. Part A Civ. Eng.* **2023**, *9*, 04023038. [CrossRef]
60. Li, Z.X.; Wen, D.; Shi, Y.; Wei, X.; Liu, B. A frequency-dependent variable damping control method for base-isolated structures under ground motions with different frequency characteristics. *Soil Dyn. Earthq. Eng.* **2023**, *165*, 107717. [CrossRef]
61. Bhaiya, V.; Shrimali, M.K.; Bharti, S.D.; Datta, T.K. Modified semiactive control with MR dampers for partially observed systems. *Eng. Struct.* **2019**, *191*, 129–147. [CrossRef]
62. Cruze, D.; Gladston, H.; Farsangi, E.N.; Banerjee, A.; Loganathan, S.; Solomon, S.M. Seismic Performance Evaluation of a Recently Developed Magnetorheological Damper: Experimental Investigation. *Pract. Period. Struct. Des. Constr.* **2021**, *26*, 04020061. [CrossRef]
63. Lenggana, B.W.; Ubaidillah, U.; Imaduddin, F.; Choi, S.B.; Purwana, Y.M.; Harjana, H. Review of magnetorheological damping systems on a seismic building. *Appl. Sci.* **2021**, *11*, 9339. [CrossRef]
64. Abdeddaim, M.; Djerouni, S.; Ounis, A.; Athamnia, B.; Noroozinejad Farsangi, E. Optimal design of Magnetorheological damper for seismic response reduction of Base-Isolated structures considering Soil-Structure interaction. *Structures* **2022**, *38*, 733–752. [CrossRef]
65. Hojat Jalali, H.; Fahimi Farzam, M.; Mousavi Gavgani, S.A.; Bekdaş, G. Semi-active control of buildings using different control algorithms considering SSI. *J. Build. Eng.* **2023**, *67*, 105956. [CrossRef]
66. Suthar, S.J.; Patil, V.B.; Jangid, R.S. Optimization of MR Dampers for Wind-Excited Benchmark Tall Building. *Pract. Period. Struct. Des. Constr.* **2022**, *27*, 04022048. [CrossRef]
67. Ostrowski, M.; Błachowski, B.; Popławski, B.; Pisarski, D.; Mikułowski, G.; Jankowski, L. Semi-active modal control of structures with lockable joints: General methodology and applications. *Struct. Control. Health Monit.* **2021**, *28*, e2710. [CrossRef]
68. Alexander, S.; Sumathi, P.; Panigrahi, S.K. Semi-Active Control of Structures with Magnetorheological Fluid Dampers. In Proceedings of the 2022 IEEE Global Conference on Computing, Power and Communication Technologies, GlobConPT, New Delhi, India, 23–25 September 2022; Institute of Electrical and Electronics Engineers Inc.: Piscataway, NJ, USA, 2022. [CrossRef]
69. Sadek, F.; Mohraz, B. Semiactive Control Algorithms for Structures with Variable Dampers. *J. Eng. Mech.* **1998**, *124*, 981–990. [CrossRef]

70. Choi, K.M.; Cho, S.W.; Jung, H.J.; Lee, I.W. Semi-active fuzzy control for seismic response reduction using magnetorheological dampers. *Earthq. Eng. Struct. Dyn.* **2004**, *33*, 723–736. [CrossRef]
71. Lara, L.A.V.; Brito, J.L.V.; Valencia, Y.G. Comparative analysis of semi-active control algorithms applied to magnetorheological dampers. *Ingeniare Rev. Chil. Ing.* **2017**, *25*, 39–58. [CrossRef]
72. Karami, K.; Manie, S.; Ghafouri, K.; Nagarajaiah, S. Nonlinear structural control using integrated DDA/ISMP and semi-active tuned mass damper. *Eng. Struct.* **2019**, *181*, 589–604. [CrossRef]
73. Renzi, E.; De Angelis, M. Semi-active continuous control of base-excited structures: An exploratory study. *Struct. Control Health Monit.* **2010**, *17*, 563–589. [CrossRef]
74. Troise, A.; Bruschi, E.; Rosso, M.M.; Alosio, A.; Rizzo, A.; Lagaros, N.D.; Quaglini, V.; Marano, G.C. Variable orifice damper implementation for seismic semi-active control of civil structures. In Proceedings of the COMPDYN 2023, 9th ECCOMAS Thematic Conference on Computational Methods in Structural Dynamics and Earthquake Engineering, Athens, Greece, 12–14 June 2023.
75. Weber, F. Semi-active vibration absorber based on real-time controlled MR damper. *Mech. Syst. Signal Process* **2014**, *46*, 272–288. [CrossRef]
76. Xu, Z.-D.; Shen, Y.-P.; Guo, Y.-Q. Semi-active control of structures incorporated with magnetorheological dampers using neural networks. *Smart Mater. Struct.* **2003**, *12*, 80–87. [CrossRef]
77. Saleh Al-Fahdawi, O.A.; Niedzwecki, J.M.; Jones, H.L.; Newton, H.J. Semi-Active Adaptive Control of Coupled Structures for Seismic Hazard Mitigation. Ph.D. Thesis, Texas A&M University, College Station, TX, USA, 2019.
78. Lin, C.C.; Lu, L.Y.; Lin, G.L.; Yang, T.W. Vibration control of seismic structures using semi-active friction multiple tuned mass dampers. *Eng. Struct.* **2010**, *32*, 3404–3417. [CrossRef]
79. Gattulli, V.; Lepidi, M.; Potenza, F. Seismic protection of frame structures via semi-active control: Modeling and implementation issues. *Earthq. Eng. Eng. Vib.* **2009**, *8*, 627–645. [CrossRef]
80. Mulligan, K. Experimental and Analytical Studies of Semi-Active and Passive Structural Control of Buildings. Ph.D. Thesis, University of Canterbury, Christchurch, New Zealand, 2007.
81. Iemura, H.; Igarashi, A.; Kalantari, A. Experimental verification and numerical studies of an autonomous semi-active seismic control strategy. *Struct. Control Health Monit.* **2006**, *13*, 301–323. [CrossRef]
82. Huang, Z.S.; Wu, C.; Hsu, D.S. Semi-Active Fuzzy Control of a MR Damper on Structures by Genetic Algorithm. *J. Mech.* **2009**, *25*, 1–6. [CrossRef]
83. Wang, L.; Shi, W.; Zhou, Y.; Zhang, Q. Semi-active eddy current pendulum tuned mass damper with variable frequency and damping. *Smart Struct. Syst.* **2020**, *25*, 65–80. [CrossRef]
84. Karamodin, A.; Kazemi, H.; Akbarzadeh, T. Semi-active control of structures using neuro-predictive algorithm for MR dampers. *Struct. Control Health Monit.* **2008**, *17*, 237–253. [CrossRef]
85. Chae, Y.; Ricles, J.M. Seismic Performance Enhancement of Base-Isolation Using Semi-Active Damping Devices Systems. In Proceedings of the 15th World Conference of Earthquake Engineering, Lisbon, Portugal, 24–28 September 2012.
86. Samali, B.; Djajakesukma, S.L.; Nguyen, H.; Li, J. An experimental study of a five storey steel frame using semi-active control system. *APVC Proc.* **2003**, *2*, 604–609.
87. Sun, Q.; Zhang, L.; Zhou, J.; Shi, Q. Experimental study of the semi-active control of building structures using the shaking table. *Earthq. Eng. Struct. Dyn.* **2003**, *32*, 2353–2376. [CrossRef]
88. Liu, Y.F.; Lin, T.K.; Chang, K.C. Analytical and experimental studies on building mass damper system with semi-active control device. *Struct. Control Health Monit.* **2018**, *25*, e2154. [CrossRef]
89. Pu, H.; Yuan, S.; Peng, Y.; Meng, K.; Zhao, J.; Xie, R.; Huang, Y.; Sun, Y.; Yang, Y.; Xie, S.; et al. Multi-layer electromagnetic spring with tunable negative stiffness for semi-active vibration isolation. *Mech. Syst. Signal Process* **2019**, *121*, 942–960. [CrossRef]
90. Symans, M.D.; Constantinou, M.C. Seismic testing of a building structure with a semi-active fluid damper control system. *Earthq. Eng. Struct. Dyn.* **1997**, *26*, 759–777. [CrossRef]
91. Ostrowski, M.; Jedlińska, A.; Popławski, B.; Blachowski, B.; Mikułowski, G.; Pisarski, D.; Jankowski, Ł. Sliding Mode Control for Semi-Active Damping of Vibrations Using on/off Viscous Structural Nodes. *Buildings* **2023**, *13*, 348. [CrossRef]
92. Gehb, C.M. Uncertainty Evaluation of Semi-Active Load Redistribution in a Mechanical Load-Bearing Structure. Ph.D. Thesis, Technische Universität Darmstadt, Darmstadt, Germany, 2019.
93. Wang, L.; Nagarajaiah, S.; Shi, W.; Zhou, Y. Seismic performance improvement of base-isolated structures using a semi-active tuned mass damper. *Eng. Struct.* **2022**, *271*, 114963. [CrossRef]
94. Lu, L.Y.; Lin, T.K.; Jheng, R.J.; Wu, H.H. Theoretical and experimental investigation of position-controlled semi-active friction damper for seismic structures. *J. Sound Vib.* **2018**, *412*, 184–206. [CrossRef]
95. Miah, M.S.; Chatzi, E.N.; Dertimanis, V.K.; Weber, F. Real-time experimental validation of a novel semi-active control scheme for vibration mitigation. *Struct. Control Health Monit.* **2017**, *24*, e1878. [CrossRef]

96. Sahasrabudhe, S.S.; Nagarajaiah, S. Semi-active control of sliding isolated bridges using MR dampers: An experimental and numerical study. *Earthq. Eng. Struct. Dyn.* **2005**, *34*, 965–983. [CrossRef]
97. Ferreira, F.; Moutinho, C.; Cunha, Á.; Caetano, E. Use of semi-active tuned mass dampers to control footbridges subjected to synchronous lateral excitation. *J. Sound Vib.* **2019**, *446*, 176–194. [CrossRef]
98. Bagherkhani, A.; Baghlani, A. Reliability assessment of MR fluid dampers in passive and semi-active seismic control of structures. *Probabilistic Eng. Mech.* **2021**, *63*, 103114. [CrossRef]
99. Zizouni, K.; Saidi, A.; Fali, L.; Bousserhane, I.K.; Djermane, M. Semi-active structural vibration control with magnetorheological damper based on hybrid fuzzy sliding mode controller. *IAES Int. J. Robot. Autom. (IJRA)* **2023**, *12*, 167. [CrossRef]
100. Barros, R.C. Semi-active Vibration Control of Buildings using MR Dampers: Numerical and Experimental Verification. In Proceedings of the 14th European Conference on Earthquake Engineering 2010, Ohrid, North Macedonia, 30 August–3 September 2010.
101. Zabihi-Samani, M.; Ghanooni-Bagha, M. Optimal Semi-active Structural Control with a Wavelet-Based Cuckoo-Search Fuzzy Logic Controller. *Iran. J. Sci. Technol.—Trans. Civ. Eng.* **2019**, *43*, 619–634. [CrossRef]
102. Gu, X.; Yu, Y.; Li, Y.; Li, J.; Askari, M.; Samali, B. Experimental study of semi-active magnetorheological elastomer base isolation system using optimal neuro fuzzy logic control. *Mech. Syst. Signal Process* **2019**, *119*, 380–398. [CrossRef]
103. Li, H.N.; Li, X.L. Experiment and analysis of torsional seismic responses for asymmetric structures with semi-active control by MRdampers. *Smart Mater. Struct.* **2009**, *18*, 075007. [CrossRef]
104. Ramezani, M.; Bathaei, A.; Zahrai, S.M. Comparing fuzzy type-1 and -2 in semi-active control with TMD considering uncertainties. *Smart Struct. Syst.* **2019**, *23*, 155–171. [CrossRef]
105. Shih, M.H.; Sung, W.P. Structural Control Effect and Performance of Structure Under Control of Impulse Semi-active Mass Control Mechanism. *Iran. J. Sci. Technol.—Trans. Civ. Eng.* **2021**, *45*, 1211–1226. [CrossRef]
106. Bani-hani, K.A.; Sheban, M.A. Semi-active neuro-control for base-isolation system using magnetorheological (MR) dampers. *Earthq. Eng. Struct. Dyn.* **2006**, *35*, 1119–1144. [CrossRef]
107. Wang, T.; Zhang, X.; Li, K.; Yang, S. Mechanical performance analysis of a piezoelectric ceramic friction damper and research of its semi-active control strategy. *Structures* **2021**, *33*, 1510–1531. [CrossRef]
108. Xu, Y.L.; Chen, B. Integrated vibration control and health monitoring of building structures using semi-active friction dampers: Part I-methodology. *Eng. Struct.* **2008**, *30*, 1789–1801. [CrossRef]
109. Bakhshinezhad, S.; Mohebbi, M. Multi-objective optimal design of semi-active fluid viscous dampers for nonlinear structures using NSGA-II. *Structures* **2020**, *24*, 678–689. [CrossRef]
110. Ab Talib, M.H.; Mat Darus, I.Z.; Mohd Samin, P.; Mohd Yatim, H.; Ardani, M.I.; Shaharuddin, N.M.R.; Hadi, M.S. Vibration control of semi-active suspension system using PID controller with advanced firefly algorithm and particle swarm optimization. *J. Ambient. Intell. Humaniz. Comput.* **2021**, *12*, 1119–1137. [CrossRef]
111. Kim, H.-S.; Roschke, P.N.; Lin, P.-Y.; Loh, C.-H. Neuro-fuzzy model of hybrid semi-active base isolation system with FPS bearings and an MR damper. *Eng. Struct.* **2006**, *28*, 947–958. [CrossRef]
112. Shaikh Faruque Ali, by. Semi-active Control of Earthquake Induced Vibrations in Structures using MR Dampers: Algorithm Development, Experimental Verification and Benchmark Applications. Ph.D. Thesis, Indian Institute of Science, Bangalore, India, 2008.
113. Wang, L.; Zhou, Y.; Shi, W. Seismic Response Control of a Nonlinear Tall Building under Mainshock-Aftershock Sequences Using Semi-Active Tuned Mass Damper. *Int. J. Struct. Stab. Dyn.* **2023**, *23*, 2340027. [CrossRef]
114. Şahin, Ö.; Gökhan Adar, N.; Kemerli, M.; Çaglar, N.; Şahin, İ.; Parlak, Z.; Kükrek, S.; Engin, T. A comparative evaluation of semi-active control algorithms for real-time seismic protection of buildings via magnetorheological fluid dampers. *J. Build. Eng.* **2021**, *42*, 102795. [CrossRef]
115. Altay, O.; Klinkel, S. A semi-active tuned liquid column damper for lateral vibration control of high-rise structures: Theory and experimental verification. *Struct. Control Health Monit.* **2018**, *25*, e2270. [CrossRef]
116. Brennan, M.J.; Day, M.J.; Randall, R.J. An Experimental Investigation into the Semi-Active and Active Control of Longitudinal Vibrations in a Large Tie-Rod Structure. *Vib. Acoust.* **1998**, *120*, 1–12. [CrossRef]
117. Zhao, D.; Li, H. Shaking table tests and analyses of semi-active fuzzy control for structural seismic reduction with a piezoelectric variable-friction damper. *Smart Mater. Struct.* **2010**, *19*, 105031. [CrossRef]
118. Ghaffarzadeh, H.; Dehrod, E.A.; Talebian, N. Semi-active fuzzy control for seismic response reduction of building frames using variable orifice dampers subjected to near-fault earthquakes. *JVC/J. Vib. Control* **2013**, *19*, 1980–1998. [CrossRef]
119. Bathaei, A.; Zahrai, S.M.; Ramezani, M. Semi-active seismic control of an 11-DOF building model with TMD+MR damper using type-1 and -2 fuzzy algorithms. *JVC/J. Vib. Control* **2018**, *24*, 2938–2953. [CrossRef]
120. Al-Fahdawi, O.A.S.; Barroso, L.R.; Soares, R.W. Adaptive neuro-fuzzy and simple adaptive control methods for attenuating the seismic responses of coupled buildings with semi-active devices: Comparative study. *J. Soft Comput. Civ. Eng.* **2019**, *3*, 1–21. [CrossRef]

121. Kawashima, K.; Unjoh, S. Seismic Response Control of Bridges by Variable Dampers. *J. Struct. Eng.* **1994**, *120*, 2583–2601. [CrossRef]
122. Yeh, K.; Chen, C.W.; Hsiung, T.K. *Fuzzy Control for Seismically Excited Bridges with Sliding Bearing Isolation*; Springer: Berlin/Heidelberg, Germany, 2007; pp. 473–483.
123. Liu, Y.; Gordaninejad, F.; Evrensel, C.A.; Wang, X.; Hitchcock, G. Comparative Study on Vibration Control of a Scaled Bridge Using Fail-Safe Magneto-Rheological Fluid Dampers. *J. Struct. Eng.* **2005**, *131*, 743–751. [CrossRef]
124. Panchal, V.R.; Jangid, R.S. Seismic isolation of bridge using variable curvature friction pendulum system. In Proceedings of the 14th World conference of Earthquake Engineering, Beijing, China, 12–17 October 2008; pp. 1–8.
125. Wang, Y.-P.; Chung, L.-L.; Liao, W.-H. Seismic Response Analysis of Bridges Isolated with Friction Pendulum Bearings. *Earthq. Eng. Struct. Dyn.* **1998**, *27*, 1069–1093. [CrossRef]
126. Bhuiyan, A.R.; Alam, M.S. Seismic performance assessment of highway bridges equipped with superelastic shape memory alloy-based laminated rubber isolation bearing. *Eng. Struct.* **2013**, *49*, 396–407. [CrossRef]
127. Yang, J.N.; Wu, J.C.; Kawashima, K.; Unjoh, S. Hybrid control of seismic-excited bridge structures. *Earthq. Eng. Struct. Dyn.* **1995**, *24*, 1437–1451. [CrossRef]
128. Kori, J.; Jangid, R. Semiactive control of seismically isolated bridges. *Int. J. Struct. Stab. Dyn.* **2008**, *8*, 547–568. [CrossRef]
129. Yeh, K.; Chiang, W.-L. Application of Adaptive fuzzy sliding mode control for Bridges. *J. Adv. Comput. Intell.* **2001**, *5*, 172–173. [CrossRef]
130. Jangid, R.S. Stochastic Response of Bridges Seismically Isolated by friction pendulum system. *J. Bridge Eng.* **2008**, *13*, 319–330. [CrossRef]
131. Lee, T.; Kawashima, K. Control of Seismic-Excited Nonlinear Isolated Bridges with Variable Viscous Dampers. In *Proceedings of the JSCE Earthquake Engineering Symposium*; Japan Society of Civil Engineers: Tokyo, Japan, 2005; pp. 1–7.
132. Lee, T.Y.; Kawashima, K.; Chen, P.C. Experimental and Analytical Study on A Nonlinear Isolated Bridge Under Semiactive Control. In Proceedings of the 14th World Conference on Earthquake Engineering, Beijing, China, 12–17 October 2008.
133. Ok, S.-Y.; Kim, D.-S.; Park, K.-S.; Koh, H.-M. Semi-active fuzzy control of cable-stayed bridges using magneto-rheological dampers. *Eng. Struct.* **2007**, *29*, 776–788. [CrossRef]
134. Ko, J.M.; Ni, Y.Q. Structural health monitoring and intelligent vibration control of cable-supported bridges: Research and application. *KSCE J. Civ. Eng.* **2003**, *7*, 701–716. [CrossRef]
135. Ghobarah, A.; Ali, H.M. Seismic performance of highway bridges. *Eng. Struct.* **1988**, *10*, 157–166. [CrossRef]
136. Jangid, R.S. Seismic Response of Isolated Bridges. *J. Bridge Eng.* **2004**, *9*, 156–166. [CrossRef]
137. Lee, T.-Y.; Chen, P.-C. Sliding Mode Control for Nonlinear Isolated Bridges. *J. Earthq. Eng.* **2011**, *15*, 582–600. [CrossRef]
138. Ucak, A.; Tsopelas, P. Effect of Soil–Structure Interaction on Seismic Isolated Bridges. *J. Struct. Eng.* **2008**, *134*, 1154–1164. [CrossRef]
139. Tongaonkar, N.P.; Jangid, R.S. Seismic response of isolated bridges with soil–structure interaction. *Soil Dyn. Earthq. Eng.* **2003**, *23*, 287–302. [CrossRef]
140. Jangid, R.S. Equivalent linear stochastic seismic response of isolated bridges. *J. Sound Vib.* **2008**, *309*, 805–822. [CrossRef]
141. Du, X.L.; Han, Q.; Zhan, J.D. Nonlinear seismic response of isolated bridges with lrb. In Proceedings of the 14th World Conference of Earthquake Engineering, Beijing, China, 12–17 October 2008.
142. Gordaninejad, F.; Saiidi, M.; Hansen, B.C.; Ericksen, E.O.; Chang, F.-K. Magneto-Rheological Fluid Dampers for Control of Bridges. *J. Intell. Mater. Syst. Struct.* **2002**, *13*, 167–180. [CrossRef]
143. Dicleli, M.; Albhaisi, S.; Mansour, M.Y. Static Soil–Structure Interaction Effects in Seismic-Isolated Bridges. *Pract. Period. Struct. Des. Constr.* **2005**, *10*, 22–33. [CrossRef]
144. Kim, T.M.; Kim, G.; Kim, M.K. TMD-Based Structural Control of High Performance Steel Bridges. *J. Phys. Conf. Ser.* **2012**, *382*, 012040. [CrossRef]
145. Dimitrakopoulos, E.G. Seismic response analysis of skew bridges with pounding deck–abutment joints. *Eng. Struct.* **2011**, *33*, 813–826. [CrossRef]
146. Konstantakopoulos, T.G.; Michaltsos, G.T. A mathematical model for a combined cable system of bridges. *Eng. Struct.* **2010**, *32*, 2717–2728. [CrossRef]
147. YeganehFallah, A.; Attari, N.K.A. Robust control of seismically excited cable stayed bridges with MR dampers. *Smart Mater. Struct.* **2017**, *26*, 35056. [CrossRef]
148. Lee, K.J.; Yun, C.B. Extended Kalman Filter for Identification of Nonlinear Earthquake Responses of Bridges. *IFAC Proc. Vol.* **2008**, *41*, 15768–15773. [CrossRef]
149. Amjadian, M. Influence of Seismic Pounding on Dynamic Response of Skewed Highway Bridges. In Proceedings of the 15th World Conference on Earthquake Engineering, Lisbon, Portugal, 24–28 September 2012.
150. Park, K.-S.; Koh, H.-M.; Ok, S.-Y.; Seo, C.-W. Fuzzy supervisory control of earthquake-excited cable-stayed bridges. *Eng. Struct.* **2005**, *27*, 1086–1100. [CrossRef]

151. Caicedo, J.M.; Dyke, S.J.; Moon, S.J.; Bergman, L.; Turan, G.; Hague, S. Phase II Benchmark Control Problem for Seismic Response of Cable-Stayed Bridges. *J. Struct. Control* **2003**, *10*, 137–168. [CrossRef]
152. Schemmann, A.G. Vibration control of cable-stayed bridges—Part 2: Control analyses. *Earthq. Eng. Struct. Dyn.* **1998**, *27*, 825–843. [CrossRef]
153. Dyke, S.J.; Caicedo, J.M.; Turan, G.; Bergman, L.A.; Hague, S. Phase I Benchmark Control Problem for Seismic Response of Cable-Stayed Bridges. *J. Struct. Eng.* **2003**, *129*, 857–872. [CrossRef]
154. Sheikh, M.N.; Xiong, J.; Li, W. MR Damper in Reducing Pounding Effect of Base- Isolated RC Highway Bridges. In Proceedings of the Australian Earthquake Engineering Society 2010 Conference, Perth, Australia, 26–28 November 2010.
155. Chen, C.-W.; Yeh, K.; Fong, K.; Liu, R. Adaptive Fuzzy Sliding Mode Control for Seismically Excited Bridges with Lead Rubber Bearing Isolation. *Int. J. Uncertain.* **2009**, *17*, 705–727. [CrossRef]
156. Jung, H.; Spencer, B.F.; Lee, I. Benchmark Control Problem for Seismically Excited Cable-Stayed Bridges Using Smart Damping Strategies. In Proceedings of the IABSE Conference: Cable-Supported Bridges—Challenging Technical Limits, Seoul, Republic of Korea, 12–14 June 2001; pp. 1–8.
157. Heo, G.; Kim, C.; Lee, C. Experimental test of asymmetrical cable-stayed bridges using MR-damper for vibration control. *Soil Dyn. Earthq. Eng.* **2014**, *57*, 78–85. [CrossRef]
158. Dyke, S.J.; Spencer, F.; Sain, M.K.; Carlson, J.D. An experimental study of MR dampers for seismic protection. *Smart Mater. Struct.* **1998**, *7*, 693–703. [CrossRef]
159. Fabio, C.; Georges, M.; Francesco, M. *Technology of Semiactive Devices and Applications in Vibration Mitigation*; John Wiley & Sons, Ltd.: Hoboken, NJ, USA, 2006.
160. Kim, Y.; You, K.; You, J. Active Control of Along-Wind Response of a Tall Building with AMD using LQR Controller. *Appl. Mech. Mater.* **2014**, *491*, 1063–1067. [CrossRef]
161. Wang, S.Q.; Li, N. Semi-active vibration control for offshore platforms based on LQG method. *J. Mar. Sci. Technol.* **2013**, *21*, 562–568. [CrossRef]
162. Neethu, B.; Das, D. A comparative study on the performance of semi- active and passive control systems for multi-span bridges. In Proceedings of the 16th European Conference of Earthquake Engineering, Thessaloniki, Greece, 18–21 June 2018; pp. 1–12.
163. Moon, S.J.; Bergman, L.A.; Voulgaris, P.G. Sliding Mode Control of Cable-Stayed Bridge Subjected to Seismic ExcitationSeismic Excitation. *J. Eng. Mech.* **2003**, *129*, 71–78. [CrossRef]
164. Das, D.; Khan, M. Hybrid Control of Offshore Jacket Platforms Using Decentralized Sliding Mode Control and Shape Memory Alloy-Rubber Bearing Under Multiple Hazards. *J. Offshore Mech. Arct. Eng.* **2020**, *142*, 011603. [CrossRef]
165. Narasimhan, S.; Nagarajaiah, S. Smart base isolated buildings with variable friction systems: H∞ controller and SAIVF device. *Earthq. Eng. Struct. Dyn.* **2006**, *35*, 921–942. [CrossRef]
166. Soleymani, M.; Khodadadi, M. Adaptive fuzzy controller for active tuned mass damper of a benchmark tall building subjected to seismic and wind loads. *Struct. Des. Tall Spec. Build.* **2013**, *23*, 781–800. [CrossRef]
167. Wasilewski, M.; Pisarski, D.; Bajer, C.I. Adaptive optimal control for seismically excited structures. *Autom. Constr.* **2019**, *106*, 102885. [CrossRef]
168. Saidi, A.; Zizouni, K.; Kadri, B.; Fali, L.; Bousserhane, I.K. Adaptive sliding mode control for semi-active structural vibration control. *Stud. Inform. Control* **2019**, *28*, 371–380. [CrossRef]
169. Lavretsky, E.; Wise, K.A. *Robust and Adaptive Control*, 2nd ed.; Springer: Berlin/Heidelberg, Germany, 2024. [CrossRef]
170. Darbanian, R.; Mousavi Gavgani, S.A.; Kalatjari, V.R. Robust Control of Structures with Active Tendon and Fuzzy-LQR Algorithm Considering Uncertainties of Structural Parameters. *J. Vib. Eng. Technol.* **2023**, *12*, 1635–1647. [CrossRef]
171. Fazaeli, H.; Nejad, H.; Karamodin, A. Semi-active control of three-story benchmark structure using LQG algorithm with a fuzzy-genetic system. *J. Struct. Constr. Eng.* **2022**, *9*, 107–123. [CrossRef]
172. Sabetahd, R.; Arash, S.; Ghasemi, M.; Poursorkhabi, R.V.; Mohammadzadeh, A.; Zandi, Y. Response Attenuation of a Structure Equipped with ATMD under Seismic Excitations Using Methods of Online Simple Adaptive Controller and Online Adaptive Type-2 Neural-Fuzzy Controller. *Comput. Intell. Neurosci.* **2022**, *2022*, 5832043. [CrossRef]
173. Wang, D.H.; Liao, W.H. Modeling and control of magnetorheological fluid dampers using neural networks. *Smart Mater. Struct.* **2005**, *14*, 111–126. [CrossRef]
174. Caterino, N.; Spizzuoco, M.; Londoño, J.M.; Occhiuzzi, A. Experimental issues in testing a semiactive technique to control earthquake induced vibration. *Model. Simul. Eng.* **2014**, *2014*, 535434. [CrossRef]
175. Laflamme, S.; Asce, S.M.; Slotine, J.J.E.; Connor, J.J.; Asce, F. Wavelet Network for Semi-Active Control. *J. Eng. Mech.* **2011**, *137*, 462–474. [CrossRef]
176. Wang, L.-X. Stable adaptive fuzzy control of nonlinear systems. *IEEE Trans. Fuzzy Syst.* **1993**, *1*, 146–155. [CrossRef]
177. Fujino, Y. Vibration, control and monitoring of long-span bridges—Recent research, developments and practice in Japan. *J. Constr. Steel Res.* **2002**, *58*, 71–97. [CrossRef]

178. Pourzeynali, S.; Datta, T.K. Semiactive Fuzzy Logic Control of Suspension Bridge Flutter. *J. Struct. Eng.* **2005**, *131*, 900–912. [CrossRef]
179. Jung, H.; Spencer, B.F., Jr.; Asce, M.; Lee, I. Control of Seismically Excited Cable-Stayed Bridge Employing Magnetorheological Fluid Dampers. *J. Struct. Eng.* **2003**, *129*, 873–883. [CrossRef]
180. Lin, X.; Lin, W. Optimal Allocation and Control of Magnetorheological Dampers for Enhancing Seismic Performance of the Adjacent Structures Using Whale Optimization Algorithm. *Shock Vib.* **2021**, *2021*, 1218956. [CrossRef]
181. Lee, S.H.; Min, K.W.; Lee, Y.C. Modified sliding mode control using a target derivative of the Lyapunov function. *Eng. Struct.* **2005**, *27*, 49–59. [CrossRef]
182. Sayed, S. Modified Sliding Mode Controller with Extended Kalman Filter for Stochastic Systems. In Proceedings of the 2007 IEEE International Conference on Control and Automation, Guangzhou, China, 30 May–1 June 2007; pp. 630–635.
183. Liang, C.-Y.; Su, J.-P. A new approach to the design of a fuzzy sliding mode controller. *Fuzzy Sets Syst.* **2003**, *139*, 111–124. [CrossRef]
184. Amini, F.; Ghaderi, P. Seismic motion control of structures: A developed adaptive backstepping approach. *Comput. Struct.* **2013**, *114–115*, 18–25. [CrossRef]
185. Bagheri, A.; Amini, F. Control of structures under uniform hazard earthquake excitation via wavelet analysis and pattern search method. *Struct. Control Health Monit.* **2013**, *20*, 671–685. [CrossRef]
186. Amiri, E.; Keshavarz, H.; Alizadeh, M.; Zamani, M.; Khodadadi, T. Energy efficient routing in wireless sensor networks based on fuzzy ant colony optimization. *Int. J. Distrib. Sens. Netw.* **2014**, *10*, 768936. [CrossRef]
187. Bitaraf, M.; Hurlebaus, S.; Barroso, L.R. Active and Semi-active Adaptive Control for Undamaged and Damaged Building Structures Under Seismic Load. *Comput. -Aided Civ. Infrastruct. Eng.* **2012**, *27*, 48–64. [CrossRef]
188. Oliveira, F.; Morais, P.; Suleman, A. Predictive control for earthquake response mitigation of buildings using semiactive fluid dampers. *Shock Vib.* **2014**, *2014*, 670683. [CrossRef]
189. Nguyen, S.D.; Kim, W.; Park, J.; Choi, S.B. A new fuzzy sliding mode controller for vibration control systems using integrated-structure smart dampers. *Smart Mater. Struct.* **2017**, *26*, 045038. [CrossRef]
190. Ghaderi, P.; Amini, F. Adaptive block backstepping control for civil structures with unknown parameters subjected to seismic excitation. *Struct. Control Health Monit.* **2017**, *24*, e1875. [CrossRef]
191. Shan, J.; Ouyang, Y.; Shi, W. Adaptive control of earthquake-excited nonlinear structures with real-time tracking on prescribed performance criteria. *Struct. Control Health Monit.* **2018**, *25*, e2247. [CrossRef]
192. Loh, C.H.; Lynch, J.P.; Lu, K.C.; Wang, Y.; Chang, C.M.; Lin, P.Y.; Yeh, T. Experimental verification of a wireless sensing and control system for structural control using MR dampers. *Earthq. Eng. Struct. Dyn.* **2007**, *36*, 1303–1328. [CrossRef]
193. Di Paola, M.; Pirrotta, A. Time delay induced effects on control of linear systems under random excitation. *Probabilistic Eng. Mech.* **2001**, *16*, 43–51. [CrossRef]
194. Nguyen, T.H.; Kwok, N.M.; Ha, Q.P.; Li, J.; Samali, B. Adaptive sliding mode control for civil structures using magnetorheological dampers. In Proceedings of the 23rd ISARC, Tokyo, Japan, 3–5 October 2006; pp. 636–641. [CrossRef]
195. Yu, W.; Thenozhi, S. *Active Structural Control with Stable Fuzzy PID Techniques*; Springer: Cham, Switzerland, 2016. [CrossRef]
196. Das, D. Semiactive and Hybrid Seismic Control of Building Frames Using MR Dampers. Ph.D. Thesis, Indian Institute of Technology, Delhi, India, 2009. Available online: https://libcat.iitd.ac.in/bib/164882 (accessed on 25 May 2025).
197. Ribakov, Y.; Agranovich, G. Control of structural seismic response by a limited set of active dampers. *Struct. Des. Tall Spec. Build.* **2011**, *20*, 594–611. [CrossRef]
198. Kim, H.; Adeli, H.; Asce, F. Hybrid Feedback-Least Mean Square Algorithm for Structural Control. *J. Struct. Eng.* **2004**, *130*, 120–127. [CrossRef]
199. Cha, Y.-J.; Zhang, J.; Agrawal, A.K.; Dong, B.; Friedman, A.; Dyke, S.J.; Ricles, J. Comparative Studies of Semiactive Control Strategies for MR Dampers: Pure Simulation and Real-Time Hybrid Tests. *J. Struct. Eng.* **2013**, *139*, 1237–1248. [CrossRef]

applied sciences

MDPI

Review

An In-Depth Analysis of the Seismic Performance Characteristics of Steel–Concrete Composite Structures

Panagiota Katsimpini *, George Papagiannopoulos and George Hatzigeorgiou

Laboratory of Structural Technology and Applied Mechanics, Hellenic Open University, GR26335 Patras, Greece; hatzigeorgiou@eap.gr (G.H.)
* Correspondence: katsimpini.panagiota@ac.eap.gr

Abstract: This review article provides an in-depth exploration of the recent advancements in the seismic analysis and design of steel–concrete composite structures, as reflected in the literature from the last ten years. It investigates key factors, such as material behavior, connection detailing, analytical modeling techniques, and design methodologies. The article highlights the synergistic benefits derived from the combination of steel and concrete components to improve seismic performance. Various composite systems, including composite beams, beam-columns, frames, shear walls, foundations, and beam–column joints, are analyzed through experimental studies to assess their dynamic response characteristics under extreme earthquake conditions. The article evaluates advanced numerical modeling methods, including finite element analysis and fiber-based models, for their capability to predict the nonlinear behavior of composite buildings and bridges. A comparative analysis of modern seismic isolation and energy dissipation techniques is also included. Furthermore, the optimization of composite structures in seismically active regions is discussed. The article concludes by identifying areas where additional research is necessary to enhance the seismic resilience of steel–concrete composite structures.

Keywords: steel–concrete composite structures; composite action; nonlinear behavior; composite beams and columns; cyclic loading; seismic isolation; energy dissipation devices; progressive collapse

1. Introduction

The seismic design and analysis of steel–concrete composite structures encounter numerous significant challenges that necessitate in-depth exploration. Although these structures are commonly utilized, considerable uncertainties persist regarding the intricate interactions between steel and concrete elements when subjected to intense seismic forces. Existing design codes frequently fall short in adequately addressing the nonlinear behavior exhibited by composite systems, particularly at the junctions of the materials. Moreover, the variety of composite configurations complicates the development of standardized design methodologies, while the absence of extensive long-term performance data introduces further uncertainty in durability assessments. Conventional analytical models may overly simplify the dynamic response features of composite structures, which could result in designs that are either excessively conservative or insufficiently robust. Additionally, the construction industry grapples with challenges in evaluating the cost-effectiveness of innovative composite solutions in comparison to traditional building methods, particularly when factoring in life-cycle performance during seismic events. These critical concerns highlight the necessity for a comprehensive review of the existing knowledge and practices related to steel–concrete composite structures.

As urban areas continue to expand and face increasing seismic risks, understanding the behavior of these composite systems under earthquake loading has become crucial for ensuring structural safety and resilience. This review paper seeks to offer a thorough examination of the existing body of research concerning the seismic analysis and design of steel–concrete composite structures. This article examines key research developments, design methodologies, and practical applications that have emerged mainly over the past decade. The paper covers various composite systems, including composite beams, columns, connections, and innovative hybrid structural solutions. This review encompasses the following areas:

- The fundamental principles of steel–concrete composite action under seismic loading
- Experimental studies on the seismic performance of composite elements and systems
- Analytical and numerical modeling techniques for predicting composite behavior
- Current design codes and standards for seismic design of composite structures
- Emerging technologies and novel composite systems for enhanced earthquake resistance
- Case studies of notable composite structures in seismic regions

By synthesizing the latest research findings and practical insights, this paper seeks to identify knowledge gaps, highlight promising research directions, and provide valuable guidance for engineers and researchers working on the seismic design and analysis of steel–concrete composite structures. To provide a comprehensive review, the following 12 topics are examined in depth:

1. Beam–column joints of composite frames under seismic loads: the behavior of composite beam–column connections under seismic loading is examined, including moment-resisting frames and various joint configurations. Research on load transfer mechanisms and failure modes is discussed.
2. The seismic behavior of steel–concrete composite shear walls: composite concrete/steel shear wall systems, including steel-plate shear walls with concrete infill and composite coupling beams, are reviewed. Their effectiveness in lateral load resistance and energy dissipation is analyzed.
3. Modeling techniques for composite members and structures under earthquake excitation: advanced numerical modeling techniques for composite structures, including finite element analysis, multi-scale modeling approaches, and the challenges in accurately representing composite behavior are discussed.
4. The behavior of composite beams under cyclic loads: the hysteretic response of composite beams under repeated loading, including local buckling phenomena and the influence of concrete encasement on steel beam performance, is investigated.
5. The behavior of composite beam-columns under cyclic loads: the performance of composite columns under cyclic loading, considering various cross-section types (e.g., concrete-filled steel tubes, encased steel sections) and their impact on ductility and energy dissipation is analyzed.
6. The seismic performance of composite buildings: the advances in the seismic design of composite building structures are analyzed. The review focuses on the inelastic behavior of composite buildings where the evaluation of inelasticity at various levels of seismic intensity is significantly enhanced by seismic design approaches that allow for the control of structural deterioration at both the elemental scale (for instance, crucial structural parts) and the overall building scale (such as individual floors).
7. Composite construction and foundation seismic design: seismic design considerations for foundations supporting composite structures, including pile-to-pile cap connections and soil–structure interaction effects, are discussed.
8. Seismic isolation and energy dissipation devices for composite structures: the integration of base isolation systems with composite structures, focusing on their effec-

tiveness in reducing seismic demands and improving overall structural performance, is examined. Additionally, the implementation of various damping devices (e.g., viscous dampers and friction dampers) in composite structures and their contribution to energy dissipation during seismic events is reviewed.

9. The progressive collapse resistance of composite structures: the resistance of composite structures to progressive collapse under extreme loading conditions, including the role of composite action in enhancing structural robustness, is evaluated.
10. Optimal design strategies for earthquake-resistant composite buildings: optimization strategies for composite structural systems, considering both material usage and seismic performance objectives is explored.
11. The seismic performance of steel–concrete composite bridges: the application of composite construction in bridge engineering, focusing on the seismic performance of composite deck systems, piers, and abutments, is explored.
12. Smart materials and sensors: new smart materials and sensors enhance the seismic performance of steel–concrete structures, offering self-centering and real-time monitoring capabilities.

The flowchart of the examined topics is shown in Figure 1.

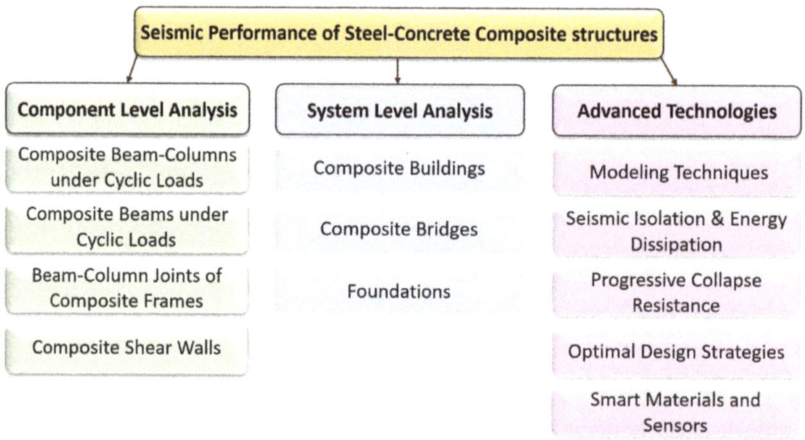

Figure 1. Comprehensive flowchart illustrating the systematic review methodology and key research topics in steel–concrete composite structural analysis.

The twelve key areas selected for this comprehensive review were carefully chosen to address the critical challenges in understanding and improving the seismic performance of steel–concrete composite structures. Beginning at the component level, the examination of beam-columns, beams, and beam–column joints under cyclic loads is essential for understanding the fundamental behavior of composite elements during seismic events. The analysis of composite shear walls complements this by addressing lateral load resistance systems. Moving to a broader scale, the review of composite buildings and bridges allows for the evaluation of system-level responses, while the inclusion of foundation design ensures a complete understanding from the ground up. The sections on modeling techniques and optimal design strategies address the crucial need for accurate prediction and efficient design methodologies. The coverage of seismic isolation and energy dissipation devices, along with progressive collapse resistance, tackles the critical aspects of enhancing structural resilience. The smart materials and sensors section reflects the cutting-edge developments in monitoring and adaptive response systems. This carefully structured approach ensures that current practices and emerging technologies are thoroughly examined, providing a comprehensive framework for understanding seismic behavior in composite structures.

The methodological approach for this comprehensive review was carefully designed to ensure both scholarly rigor and comprehensive coverage. Thus, 205 scholarly works published over the past decade are systematically analyzed, employing a multi-stage selection process that prioritized peer-reviewed publications with demonstrable methodological depth. The selection criteria focused on works that provided empirical insights, innovative analytical approaches, or significant advancements in understanding the seismic performance of steel–concrete composite structures.

By intentionally incorporating diverse research perspectives, we aimed to create a narrative that captures the complexity of composite structural systems. This approach prioritizes a holistic representation of current research, acknowledging that scientific understanding emerges from varied methodological traditions and nuanced investigative approaches.

2. Beam–Column Joints of Composite Frames Under Seismic Loads

2.1. Beam–Column Joints of Composite Frames Under Seismic Loads: State-of-the-Art

Beam–column joints play a crucial role in steel–concrete composite structures, particularly during seismic events, as they transfer forces between beams and columns and affect the structure's earthquake resistance. Composite structures have gained popularity due to their combined strength, stiffness, and ductility. This scholarly paper provides a comprehensive review of the seismic analysis and design of beam–column joints in these structures. It covers recent advancements in experimental studies, analytical methods, and design approaches, focusing on failure modes, influential parameters, experimental investigations, modeling techniques, existing guidelines and limitations, and emerging joint configurations. By consolidating recent research and identifying gaps, this review aims to offer valuable insights to earthquake engineering and structural design professionals.

Thus, Ataei et al. (2017) [1] examined a detailed finite element (FE) model for studying the performance of deconstructable beam-to-column composite joints. Their model was validated through experiments, and a parametric study was conducted to analyze various parameters. Design models were proposed to predict performance parameters. Key findings included the impact of shear connection ratio, reinforcement ratio, bolt size, slab thickness, column flange thickness, and steel grade on joint behavior. It was found that the design models could accurately predict performance parameters, even for joints with high-strength steel components. Furthermore, Van-Long et al. (2015) [2] presented a solution for connecting I-shaped beams to concrete-filled RHS columns using long bolts. The joint configuration improved rigidity and resistance, making it suitable for seismic-resistant moment frames, while experimental tests and analytical models were used to validate the joint design. Moreover, the dynamic response and failure mechanism of a new assembly of blind-bolted CFST composite structures under simulated seismic loading were studied by Wang et al. (2017) [3]. The tests showed excellent seismic performance, including good ductility, energy dissipation capacity, and damage control, while analytical modeling using OpenSees software 3.7.1 (https://opensees.berkeley.edu/ accesed on 10 September 2024) validated the results.

Peng et al. (2018) [4] presented a study on the seismic performance of an innovative end-plate connection between T-shaped CFT columns and reinforced concrete beams. Seven specimens were tested, one under monotonic loading and six under cyclic loading. The proposed end-plate connections exhibited good seismic performance, achieving the desired "strong column and weak beam" and "strong connection" objectives in seismic design. Peng et al. (2018) [5] found that the diameter of the bolts and the length of the (H-shaped) steel corbel influenced the yield strength, ultimate strength, and initial stiffness. A FEM model was also constructed and subsequently verified, and parametric studies revealed the impact of beam reinforcement ratio and concrete strength on seismic performance. Moreover, Peng et al. (2021) [6] examined the shear strength and seismic damage performance of H-section beams with unequal-depth CFT column composite joints. Four specimens with varying height ratios of the beams were subjected to lateral cyclic loading. The findings indicated that shear failure occurred primarily in the joint core area, with the joint core area reaching yield strain earlier as the beam height ratio increased.

A mathematical formula was developed to calculate the shear strength of the CFT columns with a H-shaped beam joint. This equation takes into account the combined effects of the restrained compression struts in the joint central region, concrete compression struts, and steel tube web.

Mou et al. (2021) [7] explored a new type of joint between a reinforced concrete-filled steel tube column and a beam. Four specimens with varying characteristics were tested under cyclic loading. The results showed that slipping between the concrete and steel caused pinch phenomena in the joint's hysteretic curves. Increasing the size of the middle steel tube improved stiffness and energy dissipation but had little effect on strength. Transfer sleeves played a crucial role in limiting stiffness and strength degradation while enhancing energy dissipation. Yu et al. (2022) [8] focused on evaluating the seismic behaviors of different connections in steel residential buildings, including endplate, side plate, and hot-rolled T-shaped connector (HTC) connections. The experimental findings revealed that both the connections of lateral plate and HTC connections exhibited favorable cyclic response characteristics. These included consistent force-displacement patterns, a gradual reduction in rigidity, adequate ductility, and effective energy absorption capacity. In contrast, the connection of the endplate suffered an untimely rupture at the welded interfaces between the endplate and the side plates. Furthermore, the suggested HTC connections not only demonstrated satisfactory seismic performance but also required no additional stiffeners and had a minimal impact on indoor space, making them easily concealable within partition walls.

Cheng et al. (2022) [9] presented an experimental study on the behavior of connections between T-shaped CFT columns and U-shaped steel–concrete composite beams. Five different connection details were tested, with all showing favorable seismic performance with plastic hinge failure occurring at the beam-ends. The joints demonstrated good ductility and energy dissipation capacity, meeting Eurocode 3 requirements for rigid joints in braced frames. They concluded that their proposed joint design with vertical ribs is seismically feasible and provides recommendations for practical applications. Li et al. (2023) [10] introduced a new manufactured steel beam (H-shaped) to a CFT column joint design. This system aims to improve construction efficiency and structural performance by eliminating welding in column connections. The joint employs embedded threaded steel bars for column segments and outer annular stiffeners for beam connections. They tested four joint specimens under cyclic loading, examining parameters, such as column tube thickness, sleeve distance, and steel strength, and their results showed good flexural capacity, deformability, and energy dissipation.

Chen et al. (2014b) [11] examined the performance of a new through-beam connection between reinforced concrete beams and CFT columns. Tests showed different failure modes and demonstrated good seismic performance, load capacity, and ductility. Finite element modeling validated the results, and this new connection type simplifies construction without sacrificing structural integrity and offers valuable considerations for future CFST column-to-RC beam connections. Moreover, Jeddi et al. (2017) [12] introduced and tested a novel moment-resisting connection called a "through rib stiffener beam connection" for I-beams and concrete-filled circular steel tubular columns. Four cruciform specimens were experimentally examined to evaluate the behavior of the connections under seismic loading. The new connection involved vertical rib stiffeners welded to the beam flanges, passing through pre-slotted holes in the column. It was found that the new connection exhibited rigid moment-resisting behavior, good ductility, and energy dissipation capacity. The connection appears to satisfy seismic design requirements and is practical and cost-effective.

Amadio et al. (2017) [13] introduced a refined finite element model for predicting the behavior of composite steel–concrete welded joints under earthquake loads. The model underwent validation through comparison with experimental data, successfully depicting both global and local behaviors, stress distributions, and failure mechanisms. This was achieved by incorporating the Gurson model for steel, accounting for concrete behavior and damage, and ensuring an accurate representation of connections. The validated model can be used for extensive parametric studies and can lead to suggested improvements to Eurocode 8 design provisions. Similarly, Liu et al. (2017) [14] investigated the performance of special-shaped CFT columns—steel beam connections. Four specimens were tested with different shapes and joint stiffeners. Different failure modes were observed, and it was found that exterior diaphragms performed better than vertical ribs in transferring forces. The study also developed finite element models and proposed design formulas based on internal force transfer mechanisms. The findings suggest that exterior diaphragm connections are suitable for engineering applications, while vertical rib connections require improvement.

Li et al. (2021) [15] investigated concrete-encased CFST columns/steel beam joints using experimental tests. An RC slab was installed for each joint, and an external diaphragm design was used to connect the beam flange and the CFST tube. The joints were subjected to constant axial load and reverse cyclic loading to provide test data for composite joints, evaluate their seismic performance, and analyze the effects of key parameters, specifically the joint type. Additionally, Wu et al. (2020) [16] designed three specimens of joints using an integrated combination scheme and tested them with low-cycle reciprocating loading. The results analyzed the strength degradation, stiffness degradation, failure modes, energy dissipation, and ductility, clarifying the effect of the beam-to-column stiffness ratio on seismic performance of modular composite joints.

Wang et al. (2016) [17] executed an experimental program on two blind-bolted composite joints to CFTST columns in full-scale under short cyclic loading to understand their seismic performance and failure modes. The results were used to assess various aspects, like hysteresis, strength degradation, ductility, and energy dissipation capacity, with implications for enhancing the design of composite joints in practice. Furthermore, Peng et al. (2018) [4] introduced new connections, of a ring-bar-reinforced type, for connecting reinforced concrete slabs, beams, and T-shaped CFST columns. They conducted tests on 34 columns and derived a formula to determine the axial compression bearing capacity. Additionally, an improved composite column was presented, which is more cost-effective and exhibits better seismic behavior. The connection between the T-shaped CFST column and the RC beams is strengthened by the new ring-bar-reinforced connection, enhancing the stiffness and bearing capacity.

The concept and benefits of an innovative prefabricated composite column–beam joint, combining concrete-filled steel tubular (CFST) columns with reinforced concrete (RC) beams, were first presented by Zhang and Li (2021) [18]. To evaluate the earthquake resistance of these connections, which employ bolts and non-adherent post-stressed cables, the researchers conducted alternating load experiments on eight reduced-scale prototypes (one-third of full size). The study examined how various factors, including the vertical load ratio, bolt size, steel tube thickness, and overall structural configuration, influenced the seismic response characteristics of these joints. The resulting data were thoroughly analyzed and interpreted to draw meaningful conclusions about the connections' performance under simulated earthquake conditions. A novel joining technique for linking a steel–concrete composite column (concrete-encased CFST) with a metallic beam was introduced by Wang et al. (2020) [19]. This method utilizes concealed fasteners. The composite column, which consists of a concrete-filled steel tube further encased in concrete, is manufactured off-site in a controlled environment, with bolt-holes drilled in the steel tube and concrete poured around it. Blind bolts are then installed, and the core concrete is poured on site. Numerical models and experimental results were used to analyze the connection, considering material models and contacts. Component models for stiffness and moment resistance were developed, considering the load transfer mechanism.

The seismic behavior of vertical rib connections for the joint of an H-shaped steel beam to a special-shaped CFT column was tested by Yang et al. (2021) [20]. Frames were tested under vertical and horizontal loads to examine stiffness degradation, failure mode, energy dissipation, strength-bearing capacity, and ductility. Finite element analysis was used to investigate the plastic hinge formation and failure mode. Furthermore, Xu et al. (2021) [21] explored the seismic and shear behavior of joints connecting U-shaped steel–concrete composite beams to T-shaped concrete-filled steel tubular columns. A new joint design with C-shaped slots allowing double-C channels to pass through the joint panel zone was proposed and tested.

The seismic performance of composite steel–concrete frame systems using the URSP (uplift-restricted/slip-permitted) connector was studied by Duan et al. (2022) [22]. A numerical model was used to analyze the dynamic behavior of structures with URSP connectors under seismic loads. The results were compared to traditional structures, revealing that the arrangement of full-span URSP connectors had a significant impact on the system's behavior, increasing inter-story displacement and causing adverse effects. It was also found that the partial arrangement of URSP connectors had little influence and still maintained good seismic performance. On the other hand, Wang et al. (2022) [23] analyzed the dynamic response of the beam to CFDST column joints for this connection, using established analytical models and pseudo-dynamic tests. The tests explored the effects of end plate type, column hollow ratio, and concrete slab on dynamic responses and failure mechanisms. Seismic behavior, hysteresis curves, viscous damping coefficients, and energy dissipation capacities were also examined, and analytical models of the joints were developed using OpenSees software (https://opensees.berkeley.edu/).

Zhou et al. (2022) [24] conducted experiments on 3D beam to L-shaped steel–concrete column joints subjected to dynamic loading, considering variables, such as encased steel configuration, axial compression ratio, loading angle, and beam type, to provide test data, evaluate seismic behavior, and analyze the effects of key parameters. Additionally, Chen et al. (2023) [25] introduced a new approach to designing PC joints with distinct advantages, including an efficient and eco-friendly assembly method, simple structure, concentrated failure range, and satisfactory seismic performance. The tests were conducted on full-scale joint specimens, considering various factors, like steel tube thickness and internal structure, and analyzed using an ABAQUS (https://www.3ds.com/products/simulia/abaqus) finite element model.

Gan et al. (2019) [26] developed a new joint system for RC beams and square-reinforced concrete-filled steel tube (RCFST) columns. The system utilizes openings on the square hollow section (SHS) for continuous beam reinforcement and interior diaphragms welded to the cold-formed SHS. These diaphragms confine the steel tube and concrete and can be used to position column reinforcement. The system eliminates the need for internal diaphragms to transfer forces from the RC beam, resulting in improved seismic performance. This joint system is shown in Figure 2.

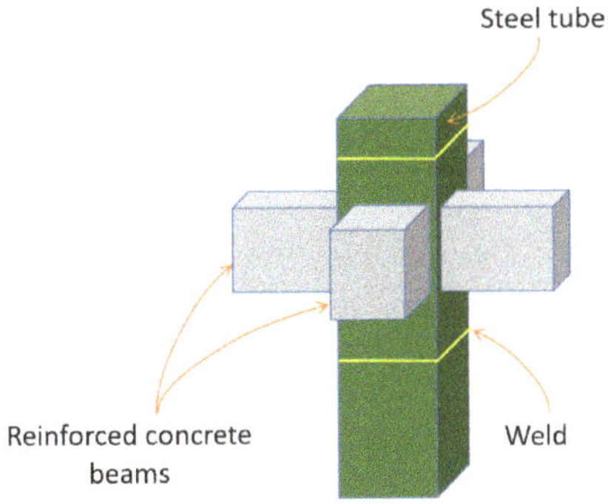

Figure 2. Detailed joint system configuration for a reinforced concrete-filled steel tube (RCFST) column connected to reinforced concrete beams, highlighting interface characteristics (adapted from Gan et al. (2019) [26]).

Additionally, Lai et al. (2019) [27] reviewed experimental research on composite special moment frame connections and presented a database of 165 tests conducted on these connections. They summarized connection behavior, evaluated them based on standard requirements, and recommended different connection types for specific frame configurations for seismic applications. Moreover, Fan et al. (2019) [28] examined the mechanical behavior of connections between composite steel–concrete beams and concrete-filled square steel tubular columns. Tests on six specimens under cyclic loading were conducted to evaluate the damage patterns, hysteretic behavior, joint ductility, and energy dissipation capacity. The result of joint geometries, connection details, and loading path on connection behavior were also analyzed.

Wu et al. (2020) [29] analyzed the dynamic behavior of three modular composite joints with various beam–column connections through quasi-static tests. Factors, such as failure mode, strain distribution, hysteresis curve, strength degradation, and energy consumption, were considered. Furthermore, Liu et al. (2020) [30] examined the seismic behavior of steel beams in large-diameter composite column connections. They found that there are no unified design provisions for these connections in high-rise buildings, while the pertinent Chinese code GB 50936–2013 is inadequate for these connections. To investigate the seismic behavior and validate the code-specified calculation method, they tested five connection specimens, examining their energy dissipation stiffness, failure mode, deformation capacity, and ductility.

Amadio et al. (2017) [31] discussed the effects of slab isolation on the structural performance of joints. Properly isolating the slab from the column, even with a small gap, prevents any over-strengthening outcome on the behavior of the joint, improving control of the seismic response of the braced frame. They also found that under lateral loads, the joint acts as a hinge and transfers loads to the bracing system, while under gravitational loads, the continuity of longitudinal rebar leads to a mostly clamped joint performance. Additionally, recent research on composite steel–concrete beam-to-column joints was discussed by Demonceau and Ciutina (2019) [32]. They examined various loading conditions, such as elevated temperatures, combined bending and axial loads, sagging bending moments, and cyclic loading. They also proposed methods to characterize the "concrete slab in compression" component for sagging moments and presented an improved analytical procedure for M-N loading.

2.2. Beam–Column Joints of Composite Frames Under Seismic Loads: Key Findings

The key findings regarding beam–column joints in steel–concrete composite frames subjected to seismic loads are as follows:

- Composite action enhances the strength and stiffness of joints when compared to bare steel joints.
- Proper reinforcement detailing in the joint area is essential for achieving ductile behavior.
- The presence of a concrete slab significantly affects the moment capacity and failure mode of the joint.
- Panel zone deformation has a noteworthy role in the overall rotation of the joint.
- Under cyclic loading, concrete crushing and bond deterioration of reinforcement can occur.
- The shear strength of joints is affected by the depth of the beam and the axial load on the column.
- Incorporating through-beam details can enhance joint performance in comparison to external beam–column connections.
- Composite joints generally demonstrate good energy dissipation capacity when subjected to seismic loading.

3. Seismic Behavior of Steel–Concrete Composite Shear Walls

3.1. Seismic Behavior of Steel–Concrete Composite Shear Walls: State-of-the-Art

Steel–concrete composite shear walls are structural elements that leverage the strengths of both steel and concrete to create efficient lateral force-resisting systems in buildings. These walls are typically composed of steel plates or sections integrated within or attached to reinforced concrete walls. The synergy between the two materials results in a system that offers superior strength, stiffness, and ductility when compared to conventional reinforced concrete or steel-only shear walls. Composite shear walls offer several advantages for seismic loading. They have increased strength and stiffness, improved ductility, better crack control, and can dissipate seismic energy. Additionally, they are lighter and have quicker construction times compared to traditional reinforced concrete walls. The walls also exhibit stable hysteretic behavior, and understanding failure modes is crucial for assessing safety.

Mo et al. (2021) [33] reviewed the behavior of composite shear walls with different connection configurations and loading conditions, highlighting the lack of previous research in the field. They discussed the motivation, structural performance, and development of these important structural elements. Furthermore, Lan et al. (2023) [34] examined a high-rise structure with irregular components that utilizes a steel frame–shear wall system, with the core wall using a double steel plate–concrete composite and the outer frame consisting of steel columns and beams. This composite shear wall was tested under seismic conditions and was proven to meet design requirements, while the structure could withstand progressive collapse caused by removing a corner column.

Todea et al. (2021) [35] investigated the impact of openings, composite connections, and steel fiber-reinforced concrete on the seismic behavior of composite steel–concrete shear walls. They evaluated the performance of these walls, assessed the influence of centrally aligned openings, and explored the use of steel fiber-reinforced concrete as an alternative solution to common concrete. Additionally, innovative composite shear walls with CFST columns and reinforced concrete walls embedded with multiple steel plates were studied by Qiao et al. (2018) [36]. Seven specimens underwent cyclic loading tests to analyze parameters, such as axial force ratio, number of steel plates, and CFT column type. It was found that the embedded steel plates improved seismic behavior and strength. Additionally, recent research by Najm et al. (2022) [37] investigated the seismic performance of steel–concrete composite shear walls compared to traditional RC walls under cyclic loading. Using ANSYS finite element analysis, they examined how reinforcement ratio, concrete strength, shear stud layout, and steel-plate yield stress affect the walls' behavior, particularly focusing on ductility and energy absorption capabilities. Figure 3 depicts the simulation of local buckling phenomena using the finite element method for the composite steel–concrete shear walls

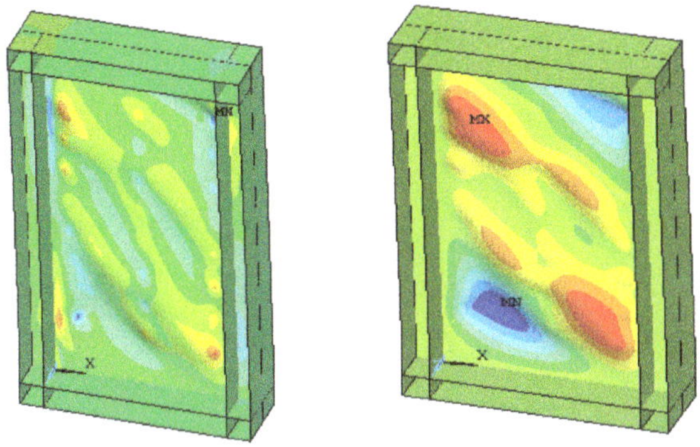

Figure 3. Finite element method simulation demonstrating local buckling phenomena in composite shear wall structures, showing stress distribution and deformation patterns (adapted from Najm et al. (2022) [37]).

Rahnavard et al. (2016) [38] presented nonlinear numerical studies on steel–concrete composite shear walls used in high-rise buildings. Five different finite element models were developed to accurately represent the complex behavior of these structures. The findings highlight the importance of parameters, such as concrete failure, hysteresis response, displacement, drift, and energy dissipation. It was observed that a steel frame with concrete on the shear plate on one side exhibited better energy dissipation. Additionally, increasing the concrete thickness reduced shear plate buckling, and decreasing the connector distance increased the energy dissipation. Furthermore, Zhang et al. (2016) [39] proposed an innovative BLC-C composite wall consisting of connected lipped channels welded together. Concrete was poured inside to enhance seismic behavior and reduce steel consumption. The wall system resisted both vertical and lateral loads and was tested for energy dissipation, failure modes, ductility, stiffness, deformation, strength, and hysteresis loops.

Hossain et al. (2016) [40] discussed the behavior of a double-skin composite wall system under cyclic shear loading. They focused on the usage of profiled steel sheets, steel sheet–concrete connections, and high-performance concrete to construct the walls. They have also examined the performance of self-consolidating concrete and engineered cementitious composite as infill materials. The latter was highlighted for its high strain capacity and ability to enhance the structural performance of the walls. They also presented data on stress–strain characteristics, failure modes, ductility, shear strength, and energy absorption, with a focus on strength and stiffness degradation. Moreover, Polat and Bruneau (2017) [41] used the finite element method to model the inelastic cyclic behavior of concrete-filled steel sandwich panel walls. The calibrated model provided insights into the design of these composite walls, addressing the lack of prescriptive guidance. The findings included the distribution of wall-to-footing forces, shear force demands, cumulative plastic strain, steel-plate yielding, and interface friction.

Zhao et al. (2016) [42] first recognized that the research on seismic behavior of steel–concrete composite shear walls for safety-related nuclear facilities lacks reliable hysteretic models for predicting structural performance near collapse. For this reason, they analyzed the experimental results of 32 specimens and developed associated hysteretic rules and a quadrilinear backbone with negative post-peak stiffness. Additionally, to reflect plasticity extension and damage accumulation, they proposed reduction factors for Young's moduli of concrete and steel. Additionally, Wang et al. (2019) [43] presented a design method for concrete-filled double steel corrugated-plate walls with a T-section. The load-bearing capacities of these walls were upgraded by the combination of infilled concrete and bolt-connected steel corrugated plates. The failure mechanism and design process were analyzed through experiments and numerical simulations, while design formulas for predicting the sectional strength were proposed.

Luo et al. (2021) [44] examined the behavior of corrugated steel-plate composite shear walls. Experimental research was conducted on six shear walls to study their failure process, hysteresis diagrams, energy consumption capacity, and stiffness degradation. It was found that the peak loads of the three shear walls increased significantly compared to conventional shear walls. Moreover, Haghi et al. (2020) [45] developed a fiber-based macro model in the PERFORM-3D program (https://www.csiamerica.com/products/perform3d) to efficiently simulate the global response of steel–concrete composite walls. The model has a simple system and decreases computational time. It was validated using test data for seventeen composite walls with and without boundary elements, considering various design variables and the presence of axial load.

Shafaei et al. (2021) [46] summarized the experimental results and discussed the development of three-dimensional FEM models for large-scale specimens that have to do with composite plate shear walls. Effective stress–strain relationships were proposed to account for various behaviors, including steel yielding, concrete cracking, and composite interaction. Two-dimensional FEM models and fiber-based FEM models were recommended for simulating seismic response, while three-dimensional FEM models were recommended for further parametric studies on structural behavior and design. Additionally, Chen et al. (2015) [47] developed a novel lateral force-resisting system comprising high-strength concrete and a steel plate. Tests were conducted to analyze the response of this composite shear wall to cyclic loading. The axial load ratio and tie bar spacing were investigated. The specimens displayed high strength, deformation capacity, and flexure-dominated behavior. The primary failure mode involved the local buckling of steel plates and compressive crushing of concrete.

An LS-DYNA (https://lsdyna.ansys.com/) finite element model was created by Epackachi et al. (2015) [48] to model the cyclic behavior of flexure-critical steel-plate–concrete composite shear walls. Validation was performed using data from tests on composite wall piers. The damage to the walls included cracking, crushing of infill concrete, and the buckling of steel faceplates. It was found that the model accurately predicted force-displacement responses, damping ratio, and damage distribution. Moreover, Huang and Liew (2016) [49] examined the behavior of composite sandwich walls, composed of two steel plates filled with ultra-lightweight cementitious material. Compression tests on different height composite walls revealed that J-hook connectors provided similar resistance and unloading behavior as overlapped-headed studs. J-hook connectors improved composite action, prevented local buckling, and showed a reasonable correlation with test results when included in predictive methods.

Hu et al. (2014) [50], based on a fiber section analysis approach, developed a program that analyzes the moment–curvature behavior of concrete-filled steel-plate composite shear walls. The program's accuracy was verified against test results, and a parametric study was conducted on 6379 configurations. Simplified formulas were developed based on the study's results, which can be used to calculate the drift capacities and ductility of composite shear walls. Furthermore, experimental and numerical analyses were conducted on low shear span ratio dovetailed profiled steel–concrete composite shear walls subjected to axial and cyclic lateral loads by Huang et al. (2022) [51]. The results showed that all specimens failed in a ductile manner with a flexure–shear mixed mode. The lateral stiffness increased with higher thickness ratios, while deformability and energy dissipation capacity decreased. It was found that simplified models accurately estimated the lateral resistance of these composite walls.

Four full-scale experiments were conducted by Wang et al. (2022) [52] to evaluate the seismic performance of double-skin composite walls under earthquakes. The experimental results showed that these specimens had similar damage and flexural failure modes. Additionally, increasing boundary column thickness improved strength but decreased displacement ductility, while increasing the axial compression ratio improved strength but reduced ductility and energy dissipation cycles. Moreover, Shafaei et al. (2021) [46] analyzed the behavior of concrete stiffened steel-plate shear walls with a reinforced concrete panel and a gap between the panel and steel frame. The reinforced concrete panel thickness greatly affects the shear capacity and ultimate strength of these composite walls, prevents elastic buckling, and allows lateral load through shear yield. The results showed that these composite walls and steel-plate shear walls exhibit different behaviors since for the latter type, the steel plate undergoes elastic buckling to resist lateral load. Furthermore, concrete-stiffened steel-plate shear walls offered higher initial elastic stiffness, shear capacity, ultimate strength, ductility ratio, and energy absorption compared to steel-plate shear walls.

Ji et al. (2016) [53] examined the seismic performance of a novel composite wall with encased steel braces. This composite shear wall consists of a steel-braced frame embedded in reinforced concrete, which increases flexural and shear strength. Two wall specimens with different types of encased braces were tested, with both failing similarly with cracked-diagonal cracking and crushing of the concrete. It was found that steel-plate braces could potentially be more efficient and improve construction quality. Formulas for assessing the shear strength of these walls were also proposed. Additionally, Kenarangi et al. (2021) [54] examined the cyclic lateral load behavior of composite plate shear wall/concrete-filled systems for core–wall structures in high-rise construction. Two large-scale walls were tested, investigating their composite behavior and plastic hinge development. The results will aid in the development of design guidelines for high-rise steel buildings with these systems as the primary lateral force-resisting systems.

3.2. Seismic Behavior of Steel–Concrete Composite Shear Walls: Key Findings

The key findings regarding the seismic behavior of steel–concrete composite shear walls are as follows:

- Composite shear walls typically demonstrate enhanced strength and stiffness when compared to conventional reinforced concrete or steel-plate shear walls alone. This is due to the combination of materials, such as steel and concrete, which work together to provide increased structural integrity.
- Furthermore, the use of steel and concrete in composite shear walls often results in improved ductility. This means that the structure is able to endure larger displacements without failure during seismic events. This is crucial for ensuring the safety and resilience of the composite building structures during earthquakes.
- Another advantage of composite shear walls is their ability to dissipate energy. This enhanced energy dissipation capability is essential for seismic resistance, as it helps to absorb and dissolve the energy generated by earthquake forces, plummeting the potential for damage.

- Composite designs also offer the advantage of reduced wall thickness. Compared to traditional reinforced concrete walls, composite designs can often achieve the same level of strength and stiffness with thinner wall sections. This not only saves space but also reduces the amount of material required for construction.
- Additionally, the use of steel elements in composite shear walls can contribute to faster construction. Compared to conventional reinforced concrete systems, the incorporation of steel elements can expedite the construction process, resulting in time and cost savings.
- The presence of steel elements in composite shear walls also improves crack control. Steel helps to control crack formation and propagation in the concrete, enhancing the overall durability and longevity of the structure.
- Composite shear walls generally exhibit improved cyclic performance under cyclic loading conditions, which are representative of seismic actions. This means that they are better able to withstand repeated loading and unloading cycles without significant damage or failure.
- Proper design of composite shear walls can also lead to more predictable and desirable failure modes during extreme seismic events. By carefully considering the design and configuration of the walls, engineers can ensure that the structure behaves in a controlled manner, minimizing the risk of catastrophic failure.
- Furthermore, composite shear walls offer versatility in design. They provide various configuration options, allowing designers to optimize the performance of the walls for specific seismic demands. This flexibility enables engineers to tailor the design to meet the specific requirements of each project.

4. Modeling Techniques for Composite Members and Structures Under Earthquake Excitation

4.1. Modeling Techniques for Composite Members and Structures Under Earthquake Excitation: State-of-the-Art

Accurate simulation of composite members and structures is essential for forecasting performance and guaranteeing safety in seismic analysis. Critical elements encompass material simulation, capturing nonlinear behavior, such as plasticity and cracking, and interface simulation, to depict the interaction between steel and concrete. Member simulation encompasses beams, columns, and floor systems, while joint simulation replicates the behavior of connections. Global structural simulation integrates all elements, while dynamic analysis applies suitable methods. It is also crucial to integrate damping mechanisms and realistic energy dissipation. The range of models varies from simple 2D to complex 3D nonlinear models, depending on design requirements, building complexity, and available computational resources. Thus, a concise and effective analytical simulation for the strength capacity and cyclic behavior of circular CFTs was developed by Serras et al. (2016) [55]. Finite element analysis was used to validate the model and create a database of CFT behavior. Empirical expressions were developed to analyze the Ramberg–Osgood model and to provide a representation of ultimate strength. The proposed comprehensive model reliably captured the cyclic behavior of circular CFT columns.

Sahin et al. (2022) [56] explored the inelastic response of composite steel–concrete beam members, focusing on cyclic degradation phenomena. A continuum simulation was examined to model the hysteretic behavior of concrete slab and composite steel beam assemblages. Parametric assessments were conducted to understand key response characteristics, such as stiffness, capacity, and ductility. The results showed that the deterioration effects depend on various factors, like composite beam depth and steel cross-section slenderness, and that generally, under cyclic loading, composite members show 20% additional deterioration compared to bare steel counterparts. Moreover, Liu et al. (2016) [57] presented a study on high-strength friction-grip bolts as shear connectors for steel–concrete composite beams. A 3D FEM model was constructed to analyze the structural behavior of these beams, including the material nonlinearities and interactions between components. They also explored the differences in mechanical performance compared to traditional headed stud connectors and examined the effects of various factors on beam behavior for design purposes.

Katwal et al. (2018) [58] found that most of the finite element (FE) models simplified complex interactions, limiting their ability to capture failure modes, and for this reason, they developed a detailed FE model for composite beams, considering realistic component interaction and concrete damage. Their model accurately predicted the load–deformation curves and shear force–slip relationships of embedded studs.

Peng et al. (2022) [59] examined the seismic damage assessment process for recycled aggregate concrete-filled square steel tube columns using test results from numerous structures. By considering deformation, energy, and mechanical characteristics, a modified Park–Ang model was proposed, and a performance-based seismic design index as well as damage performance criteria for different failure states were suggested. The damage levels were categorized as intact, slightly damaged, moderately damaged, severely damaged, and collapsed, while by controlling the damage threshold, the assessment basis for different performance levels was established. Moreover, Sahin et al. (2023) [60] modeled and assessed the cyclic behavior of composite members with steel beams and concrete slabs. They proposed nonlinear relationships to simulate their response, showing that degradation is influenced by steel cross-section slenderness and composite beam depth. They also provided important data for seismic design and proposed expressions for computationally efficient frame-level analysis.

El Jisr et al. (2022) [61] presented a macro-model for modeling the response of composite steel beams in completely restrained types of beam-to-column connections. The proposed model accurately captured the asymmetric hysteretic response, cyclic deterioration, and force transfer mechanisms and can be used to assess the seismic collapse risk of composite steel buildings in Europe and to quantify slip demands. The results showed a system over-strength, a low probability of collapse, and that beam–slab connections with partial composite action experienced minimal damage. Additionally, Papavasileiou (2017) [62] introduced a mathematical framework for simulating steel–concrete composite columns using equivalent steel columns. Three simulation methods were presented for circular and rectangular concrete-filled hollow sections and concrete-encased I-shaped sections. The simulation was achieved by satisfying three equations for axial resistance and flexural stiffness and the proposed methods provided the dimensions of equivalent steel sections.

Skalomenos et al. (2015) [63] investigated the seismic performance of planar framed structures made of CFT columns and I steel beams. The impact of modeling details, such as panel zones, beam–column connections, steel I-beams, and composite CFT columns on seismic response, was examined. Their modeling of panel zones of beam–column joints is shown in Figure 4.

Skalomenos et al. (2015) [63] also created fragility curves for three composite frames designed according to European codes, allowing for the selection of the appropriate modeling level for desired seismic behavior. Moreover, Lin and Zhang (2021) [64] highlighted the importance of the slab spatial composite effect in the load-bearing capacity and load resistance of composite frame structures. They modified existing material models in OpenSees (https://opensees.berkeley.edu/) to capture this effect accurately. They found that the conventional fiber model overestimated the probability of the structure being in a specified limit state, while their improved fiber model provided results that are more accurate.

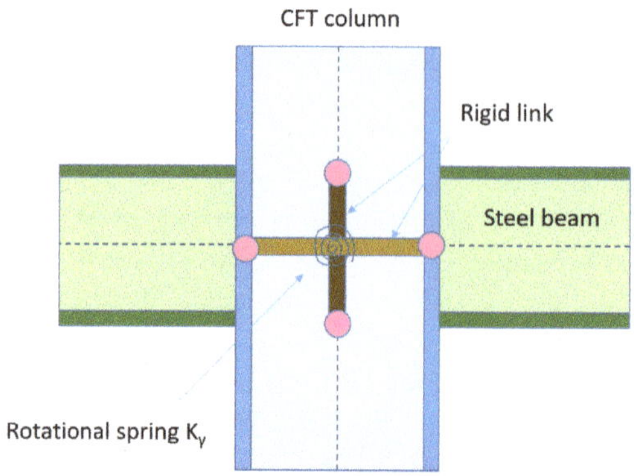

Figure 4. Panel zone model (adapted from Skalomenos et al. (2015) [63]).

4.2. Modeling Techniques for Composite Members and Structures Under Earthquake Excitation: Key Findings

Some of the key findings regarding the seismic analysis of steel–concrete composite buildings are as follows:

- The seismic response of composite buildings is significantly influenced by the modeling of composite action, which alters the stiffness, strength, and energy dissipation capacity of the structure compared to non-composite systems.
- Accurate modeling of interfaces, such as the steel–concrete interface behavior, is crucial for overall structural performance, including slip and partial interactions.
- Nonlinear behavior, including inelastic deformations in steel and concrete components, as well as their composite action, must be taken into account for realistic seismic analysis.
- The performance of beam–column joints, especially in moment-resisting frames, is greatly influenced by connection details, affecting energy dissipation and overall ductility.
- The modeling of floor diaphragms is important, as composite floor systems affect load distribution and structural dynamics, requiring careful consideration in the modeling process.
- Material degradation due to cyclic loading should be considered, as it can lead to strength and stiffness degradation in both steel and concrete, affecting the seismic response.
- Three-dimensional modeling is often necessary to capture effects, such as torsion and bidirectional loading, which can be significant in composite structures.
- Utilizing sophisticated numerical techniques, like finite element analysis incorporating intricate material models, has the potential to offer precise assessment of the local and global performance of composite structures.
- Future research should focus on creating standardized protocols for assessing the durability of composite structures under different seismic intensities and frequencies.

5. Behavior of Composite Beams Under Cyclic Loads

5.1. Behavior of Composite Beams Under Cyclic Loads: State-of-the-Art

Composite beams, which consist of steel beams and concrete slabs, are commonly utilized in the construction industry due to their effectiveness. It is imperative to comprehend how these beams behave when subjected to cyclic loads to ensure safe design practices. Important factors to consider include the transmission of loads via shear connectors, the level of ductility needed for seismic occurrences, the reduction in stiffness over time, the decrease in strength after multiple cycles, the dissipation of energy through inelastic deformation, the role of the concrete slab, the occurrence of local buckling in steel components, and the residual deformations. This knowledge is essential for the development of composite structures that are capable of withstanding seismic forces. Current research is concentrated on enhancing design approaches and connection specifics to improve seismic performance.

Fa-xing et al. (2018) [65] investigated the seismic behavior of steel–concrete composite beams through experiments. Parameters, such as shear connection degree, reinforcement ratio, and section type, were examined. The results showed that composite beams had favorable seismic performance, with a high displacement ductility and damping ratio. Increasing the shear connection, reinforcement, and transverse reinforcement enhanced the bearing and energy dissipation capacity. Transverse reinforcement ratio, shear connection degree, and stud spacing were important considerations. Welding quality and additional reinforcement were advised for practical engineering. Additionally, Han et al. (2015) [66] investigated the bending behavior of large-scale concrete-encased concrete-filled steel tube box members using experimental tests and finite element analysis. Variations in steel tube diameter and sectional height were examined, comparing these composite box members with reinforced concrete box members. A finite element analysis model analyzed the flexural performance and a simplified model for predicting the capacity of composite box members was proposed.

Based on previous tests, Suzuki and Kimura (2019) [67] showed that cyclically loaded stud connections in composite beams degrade faster than in monotonic push-out tests during earthquakes. Then, they used finite element analysis to assess the mechanical performance of the composite connection, considering various factors, such as stud and reinforcement sizes and configurations, material properties, slab widths, and loading protocols. The results were compared with existing equations, and a new mathematical model evaluated stud connection stiffness. Moreover, Zhu et al. (2023) [68] highlighted the application of super-elastic NiTi shape memory alloy (SMA) bolts, and brass-based friction-energy dissipation devices were utilized in self-centering connections. This study involved the cyclic testing of connections between concrete-filled square steel tubular columns and steel beams, showing that the SMA connection exhibited excellent recentering ability and moderate energy dissipation. The pre-strain of the SMA bolt was also modeled, and it was found that its increment improved self-centering capacity, while increasing the pre-tension force in the friction bolts enhanced energy dissipation.

Huang et al. (2014) [69] discussed the behavior of reduced beam section connections with composite beams in steel beam composite frames. A mathematical model was derived, and an amplification factor was proposed to account for the increased potential of the fracture at the beam bottom flange. Parametric analyses were conducted to investigate the effects of beam dimension, reinforcement, and slab dimension. A simplified design formula was proposed based on theoretical data, and the mathematical models were verified with finite element models. Additionally, Di Cesare et al. (2023) [70] conducted experimental and numerical analyses on ductile beam–column connections between composite reinforced concrete truss beams and RC/CFT columns under cyclic loading. This system, entitled the MTR-A beam (see also https://www.metalri.it/en/), is shown in Figure 5. Their study examined two experimental models with different testing schemes, validating the connections' ductile behavior and demonstrating that the results could be accurately reproduced through simple nonlinear modeling approaches.

Furthermore, Chen et al. (2017) [71] examined partially encased composite beams subjected to monotonic and cyclic loading conditions. More specifically, nine specimens with varying link details, beam lengths, and loading schemes were tested. The results indicated that failure typically occurs through beam flange fracture, local flange buckling, and damage to links and concrete. Moment capacities were higher when specimens experienced combined bending–shear action with an adequate shear span ratio. They also found that Eurocode 4 provides generally reasonable predictions for stiffness but overestimates its initial values. Additionally, El Jisr et al. (2019) [72] presented the construction of a freely available databank on steel beam-to-column connections. They used this databank to offer suggestions for nonlinear performance assessment seismic and design of steel and composite steel–concrete frames, and they concluded that the Eurocode 8-Part 3 overvalues the capacity of plastic rotation for composite beams. Empirical relationships were developed to predict plastic rotation capacity. They also highlighted that the composite steel connections experience higher shear demands in the web panel zone compared to non-composite connections.

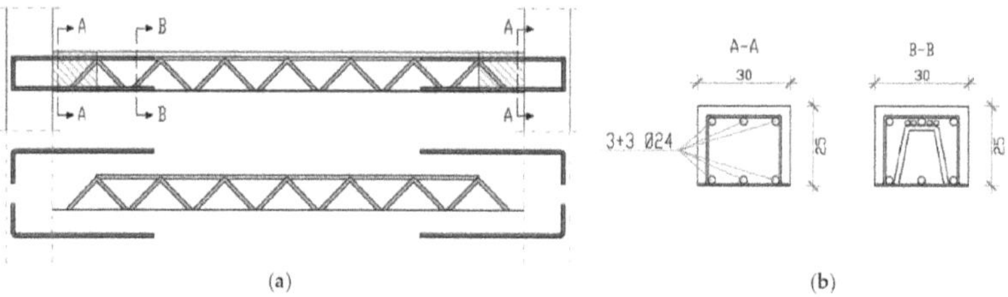

Figure 5. Ductile beam–column connections between composite reinforced concrete truss beams and RC/CFT columns: (**a**) longitudinal section and (**b**) cross-sections A-A and B-B (adapted from Di Cesare et al. et al. (2023) [70] and https://www.metalri.it/en/).

Suzuki and Kimura (2021) [73] found that the composite effects of stud shear connectors in composite beams' design are conventionally measured through push-out tests. However, these tests did not account for the reversed stress on concrete slabs during earthquakes. For this reason, they introduced a component model for composite beams that considers the stress in real structures. Cyclic loading tests on fourteen specimens revealed that the ultimate shear strength was significantly lower under compressive stress. To accurately evaluate performance in composite structures, they presented improved pertinent equations. Furthermore, Suzuki et al. (2023) [74] documented that novel shear connectors, specifically clothoid-shaped connectors, have gained interest in Europe for bridge engineering, although their mechanical performance evaluation has been based on pushout tests, which differ from the stress history experienced during an earthquake. To address this, cyclic loading tests were conducted on 14 specimens, revealing the stress transfer mechanism and confirming the dependence of mechanical capacity on stress orientation. They presented a new formula for evaluating ultimate shear strength and load–displacement relationships, considering the stress history.

An experimental study on concrete-filled tube specimens under cyclic bending loads was conducted by Montuori et al. (2024) [75]. Precise strain measurements using gauges generated load–displacement and moment–curvature curves. A new analytical model incorporating stress–strain relationships for concrete and steel was calibrated to match experimental results. They also provided insights into concrete-filled tube behavior under bending, supporting future modeling and design optimization for engineering applications. Moreover, Kim et al. (2016) [76] introduced a new hybrid composite beam design and assessed its performance through monotonic and cyclic tests. The study fabricated two beam-to-column connection specimens and six bending specimens for testing. The results demonstrated that the beam's capacity consistently increased with greater beam depth and steel-plate thickness. Additionally, the beam exhibited a satisfactory maximum moment compared to the nominal moment.

5.2. Behavior of Composite Beams Under Cyclic Loads: Key Findings

Some of the key findings regarding the behavior of composite beams under cyclic loads are as follows:

- The beam–column connection performance is essential in composite structures under seismic loads, as these are often critical points for energy dissipation and load transfer.
- Composite beams generally possess excellent energy dissipation capacity under cyclic loading, a crucial factor for seismic performance. This behavior is attributed to the interaction between steel and concrete components.
- Composite beams typically exhibit high ductility when properly designed, allowing for significant deformation without sudden failure. This characteristic is essential for maintaining structural integrity during earthquakes.
- Composite beams may experience stiffness degradation under repeated cyclic loading, often due to concrete cracking, steel yielding, and slip at the steel–concrete interface.
- The strength of composite beams can deteriorate with increasing load cycles, primarily due to local buckling of steel components and concrete crushing.
- The performance of shear connectors, such as headed studs, is critical in composite action. Inadequate shear connection may result in heightened slippage and diminished composite behavior when subjected to cyclic loading.
- Steel components in composite beams, particularly in the beam flanges and web, may experience local buckling under high cyclic loads, affecting the overall performance of the composite beam.
- The concrete slab significantly influences the behavior of composite beams, providing additional stiffness and strength. However, concrete damage and cracking can occur under severe cyclic loading.
- Composite beams may accumulate residual deformations after cyclic loading, which can affect the post-earthquake serviceability of the structure.
- Maintaining composite action throughout cyclic loading is essential for optimal seismic performance. The proper detailing and design of shear connectors are crucial to achieve this.

6. Behavior of Composite Beam-Columns Under Cyclic Loads

6.1. Behavior of Composite Beam-Columns Under Cyclic Loads: State-of-the-Art

Composite beam-columns are widely used in construction due to their strength, stiffness, and ductility. They typically consist of steel sections encased in or filled with concrete. During seismic events, they exhibit complex behavior that is crucial to understand for earthquake-resistant building design. Important features include load transfer mechanisms, ductility, stiffness degradation, strength deterioration, bond–slip behavior, local buckling, and concrete confinement. Understanding these behaviors is crucial for accurately modeling and designing steel–concrete composite buildings to withstand seismic loads. Researchers and engineers use this knowledge to develop design guidelines, analytical models, and performance-based seismic design approaches for composite structures. The current practices and future trends for the behavior of composite beam-columns under cyclic loads are examined in the following section.

Wang et al. (2021) [77] investigated the damage behavior and bearing capacity of circular high-strength CFT columns using quasi-static experimental tests. The moment–axial force relationship was analyzed, and the effects of key parameters were studied to propose a design method. Furthermore, they established a fracture criterion for high-strength steel and a restoring force model. The results showed that steel yield strength, concrete strength, and D/t ratio affected the moment–axial force curves. The proposed model aligned with tested bearing capacity, and the HS steel tube exhibited ductile fracture. Moreover, Qian et al. (2016) [78] discussed the analytical behavior of concrete-encased concrete-filled steel tubular columns during cyclic lateral loading. A finite element analysis model accurately predicted failure modes, load–displacement relationships, and ultimate strength. The study compared the behavior of these composite columns with conventional concrete-filled steel tubular columns and reinforced concrete columns, and a simplified hysteretic model for the M–φ relationship of the examined composite columns was proposed.

Tang et al. (2017) [79] conducted low cyclic loading tests on CFT columns filled with normal concrete and recycled aggregate concrete (RAC) to analyze their seismic performance. Findings showed that RAC-filled columns had a similar seismic performance to normal concrete-filled columns. However, RAC-filled columns exhibited better lateral bearing capacity, improved ductility, and slightly lower energy dissipation ability. Similarly, Ma et al. (2022) [80] conducted cyclic loading tests on eleven composite columns filled with RAC to investigate their seismic performance. Factors, such as the diameter-to-thickness ratio of circular steel tube, profile steel ratio, axial compression ratio, replacement percentage of recycled coarse aggregate, and section form of profile steel, were considered. The results showed that increasing the recycled coarse aggregate replacement percentage led to a decrease in ductility and energy dissipation capacity but an initial reduction and subsequent increase in bearing capacity and stiffness. Increasing the wall thickness of the circular steel tube and profile steel ratio improved seismic performance, while increasing the axial compression ratio had a negative effect. Modified formulas for the horizontal bearing capacity were proposed and verified with the test results. In the same way, cyclic load tests were performed on nine steel-reinforced RAC-filled square steel tube composite columns by Ma et al. (2024) [81]. Factors, such as steel ratio, width thickness ratio, recycled coarse aggregate replacement rate, and axial compression ratio were studied. The results showed that columns exhibited a compression-bending failure mode and good energy dissipation capacity. The use of recycled coarse aggregate had minimal impact on seismic performance, and decreasing the width thickness ratio or increasing profile steel ratio improved ductility. The axial compression ratio enhanced the bearing capacity but not the energy dissipation. A modified formula for horizontal bearing capacity was proposed and verified. Similarly, Luo et al. (2022) [82] investigated the behavior of composite columns filled with RAC containing ferronickel slag. Nine specimens were designed and tested, with parameters, such as coarse aggregate replacement ratios, diameter–thickness ratio, and axial load ratios, varied. Damage patterns, hysteresis curves, and other seismic parameters were analyzed. Numerical models were also established and compared with the test results. It was found that the location of damage was in the plastic hinge area, the coarse aggregate replacement ratios had a minimal impact on behavior, the axial load ratio affected ultimate strength and stiffness degradation, and the diameter–thickness ratio improved ultimate strength and ductility. The slenderness ratio, axial load ratio, and steel ratio had a significant effect on strength and ductility, while the RAC strength had a smaller effect. Moreover, Zeng et al. (2020) [83] focused on the behavior of a new composite column that involved a hybrid FRP-recycled aggregate concrete–steel tubular column. This composite column consists of an outer tube made of environmentally friendly basalt fibers and recycled coarse aggregates for the concrete. Quasi-static tests were conducted on five full-scale basalt composite columns to understand their behavior and potential for use in earthquake-prone areas. Various parameters were examined, including RAC replacement percentages, axial load ratios, and steel reinforcement ratios, to determine their influence on the composite columns' performance. The results showed that RAC replacement percentage has little impact, while axial load ratio significantly affects the seismic performance. These composite columns have better bearing capacity and energy dissipation compared to traditional composite columns with natural aggregate concrete.

The nonlinear response of square concrete-filled steel tubes under axial and flexural loads was studied using computational analysis by Skalomenos et al. (2014) [84]. A reliable finite element model was developed, considering factors, like local buckling, concrete behavior, cyclic softening, and interface interaction, as shown in Figure 6.

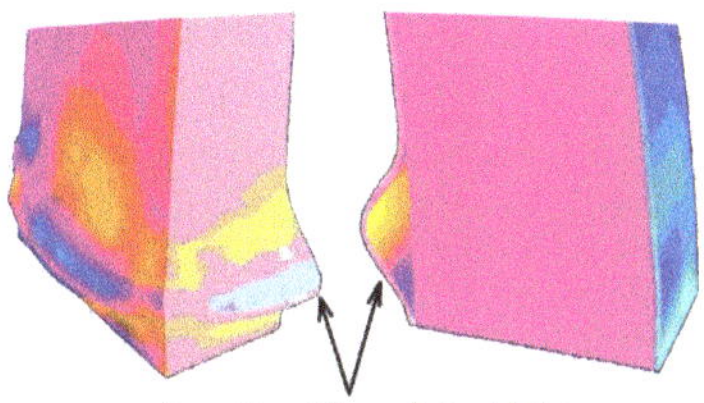

Figure 6. Advanced finite element modeling techniques for concrete-filled tube (CFT) columns, visualizing complex material interaction and structural response (adapted from Skalomenos et al. (2014) [84]).

Parametric analysis yielded expressions for three hysteretic models. Test results were used to validate these models, which are capable of simulating the response of concrete-filled steel tube columns or composite MRFs to cyclic loading.

Zhang et al. (2019) [85] experimentally examined the behavior of steel tube-reinforced, high-strength concrete columns with high-strength steel bars under cyclic loads. They found that the combination of C90 core concrete with C70 outer concrete provided superior composite effects compared to using C80 concrete for both. Additionally, adding steel fibers to the outer concrete improved ductility and energy dissipation. A calculation model for predicting maximum load-carrying capacity was also proposed and verified. Furthermore, Zheng et al. (2020) [86] investigated the behavior of multi-cell L-shaped concrete-filled steel tubular columns subjected to earthquake loading. Four tests were conducted on columns with different loading angles and axial load levels. The results obtained from experimental tests were compared with finite element analysis. Parametric analysis was also performed to study the effects of different parameters on column performance.

Kim et al. (2022) [87] examined composite columns, featuring a prefabricated steel cage, to enable the rapid construction of large industrial buildings. The cage consists of bolt-connected longitudinal steel angles and transverse steel plates, with integrated steel forms for concrete casting. Through seismic performance tests, it was observed that the steel angles improved flexural strength and stiffness, while the transverse plates reduced spalling and increased deformation capacity. The results aligned with existing models' predictions. Moreover, a concrete-filled steel tubular column and a metamaterial concrete-filled steel tubular column using rubber, lead, steel, and high load-bearing capacity concrete were designed by Xiong et al. (2024) [88] for vibration mitigation. The flexural bandgap theoretical model based on PWEM and a Timoshenko beam was established and analyzed. Shaking table tests confirmed its effectiveness in reducing vibrations, making it a potential solution for protecting structures from earthquakes and low-frequency lateral vibrations.

Hassan and Farag (2021) [89] recognized that, due to a lack of data, the seismic behavior of steel-reinforced-concrete composite columns having non-seismic details in older buildings is poorly understood. Furthermore, the influence of axial loads on the deformation capacity of steel-reinforced-concrete composite columns is uncertain, and there are no formal guidelines for modeling parameters or acceptance criteria for columns with modern seismic details. For these reasons, they addressed these gaps through experimental and analytical methods, examining these composite columns under cyclic loads. The test results indicate that tension-controlled steel-reinforced-concrete composite columns with non-seismic details meet the seismic assessment criteria, while compression-controlled columns with high axial loads exhibit premature failure. They also proposed new expressions and parameters for assessing steel-reinforced-concrete composite columns with non-seismic details. Mostafa et al. (2019) [90] also discussed the pros and cons for steel-reinforced-concrete composite columns under axial and seismic loads. Additionally, Hassan et al. (2021) [91] developed fragility and resilience functions, and assessed ASCE 41-17, examining the seismic performance of steel-reinforced-concrete composite columns. The methodology used Monte Carlo simulation techniques to establish economic vulnerability functions, and the results showed that ASCE 41-17 criteria underestimate structural capacity and resilience, while overestimate fragility and vulnerability.

Campian et al. (2015) [92] examined the seismic performance of composite steel–concrete columns with steel-encased profiles. It was found that composite columns with high strength concrete experienced an increase in lateral force and maximum lateral loading, while columns with normal strength concrete showed a gradual decline in bearing capacity. HSC columns exhibited a brittle failure mode but have higher energy absorption capacity, making them suitable for seismic areas. Composite columns with class C70/85 concrete demonstrated better structural performance. The fully encased composite column solution is competitive for both seismic and non-seismic zones, offering improved seismic performance and fire protection. Additionally, Chen et al. (2014) [93] examined, under low cyclic reversed loading tests, twenty-six steel–concrete composite columns simulating seismic conditions. An analysis of ductility, energy dissipation, hysteresis loops, and failure patterns highlighted the influence of steel section shape, stirrup ratio, axial compression ratio, and embedded depth ratio on seismic performance.

Chen et al. (2021) [94] conducted an experimental and computational analysis of the seismic performance of fiber-reinforced polymer (FRP)—confined high-strength rectangular CFT columns employing high-strength thin-walled steel tubes and concrete. Aspect ratio, axial compression ratio, number of FRP layers, and fiber direction are among the factors taken into account. Failure modes, hysteretic behaviors, capacity degradation, stiffness degradation, ductility, and stresses were used to assess seismic performance. By raising constraint stress in the effective confinement area, CFRP confinement increased bearing capacity by reducing local buckling and improving the confinement impact of steel tubes on core concrete, according to numerical analysis. Additionally, Gautham and Sahoo (2021) [95] investigated the behavior of structural concrete columns reinforced with steel under combined axial and lateral cyclic loadings. Energy dissipation potential, stiffness degradation, hysteresis response, failure mode, and lateral strength were among the parameters that were assessed in the study. A parametric study was carried out to evaluate the validity of design guidelines from multiple international codes and to predict the flexural capacity of structural steel-reinforced concrete columns under varying axial load levels using the ABAQUS program (https://www.3ds.com/products/simulia/abaqus).

6.2. *Behavior of Composite Beam-Columns Under Cyclic Loads: Key Findings*

Composite beam-columns have improved ductility, allowing for greater energy dissipation during seismic events. The composite action between steel and concrete results in higher strength and stiffness, improving overall structural performance. Concrete encasement provides confinement, enhancing compressive strength and ductility. The interaction between steel and concrete through shear connectors plays a crucial role in load transfer and overall structural behavior. Under repeated cyclic loading, there may be strength and stiffness degradation, and concrete encasement can help mitigate local buckling of steel sections. The bond between steel and concrete can deteriorate under cyclic loading, affecting overall performance. Composite beam-columns may accumulate residual deformations after severe seismic events, affecting post-earthquake serviceability. Understanding various failure modes (concrete crushing, steel yielding, and shear connector failure) is crucial for proper design. The performance of beam–column connections in composite structures is critical for seismic resistance and energy dissipation.

7. Seismic Performance of Composite Building Structures

7.1. Seismic Performance of Composite Buildings: State-of-the-Art

Considering that steel–concrete composites combine the advantages of both materials and improve seismic resistance, they are common in earthquake-prone areas. These structures consist of steel frames with concrete slabs, composite columns, or composite shear walls. The interaction between steel and concrete components enhances the overall response of the structure. Steel elements provide ductility and energy dissipation capacity, while concrete components enhance stiffness and mass distribution. Composite structures also have higher strength-to-weight ratios. Various factors, like connection design, stiffness and mass distribution, and composite systems, influence seismic performance. Challenges include predicting behavior and considering bonds, concrete cracking, and nonlinear material behavior. Ongoing research aims to optimize seismic performance and create resilient structures.

A displacement/damage-controlled seismic design approach for composite frames—consisting of composite beams and steel tube columns filled with concrete—was created by Serras et al. (2021) [96]. This method directly controls displacement and damage for all seismic performance levels, even near collapse. Empirical expressions were used to estimate the inter-story drift ratio and evaluate the damage index of critical members. By analyzing a large response databank, the necessary expressions were derived. The displacement/damage-controlled method reduced design iterations and eliminated the need for nonlinear time-history analysis. Design examples showed that the method successfully estimated inter-story drift ratio and controlled the damage index, resulting in an improved seismic performance procedure. Thus, Figure 7 shows the response (IDRs, inter-story drift ratios) of a 10-story composite building, using incremental dynamic analysis.

Furthermore, Serras et al. (2017) [97] created a computational process to ascertain how circular steel tube columns filled with concrete would react to constant loading. Experiments were used to validate the creation of precise 3D nonlinear finite element models. A total of 192 CFT specimens with varying diameter-to-thickness ratios, steel tube yield stress, compressive strength of the concrete core, and axial load levels were subjected to a parametric investigation. Empirical expressions were derived from the study to estimate force-displacement behavior. Then, these expressions were applied for the assessment of the seismic performance of composite buildings, examining their behavior through nonlinear time-history analysis. The material models for steel and concrete examined by Serras et al. (2017) [97] are shown in Figure 7.

Bai et al. (2022) [98] examined the impact of strong earthquakes on high-rise steel moment-resisting frames and concrete-filled tubular column frames. Over-design ground motions from large earthquakes can cause severe damage to structural components. The study used synthetic earthquake waves to assess the collapse criteria and deterioration margins of 40-story steel and composite building structures. The results showed that steel building experience collapse mechanisms in lower stories during very rare earthquakes, while the composite frames have a higher margin against overall collapse.

Additionally, the behavior of steel-reinforced concrete structures prestressed by bonding tendons was investigated by Ji et al. (2023) [53]. Fifteen groups of specimens were designed with various parameters. Finite element models were created using ABAQUS software (https://www.3ds.com/products/simulia/abaqus) to analyze the structures under axial forces and horizontal loads. The modeling method was validated by comparing it with the experimental results. The examined composite structures satisfied the strong columns, weak beams principle. The impact of several characteristics, such as ductility, hysteretic curves, skeleton curves, energy dissipation capabilities, and stiffness degradation, on seismic behaviors was investigated.

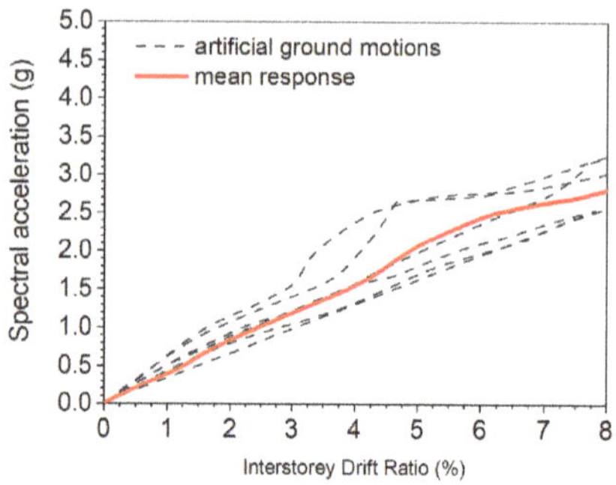

Figure 7. Dynamic inelastic analysis results showing maximum inter-storey drift ratios for a 10-storey composite structure under seismic loading conditions (adapted from Serras et al. (2021) [96]).

Wu et al. (2023) [99] found that slip-permitted and uplift-restricted connectors significantly enhanced the anti-cracking capabilities of reinforced concrete slabs. They used numerical analysis and a sophisticated finite element model to analyze the seismic performance of steel–concrete composite frames using these connectors. However, the lack of research has limited their applicability in composite frames. The study emphasized how crucial design elements are for maintaining structural integrity during seismic events, including steel beam height, flange thickness, and connector arrangement length. Furthermore, Zhang et al. (2024) [100] proposed a composite frame with a high-strength steel-plate wall core tube resilient structural system as a new high-performance structural system. It is composed of a core tube with composite frames, interchangeable energy-dissipation coupling beams, and double steel-plate–concrete composite shear walls. They showed how the benefits of this novel system include decreased wall thickness, greater space utilization, enhanced control over the drift ratio, less damage and stiffness degradation, and higher seismic resistance to safeguard people and property.

Taking into account the impact of floor loads, Zhao et al. (2020) [101] examined the seismic behavior of steel–concrete composite frames with large floor slabs. Three load cases were tested on a two-story, two-bay composite frame: pushover, longitudinal cyclic loading, and vertical floor loading. They examined strain distribution, mid-span deflection, and the emergence of cracks in the floor slabs. They also covered force processes of the composite frame, strength and stiffness degradation, load–displacement curves, energy dissipation capacity, and failure events. They also showed that fractures in the welds close to the external joint and the shear lag effect in the wide floor slabs affected the deformation pattern and force mechanism of the composite frame. Additionally, in order to evaluate the seismic risk and calculate the financial damages, Tondini et al. (2018) [102] examined the seismic demand of a steel–concrete composite structure utilizing a high-strength steel moment-resistant frame. Due to the intricate design of the building, a thorough 3D probabilistic seismic demand study that took the angle of the earthquake into account was necessary. Mild steel beams and round, high-strength steel columns filled with concrete made up the construction. There were several incremental dynamic analyses and nonlinear 3D FE models. Taking into account a variety of factors and the randomness of the impact angle, seismic fragility functions were constructed for damage and collapse limit states. The significance of peak ground displacement in the probabilistic model was demonstrated by the results.

Braconi et al. (2015) [103] discussed the results of a performance analysis on various case studies, which are included in a final project report of a research project entitled OPUS. The seismic performance of the structures was assessed using 2D nonlinear models, and the structural behavior and collapse modalities were determined using pushover analysis and incremental dynamic analysis. The study also used the incremental dynamic analysis technique to identify the level of peak ground acceleration that activates collapse, and the Ballio–Setti procedure was modified to account for frequency discrepancies. This allowed for a comparison of the actual q-factor with the design standards. For the same research project, namely OPUS, Badalassi et al. (2017) [104] analyzed the impact of material properties on the seismic performance of steel and steel–concrete structures. Various lateral-resisting systems and steel qualities were considered, with 15 structures designed. An accurate probabilistic procedure was developed to estimate failure probability and a model for the mechanical properties of European structural steel products was calibrated.

The seismic behavior of steel–concrete composite buildings designed with Turkish design codes was investigated by Etli (2022) [105]. Composite moment-resisting frame buildings with different numbers of stories were analyzed at high ductility levels. Nonlinear static and dynamic analyses were conducted to assess the lateral response, over-strength factors, and ductility factors. The examined composite frames exhibited excellent performance, surpassing design expectations. Section deformation capacities were also evaluated during dynamic and static analyses. Similarly, Etli (2023) [106] examined the seismic behavior of steel concrete composite buildings where 5-, 10-, 15-, and 20-story framed structures were designed using concrete-filled steel tube columns. The buildings were designed with high ductility based on Turkish design codes regulations, and nonlinear static pushover and incremental dynamic analyses were applied. The nonlinear analyses showed that the composite framed structures had good performance, with high ductility and the ability to absorb seismic energy through inelastic deformations.

Denavit et al. (2016) [107] examined structural system performance factors, such as strength and ductility for steel–concrete composite moment frames under seismic loads. They used new finite element formulations to analyze the behavior of composite moment frames, where archetype frames were designed according to seismic codes specifications and subjected to nonlinear static pushover and dynamic response history analyses to determine performance factors. Moreover, Zhao et al. (2023) [108] investigated the seismic performance of post-earthquake composite frame structures with varying damage levels. Nonlinear time history analysis was conducted on a ten-story steel–concrete composite frame, using peak ground acceleration (PGA) as a measure of seismic intensity and damage. The structure was subjected to different PGAs in the first stage, resulting in three damage levels. The analysis focused on inter-story drift ratios and plastic hinge ratios, which show a uniform distribution pattern across the damage levels. The structure was deemed safe based on residual inter-story drift ratios. The second stage examined the residual capacity of the structure in strong earthquakes. The results revealed that pre-existing damage could increase lateral deformation and reduce resistance capacity.

The performance of composite frames with circular concrete-filled steel tube columns attached to composite beams made of steel and concrete was studied by Ding et al. (2018) [109] under axial and lateral stresses. The effects of the beam-to-column stiffness ratio, axial compression ratio, and stiffness ratio on seismic performance were examined through the testing of seven frames. Discussions were held regarding the experimental findings, which included ductility, energy dissipation capacity, load–deformation responses, damage development, and stiffness degradation. Additionally, a finite element approach, considering geometrical and material nonlinearity, was developed and validated against the experimental results. Etli and Güneyisi (2021) [110] investigated the performance of steel–concrete composite buildings under earthquake loads with different numbers of stories and different, according to Eurocode 8 regulations, ductility levels. SeismoStruct software (https://seismosoft.com/products/seismostruct/) was used for design and performance assessment, utilizing nonlinear static pushover and incremental dynamic analyses. They analyzed the effects of ductility and the number of stories on the seismic performance of the buildings, finding that all structures had behavior factors higher than the design assumptions, particularly those in the medium ductility class.

Zhao (2016) [111] proposed a simplified approach for analyzing and evaluating the performance of steel and concrete composite frames under extreme loads, such as earthquakes. The study focused on a four-story two-bay underground composite frame. The nonlinear behavior of composite members was characterized to establish a macro-model for the frame. The pushover method was applied to determine the lateral force VS. top horizontal displacement. A method for identifying damage in composite frames was proposed, and the damage evolution of the case study frame was analyzed. The failure mode was attributed to substantial damage in the bottom CFST columns, and the seismic performance of the frame with high-strength steel was compared to that with ordinary strength steel, showing improved lateral resistance and elasticity. Furthermore, the test results of two steel–concrete composite frame specimens with encased CFST columns subjected to cyclic loading were analyzed by Wang et al. (2017) [112] using fiber beam-column elements. Parameters, such as column slenderness ratio, axial compressive ratio, and reinforcement ratio, were studied. Hysteretic models for composite frames were proposed based on parameter analysis, showing favorable energy-dissipating capacity.

Skalomenos et al. (2015) [113] tested a family of 96 regular plane CFT-MRFs using different ground motions to create a response databank. Using regression analysis, they derived simple formulas to estimate the strength reduction factor, ductility demands, drift, and seismic displacements. Several factors, such as the number of stories, stiffness ratio, strength ratio, deformation, and material strengths, were studied. The formulas can be used in seismic design methods and allow for the assessment of existing structures as well as for the design of new ones based on deformation.

7.2. Seismic Performance of Composite Buildings: Key Findings

Examining the seismic performance of steel–concrete composite buildings, these structures appear to have, in comparison with steel or reinforced concrete structures, enhanced ductility, increased stiffness, improved strength, enhanced damping characteristics, effective load transfer, reduced structural weight, enhanced performance of beam–column joints, beneficial composite floor systems, and the importance of proper detailing. Composite buildings exhibit improved energy dissipation, reduced lateral deflections, higher strength capacity, superior damping properties, efficient load transfer, lighter structures, improved behavior under cyclic loading, and effective distribution of lateral loads. Proper detailing, especially in critical areas, is crucial for the seismic performance of these buildings.

The development of comprehensive performance-based design approaches for steel–concrete composite structures in seismic zones represents a critical challenge at the intersection of structural engineering, materials science, and advanced computational modeling. The complexity stems from the multifaceted nature of composite systems, which exhibit intricate material interactions and nonlinear behavioral characteristics under seismic loading conditions. Traditional design methodologies have predominantly relied on prescriptive approaches that often fail to capture the nuanced response of composite structural elements to dynamic loading scenarios [114].

The fundamental limitation of current design frameworks lies in their inability to fully incorporate the inherent variability of material performance and structural response across different seismic intensities and loading configurations. Composite structures present unique challenges due to the differential mechanical properties of steel and concrete, including their distinct elastic moduli, thermal expansion coefficients, and strain compatibility characteristics. These differences generate complex stress transfer mechanisms that are not adequately addressed by conventional linear elastic analysis methods (Duan et al., 2022) [22].

Advanced numerical modeling techniques have emerged as a promising avenue for developing more sophisticated performance-based design approaches. Finite element analysis and multi-scale modeling techniques enable more precise characterization of the nonlinear behavior of composite structural elements. These computational methods allow for more sophisticated predictions of local and global structural responses, including the complex interactions between steel and concrete components under cyclic loading conditions [115]. The integration of machine learning algorithms with advanced numerical modeling has further expanded the potential for developing more adaptive and context-specific design methodologies.

Developing performance-based design guidelines requires a comprehensive understanding of the dynamic response characteristics of composite structures across various scales. This necessitates a holistic approach integrating experimental investigations, advanced numerical modeling, and probabilistic risk assessment techniques. Researchers have increasingly focused on developing hybrid modeling approaches that combine experimental data with sophisticated computational methods to generate robust predictive frameworks (Li and Xu, 2023) [116]. These approaches aim to bridge the existing knowledge gaps in understanding the complex behavior of composite structures under seismic loading.

One of the most significant challenges in developing performance-based design approaches is the characterization of higher mode effects and torsional responses in composite structural systems. Traditional design methodologies often simplify these complex dynamic interactions, potentially leading to significant discrepancies between predicted and actual structural performance. Advanced modal analysis techniques have demonstrated the critical importance of considering multi-modal responses, particularly in tall and irregularly configured composite structures [117–119]. These investigations reveal that higher mode effects can significantly influence the overall seismic response, introducing additional complexity beyond fundamental mode considerations.

The implementation of performance-based design approaches requires a fundamental reimagining of design methodologies that move beyond prescriptive code-based requirements. This paradigm shift necessitates the development of more sophisticated performance metrics that can comprehensively evaluate structural response under various seismic loading scenarios [120]. Such metrics must account for not only ultimate limit states but also serviceability requirements, considering various factors, such as residual deformation, energy dissipation capacity, and long-term structural integrity.

Recent advances in sensing and monitoring technologies have opened new avenues for developing more dynamic and adaptive performance-based design approaches. Smart sensor systems and real-time monitoring capabilities enable more precise characterization of structural behavior, providing unprecedented insights into the actual performance of composite structures during seismic events [121,122]. These technologies facilitate a more nuanced understanding of structural response, allowing for more targeted and context-specific design interventions.

The path toward comprehensive performance-based design methodologies for steel–concrete composite structures requires a multidisciplinary approach that integrates advanced computational techniques, experimental investigations, and innovative monitoring technologies. Future research must focus on developing more sophisticated modeling frameworks that can capture the complex material and structural interactions unique to composite systems [123]. This will necessitate continued investment in advanced computational resources, experimental facilities, and interdisciplinary research collaborations that can address the multifaceted challenges inherent in the seismic design of composite structures [124].

7.3. Contextual Variability in the Seismic Performance of Composite Structures

The complexity of seismic performance in steel–concrete composite structures necessitates a critical examination of the inherent variability across different geological, geographical, and structural contexts [108]. While this review synthesizes comprehensive research findings, it is paramount to acknowledge that the behavioral characteristics of composite structures are profoundly influenced by a multitude of site-specific factors that cannot be universally generalized. Geological conditions, including ground motion characteristics, soil properties, and local seismic hazard profiles, play a crucial role in determining structural response that extends beyond standardized design parameters [125]. Regional variations in construction practices, material quality, and local building codes introduce additional layers of complexity that significantly affect the seismic performance of composite systems [126]. For instance, the same structural design may exhibit markedly different responses in regions with varying tectonic environments, such as subduction zones versus transform fault boundaries. Furthermore, micro-level variations in material composition, fabrication techniques, and construction quality can introduce substantial uncertainties that are challenging to predict through generalized analytical models [127]. The interaction between structural geometry, connection detailing, and local seismic loading conditions creates a dynamic system where small variations can lead to significantly divergent performance outcomes [128]. These contextual nuances underscore the importance of site-specific investigations and the limitations of broad, generalized conclusions. Researchers and practitioners must approach composite structural design with a nuanced understanding that recognizes the inherent complexity and site-specific nature of seismic performance. This approach demands a more sophisticated methodology that integrates local geological data, advanced numerical modeling, and comprehensive experimental validation to develop truly robust and context-sensitive design strategies for steel–concrete composite structures in seismic regions [110].

8. Composite Construction and Foundation Seismic Design

8.1. Composite Construction and Foundation Seismic Design: State-of-the-Art

Seismic design of foundations for steel–concrete composite buildings is crucial for ensuring safety during earthquakes. The process involves analyzing the dynamic response of the structure, considering factors, like frequency and resonance. Deep foundations, like piles or caissons, may be used to transfer loads to stable soil layers. Measures, like base isolation and energy dissipation devices, help mitigate ground motion effects. The stiffness and strength of the foundation are important for overall seismic performance. Design codes and standards provide guidelines, emphasizing capacity design principles to ensure plastic deformations occur in predetermined locations. Overall, a comprehensive understanding of soil–structure interaction, material behavior, and dynamic analysis techniques is necessary for creating resilient structures that can withstand seismic events.

Abbas et al. (2021) [129] developed two column-to-foundation connections for square CFT columns. Experimental and numerical assessments were conducted, comparing the connections with existing ones for their inelastic deformation capacity when subjected to cyclic loading. The results demonstrate the efficient and improved behavior of the suggested connections, with the numerical models accurately predicting the experimental response. Similarly, Khateeb et al. (2020) [130] introduced two new and efficient connections for connecting concrete-filled steel tube (CFST) columns to reinforced concrete foundations. The connections were designed to withstand seismic loading and were tested against conventional CFT and reinforced concrete column-to-foundation connections. The results showed that the proposed connections performed better. Finite element models accurately predicted the behavior of these connections, as confirmed by the experimental results.

Serras et al. (2021) [131] examined the use of concrete-filled steel tube piles in deep foundation systems under seismic and cyclic loads. They found that composite piles effectively mitigate damage in hard-to-reach areas, such as the pile heads and deep depths. Controlled loading analyses and seismic-intensity analysis confirmed the capacity margins of the composite pile system. They also discussed the damage patterns, displacement profiles, and residual displacement of composite piles compared to concrete piles. The results showed that composite piles exhibited 40% less damage on average than reinforced concrete piles. Moreover, Zhou et al. (2015) [132] examined the mechanical properties of reinforced concrete and steel–concrete composite members under seismic loads, specifically in the context of elevated pile-group foundations used in bridge and ocean engineering. Four scale-specimens were created and tested, comparing their seismic performance. The results showed that the composite specimens had higher peak strength, higher ultimate displacement, and higher energy dissipation capability than the reinforced concrete specimens.

Wang et al. (2020) [133] focused on steel-plate composite walls and their connections to foundations. Eight large-scale specimens were tested to examine the behavior of composite wall-to-foundation connections under axial compression and cyclic lateral force. The specimens were tested in different loading directions and using different connection construction details and aspect ratios. The results showed that the lap splice connection was stronger, while the embedding connection allowed for more ductility. The current expressions for load-carrying capacities were found to be underestimated. Similarly, Vakili Sadeghi et al. (2022) [134] demonstrated how a baseplate and eccentric anchors could be used to attach steel-plate composite walls to a concrete basemat. Direct force transfer from the wall to the anchors was made possible by the connection of the anchors to the bottom of the faceplates, which eliminated the baseplate from the force transmission chain. LS-Dyna software (https://lsdyna.ansys.com/) was used to verify three test walls. It was discovered that the split-baseplate connection with concentric anchors was more resilient than the other connections. Kurt et al. (2016) [135] examined the direct shear behavior of rebar-coupler anchor systems used in nuclear reactors to secure steel-plate composite walls to the concrete basemat. The anchor specimens were subjected to extensive testing up to failure. The results showed the load–slip displacement responses, direct shear strength, and failure mode. It was found that the American Concrete Institute (ACI) 349 code equation underestimated the shear strength because it assumed that failure occurred in the rebars, while in reality, it happened in the couplers. An updated design equation using the net shear area of the couplers improved the accuracy of calculating shear strength. An empirical model for the shear force vs. slip displacement response of rebar-coupler anchor systems was also proposed with the use of experimental data. Analytical models were developed by Siddiqui et al. (2024) [136] to determine the minimum depths required for connecting circular and square concrete-filled steel tubular columns. The pullout capacity of the reinforced concrete footing, which was reliant on the strength of the footing concrete, reinforcement, and cement grout, was equivalent to the pullout strength of the tube in the models. These models were used to calculate the optimal depths for composite columns using the Monte Carlo simulation technique. The second connecting scheme showed significant improvements over the first scheme, with 37.3% and 45.2% reduced embedment depth for circular and square composite columns, respectively. Additionally, to address concrete-filled steel tube column damage during seismic load, Feng et al. (2024) [137] suggested a rubberized concrete-filled corrugated steel tube composite column–foundation connection. Experimental studies confirmed improved ductility and energy dissipation. The results showed sufficient connection strength under seismic load and proposed seismic design methods using a side shear database to inform future column–foundation connections.

Li et al. (2020) [138] examined the structural behavior of double-pile foundations under cyclic loads. The effects of inclination angles and the embedment depths of unequal-height concrete-filled steel tube piles on failure modes, ductility, stiffness, strength, and energy dissipation capacity were evaluated. Battered piles increase loading capacity and energy dissipation but decrease deformation capacity and ductility. Unequal heights of battered composite piles can lead to premature failure of the shorter pile. Unequal heights also decreased the deformation capacity of the foundation. Won et al. (2020) [139] examined a steel composite hollow reinforced concrete column, characterized by an inner tube within the hollow section, which required innovative connecting techniques for the column–footing joint segment. They introduced three novel connecting methods, which were assessed through experimental and finite element analysis investigations. Subsequently, upon identification of an effective connecting method for column footing joints, it was implemented in large-scale test specimens to validate and assess its performance.

An analysis of composite pile performance was conducted by Thusoo et al. (2021) [140], comprising a comprehensive dataset of 79 bending tests, which facilitated the comparison of observed bending moment capacities against theoretical predictions derived from various design guidelines. However, the assessment of composite pile behavior necessitates consideration of drift capacity predictions for a more thorough evaluation. To address this requirement, researchers developed and validated a computationally optimized fiber-based model suitable for nonlinear static analysis. This analytical framework incorporated the primary failure mechanisms characteristic of composite piles, specifically addressing concrete compressive failure and steel casing local buckling phenomena. Furthermore, Stephens et al. (2016) [141] focused on the seismic behavior of the embedded ring concrete-filled steel tube connections, which allows for accelerated bridge construction and sustains minimal damage during large inelastic deformations caused by earthquakes. Practical design expressions and a design example were provided for implementing the research findings. Furthermore, Wang et al. (2024) [142] focused on the seismic performance of reinforced thin-walled irregular steel tube concrete frame structures after a fire, considering soil–structure interaction. Finite element models were established using ABAQUS software (https://www.3ds.com/products/simulia/abaqus). The analysis evaluated the effects of site conditions and fire duration on various structural parameters. The results showed that soil–structure interaction increased the natural vibration period by 10–30%. The acceleration for structures on softer soil with longer fire durations was higher compared to assuming a rigid foundation. Inter-story shear force and inter-story displacement angle also increased with soil–structure interaction, especially with longer fire duration, larger seismic wave amplitude, and softer soil.

8.2. Composite Construction and Foundation Seismic Design: Key Findings

The seismic design of steel–concrete composite building foundations requires consideration of soil–structure interaction. Adequate stiffness and strength in the foundation system prevent excessive deformation. Proper connection between the superstructure and foundation is crucial for effective load transfer. Foundation type should align with site conditions and building characteristics. Energy dissipation mechanisms reduce seismic demands. Composite action enhances overall performance but requires attention to connection details. Nonlinear behavior of both foundation and superstructure should be considered, and performance-based design approaches can be used to optimize foundations in highly seismic regions.

9. Seismic Isolation and Energy Dissipation Devices for Composite Structures

9.1. Seismic Isolation and Energy Dissipation Devices for Composite Structures: State-of-the-Art

Two common strategies employed to improve the performance of steel–concrete composite buildings and minimize their damage during seismic events are seismic isolation and supplementary dampers. Seismic isolation entails the introduction of a flexible layer between the composite building and its foundation to isolate the structure from ground motion. On the other hand, supplementary dampers work by dissipating energy during seismic events, thereby reducing acceleration and displacement demands on composite structural elements. The combination of these techniques offers numerous benefits for steel–concrete composite buildings, including enhanced strength, ductility, and flexibility. Extensive studies, simulations, and practical applications have demonstrated the effectiveness of seismic isolation and supplementary dampers. By optimizing the integration of these systems, engineers can create resilient and secure composite structures capable of withstanding seismic loads and improving overall performance in earthquake-prone areas.

Darwish and Bhandari (2022) [143] examined the effectiveness of high-rise composite buildings with base isolation in reducing seismic response during earthquakes. Two base-isolated steel–concrete buildings with twelve- and fifteen-story levels were analyzed, using lead rubber-bearing isolators for base isolation. Composite base-isolated buildings were compared to fixed-base composite and reinforced concrete counterparts. Seismic responses were evaluated using the response spectrum method for the highest seismic zone according to the Indian Code. The results showed that base isolation with lead rubber-bearing isolators reduces overall seismic responses of composite buildings by 50–60%. The 15-story building demonstrated higher effectiveness. Base-isolated composite buildings had approximately 50% less story drift and displacements compared to reinforced concrete buildings. Similarly, a comparative analysis of the seismic responses of three different building types, namely reinforced concrete buildings, fixed-base composite buildings, and base-isolated composite buildings using lead rubber-bearing base isolators, was investigated by Darwish and Bhandari (2022) [144]. They indicated that the integrated application of base isolation and composite building in base-isolated composite buildings results in superior overall performance compared to reinforced concrete and fixed-base composite buildings, particularly in three-, six-, and nine-story buildings. Furthermore, the study revealed that fixed-base composite buildings outperform reinforced concrete buildings across all three levels of stories, highlighting the suitability of base-isolated composite buildings for high-rise, mid-rise, and low-rise buildings.

Li and Fa (2023) [145] investigated alternative dissipative systems in composite steel–concrete frame structures. An experiment evaluated the mechanical properties of three plate flanges with different parameters and power factors under various cyclic loadings. The examined systems demonstrated high resistance and the ability to dissipate seismic vibration energy. Similarly, Kanyilmaz et al. (2019) [118] evaluated the effectiveness of a standard multi-story building with a steel–concrete composite frame designed with and without structural fuses using nonlinear transient dynamic analysis. Numerical models were created using a distributed plasticity approach. These models were adjusted based on experimental data from the literature. The study quantified global response parameters including energy dissipation, base shear, and inter-story drifts.

Li et al. (2021) [146] developed a protective system for composite buildings using buckling-restrained braces (BRBs) and viscous dampers (VDs) to address the combined threat of earthquakes and winds. They investigated the effectiveness and design parameters of these devices by employing the fragility function method. The results showed that the hybrid-damped frame (HDF) is effective in withstanding multiple hazards, and that the energy dissipation contributions of VDs and BRBs vary based on the hazard intensities. Zhuang et al. (2022) [147] experimentally studied the seismic behavior of eccentrically braced composite frames with vertical shear links. The use of low-yield-point steel in the shear links was found to enhance seismic performance. The eccentric bracing system increased the stiffness, strength, and energy dissipation capacity of the composite frame without affecting force conditions on primary beams and columns. However, strain hardening of the low-yield-point steel resulted in severe damage to the concrete slab in the middle segment of the primary beam. Similarly, Javaid and Verma (2023) [148] evaluated the impact of buckling-restrained braces and viscous dampers on the seismic performance of asymmetrical composite frames. Both devices effectively reduced seismic response, with viscous dampers being more efficient in reducing period and base shear by 65–73% and 80–90%, respectively. Buckling-restrained braces performed better in reducing maximum overturning moments. Implementing both devices reduced inter-story drift ratio and horizontal displacement, but increased compression force on columns by 20–25% for buckling-restrained braces and 15–22% for viscous dampers. Viscous dampers were found to be more effective.

Guo and Wang (2023) [149] examined a dual system consisting of a double-skin CFT frame and metallic dampers. The energy distribution along the height was analyzed using nonlinear models validated by test results. They found that the metallic damper absorbed significant energy under seismic loads, whereas a semi-rigid connection harmed the column. A method to estimate energy distribution in the dual system was developed, providing valuable insights for future applications.

Zhuang and Zhao (2022) [150] analyzed the seismic performance of an eccentrically braced composite frame with a low-yield-point steel shear link. A theoretical and finite element model is used to understand the structural mechanics and energy dissipation mechanisms. Various design parameters were investigated to determine their impact on the system's mechanical properties. Additionally, recommended design parameter values were provided.

Furthermore, Thakur and Tiwary (2023) [151] examined the seismic response of composite structures integrated with fluid viscous dampers and base isolation. They evaluated the seismic performance of composite structures using finite element analysis. The results showed that integrating fluid viscous dampers and base isolation reduces the structural response to seismic events. A synergistic combination of fluid viscous dampers and base isolation outperforms their individual use.

A new type of shear panel dampers with bent web panels, installed in composite frames to enhance stiffness and strength was examined by Zhao et al. (2022) [152]. Unlike traditional dampers, BSPDs minimize bending moments on the primary beam, reducing damage to the concrete slab. Experimental and numerical investigations confirmed their ductile behavior and energy dissipation capacities. Design recommendations included using low-yield-point steel and link flanges to increase deformation capacity and efficiency.

Li et al. (2020) [153] presented analytical studies and experiments on blind-bolted end plate CFT composite frames with buckling-restrained braces. Pseudo-dynamic tests on two 2/3-scaled two-story composite frames with buckling-restrained braces showed good hysteretic behavior and high ductility. The moment capacity, rotation capacity, and mathematical model of a composite joint were analyzed and modified. The shear force–deformation relation of a CFT panel zone was checked. A macroscopic finite element model showed that buckling-restrained braces provide lateral stiffness and resistance under small earthquakes and dissipate energy under severe earthquakes. This model is part of the OpenSees program (https://opensees.berkeley.edu/).

Kastimpini et al. (2024) [154] conducted a comprehensive investigation of a cable-stayed bridge featuring concrete-filled steel tube pylons and an innovative seesaw system for seismic protection. The geometry of the problem is shown in Figure 8. Through detailed finite element modeling and extensive parametric studies, they evaluated the system's effectiveness in mitigating seismic demands under various earthquake scenarios. Their analysis incorporated soil–structure interaction effects across different soil conditions and foundation types, demonstrating the system's capacity to reduce deck displacements, pylon base shear, and cable forces. Furthermore, Katsimpini (2025) [155] investigated the seismic performance of two-, four-, and six-story composite buildings with viscous wall dampers using concrete-filled steel tubular columns and steel beams. Their work assessed dampers' effectiveness through nonlinear time history analyses. The results showed reduced inter-story drift ratios and peak floor accelerations, validating the dampers' integration in mid-rise structures. The behavior and bearing capacity of composite members were based on Ref. [156]. Furthermore, the influence of multiple earthquakes and soil flexibility on the structural response have been examined in Ref. [157]. The results show that multiple ground motions increase demands compared to single events. Incorporating soil–structure interaction reduces drift and accelerations but increases displacements.

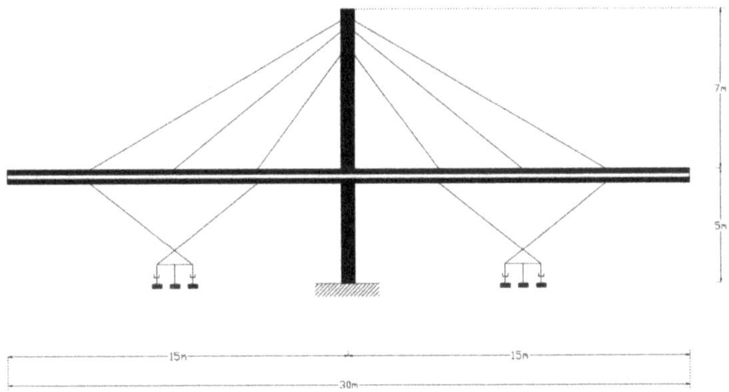

Figure 8. Innovative seismic protection system for composite bridges utilizing a seesaw-type energy dissipation mechanism, demonstrating advanced structural resilience techniques (adapted from Katsimpini et al. (2024) [154]).

Javaid and Verma (2023) [158] examined the effectiveness of buckling-restrained braces and viscous dampers in enhancing the seismic performance of composite buildings during earthquakes. Steel–concrete composite frames were analyzed with and without these devices. The results showed that viscous dampers were more efficient, especially when placed in center bays for regular buildings and corner bays for C- and L-shaped composite buildings, reducing period, maximum story displacement, base shear, and maximum inter-story drift significantly, and making them a great option for mid-rise composite buildings.

9.2. Seismic Isolation and Energy Dissipation Devices for Composite Structures: Key Findings

Seismic isolation and supplementary dampers enhance the performance of steel–concrete composite buildings during earthquakes. They reduce structural damage and improve occupant safety by decoupling the composite building's movement from ground motion and dissipating seismic energy. Base isolation systems elongate the composite building's natural period, decreasing inter-story drifts, floor accelerations, and structural demand. Supplementary dampers provide additional energy dissipation capacity. The combination of isolation and dampers offers synergistic benefits, leading to cost savings. Composite buildings with these systems exhibit improved ductility, energy dissipation, and seismic resilience. Design considerations include the interface between isolated and non-isolated portions and the proper integration of dampers. Analysis and design are crucial to maximize the benefits for each project.

10. Progressive Collapse Resistance of Composite Structures

10.1. Progressive Collapse Resistance of Composite Structures: State-of-the-Art

Progressive collapse in composite structures, especially steel–concrete composite buildings under seismic loads, is a complex phenomenon that has received considerable attention in structural engineering. This form of collapse occurs when localized damage triggers a chain reaction of failures, potentially leading to disproportionate and catastrophic structural collapse. The interaction between steel and concrete elements in composite buildings plays a crucial role in their seismic performance. During earthquakes, these structures experience intense lateral forces and dynamic loading, which can cause steel members to yield and the concrete to crack or be crushed. The composite action between steel and concrete has the potential to improve overall structural ductility and energy dissipation capacity, thereby potentially reducing the risk of progressive collapse. However, the behavior of connections between composite elements, load redistribution mechanisms, and the potential for sudden failure of key structural components are critical factors that must be carefully considered in the design and assessment of these buildings to ensure their resilience against progressive collapse under seismic conditions.

Zandonini et al. (2019) [159] focused on steel–concrete composite frames in the event of column loss. Two geometrically distinct three-dimensional composite full-scale substructures were selected from reference buildings and tested to simulate the column collapse scenario. They outlined the preparatory studies, the main features of the specimens, and the results of the initial test. The test provided insight into the need for an improved design of joints and emphasized key aspects of the response of the floor system. Additionally, static tests on six two-story three-span composite frames with CFST columns were conducted by Zheng et al. (2022) [160] under a penultimate column removal scenario. The effects of weld-bolted connection and RC slab on progressive collapse resistance were analyzed. The results showed that the torsion in steel beams at the directly affected region was reduced for frames with composite beams. Differences in failure modes were observed between the penultimate and middle-column removal scenarios. It was also found that frames with weld-bolted connection had lower resistance but multiple protection measures against progressive collapse.

Zheng et al. (2022) [161] initially presented a fiber-based model that incorporates nonlinear beam-column elements and zero-length elements to account for the composite relationship between the slab and beam, as well as the material's damage evolution for composite members. Then, the suggested modeling method was applied to an anti-collapse analysis of composite frames with various connections. Additionally, design suggestions obtained through parameter analysis were provided to prevent the development of a chain reaction. Furthermore, a multi-scale model was used by Wang et al. (2017) [162] to study the collapse performance of CFT column-to-steel beam connections. Nonlinear static and dynamic analysis methods were used to reveal resistance mechanisms, failure modes, and stress distribution of joints. The results showed that these joints could prevent progressive collapse by forming resistance mechanisms and providing alternate load paths. The adjacent framework also enhanced the anti-collapse ability of the joints.

Papavasileiou and Pnevmatikos (2017) [163] examined the use of steel cables for retrofitting steel–concrete composite buildings to prevent progressive collapse. They also investigated the impact of the building's characteristics on the overall cost of retrofitting to identify the best cost-effective approach for each scenario. To achieve this, an optimization algorithm known as Evolution Strategies was used to determine the solution that offers the desired performance at the lowest cost. Furthermore, Wang and Li (2023) [164], examining composite structures consisting of CFT columns and H-shaped steel beams, compared six beam–column connections for collapse resistance during a middle-column loss scenario. While the connections showed minimal differences under normal conditions, they vary greatly under extreme loads. A design concept to improve collapse capacity was proposed and validated through finite element analysis. Characteristic results using ABAQUS software (https://www.3ds.com/products/simulia/abaqus) are shown in Figure 9.

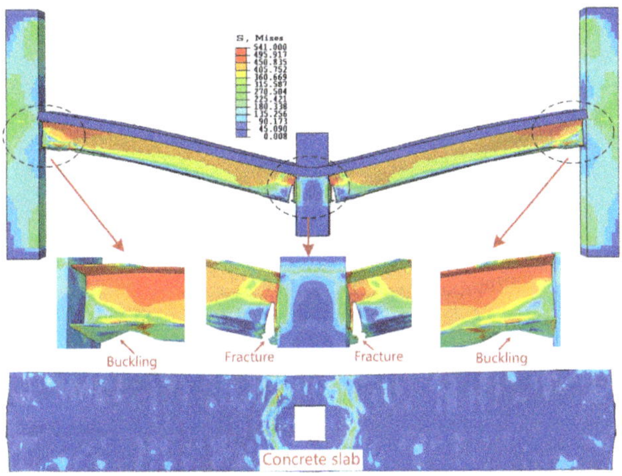

Figure 9. Finite element model of structural connections analyzing collapse resistance mechanisms in composite structural systems (adapted from Wang and Li (2023) [164]).

Bai et al. (2017) [165] discussed the seismic collapse capacity for the case of high-rise CFT moment-resisting frames subjected to extreme earthquakes beyond design levels. Ground motions with a flat velocity spectral shape were selected to minimize record-to-record uncertainty. A numerical approach using fiber elements with stiffness and strength degradation in stress–strain models was developed. Incremental dynamic analyses were conducted to assess the P-Delta and degradation effects on local and global collapse mechanisms. The results showed drift concentration at lower stories triggering side-sway collapse controlled by post-buckling strength deterioration of CFT columns. Moreover, Wang et al. (2020) [23] concentrated on the ability of composite building structures to withstand accidental loads, such as earthquakes, fires, or explosions, by depending on the beam and catenary mechanisms of steel beams to prevent collapse. They also examined a concrete-filled square steel tubular column attached to a steel beam with a bolted–welded hybrid joint in a scenario where the central column was eliminated. They computed the collapse resistance and mechanism under vertical loads, demonstrating that the steel beam can still provide resistance even after the failure of a short-span beam. They recommended the utilization of a welded haunch joint to enhance collapse resistance, offering valuable insights for engineering design to avert progressive collapse.

A numerical simulation method was proposed by Zheng and Wang (2022) [166] to integrate multi-scale element modeling with simplified beams and columns, reducing modeling elements and increasing computing efficiency. This method allowed for the accurate simulation of material damage and fractures in critical areas. Testing on CFST column composite beam frames showed that the new simplified multi-scale modeling approach was efficient and accurate. Different column-loss scenarios and the number of stories were studied to improve the composite frame's resistance to progressive collapse. Two improvement alternatives were proposed and analyzed to strengthen the frame's capacity. Wang et al. (2021) [167] simulated the hysteretic behavior on a single-story, single-span frame with concrete-filled steel tube columns and prestressed concrete-encased steel beams. Parameters for stiffness degradation, strength degradation, and pinching behavior were determined. An analytical model was developed for out-jacketing frames with single-span, multi-story layouts. Seismic response analysis was conducted for different seismic fortification levels, with the results showing that certain frames needed improvements for seismic resistance. The mechanical behavior and collapse potential of out-jacketing frames were determined from the findings of the nonlinear time-history analysis, leading to the development of pertinent seismic design recommendations.

10.2. Progressive Collapse Resistance of Composite Structures: Key Findings

Seismic loads pose a significant risk for progressive collapse in steel–concrete composite structures, making it a critical concern in structural engineering. Research has demonstrated that composite buildings generally outperform traditional steel or concrete structures due to their increased stiffness and strength. However, the interaction between steel and concrete components can result in intricate failure mechanisms. It has been found that proper detailing of connections, particularly beam-to-column joints, is essential for preventing localized failures that could initiate progressive collapse. The utilization of composite action can greatly enhance the overall system ductility and energy dissipation capacity, thereby reducing the likelihood of unbalanced failure. Furthermore, recent studies have emphasized the importance of considering dynamic amplification effects during seismic events, as they can exacerbate the potential for progressive collapse. Moreover, incorporating redundancy and alternate load paths in the design process has been shown to bolster the resilience of composite structures against progressive collapse under seismic loads.

11. Optimal Design Strategies for Earthquake-Resistant Composite Buildings

11.1. Optimal Design Strategies for Earthquake-Resistant Composite Buildings: State-of-the-Art

Achieving the optimal design of earthquake-resistant composite buildings involves a complex interplay of structural engineering principles, material science, and seismic analysis. Strategies for designing cost-effective steel–concrete composite structures aim to strike a balance between safety, performance, and economic efficiency. These strategies typically focus on optimizing the distribution of steel and concrete elements to maximize strength and ductility while minimizing material usage. Advanced analysis techniques, including nonlinear dynamic analysis and performance-based design, are often utilized to evaluate and enhance the seismic response of composite buildings. The ultimate goal is to create structures that can withstand earthquake forces while maintaining occupant safety and minimizing structural damage, all within reasonable cost constraints.

Papavasileiou et al. (2020) [168] compared the cost-effectiveness of three seismic retrofit methods for framed structures with composite columns that did not meet building codes. The first two methods involved strengthening individual composite columns with reinforced concrete jackets or steel cages. The third method upgraded the building frame by adding steel bracings. A specialized optimization procedure was used to minimize retrofit material costs while meeting the design requirements. Thirty cases of under-designed composite buildings were analyzed, revealing the most cost-effective retrofit approach for different conditions. Similarly, Papavasileiou and Charmpis (2020) [169] examined the cost-effectiveness of seismic buildings with pure steel or steel–concrete columns using a structural optimization process for unbiased comparison. An Evolution Strategies algorithm minimized material cost while meeting safety provisions from Eurocodes 4 and 3 and considering seismic behavior. The results from 154 optimization runs suggested the benefits of using concrete to partially replace steel in column designs for earthquake-resistant structures.

Moreover, Papavasileiou and Charmpis (2016) [170] focused on optimizing the design of earthquake-resistant multi-story composite buildings with steel–concrete columns. The composite columns consist of steel members fully encased in concrete, along with steel beams and optional steel bracings. The goal was to minimize materials costs while meeting design code constraints, including Eurocodes. The optimization procedure included constraints on member capacity, inter-story drifts, and top-story displacements, as well as limits on the fundamental period. The Evolution Strategies algorithm was used to solve the optimization problem and was linked with structural analysis software for evaluation purposes.

Lin and Zhang (2023) [171] introduced a practical method for life-cycle cost analysis for composite structures, combining the economic loss rate with probabilistic seismic demand modeling. A framework using a multi-objective cuckoo search algorithm was proposed for seismic design optimization. An eight-story composite frame prototype was used to compare design alternatives. The over-strength factor was identified as critical for construction, seismic damage, and life-cycle costs. Improved fiber models were developed to evaluate spatial composite effect on designs using the OpenSees software (https://opensees.berkeley.edu/). It was found that certain design parameters, like over-strength factor and flexibility, significantly affected costs.

Moreover, Kamaris et al. (2016) [172] developed a seismic optimization framework for composite structures through an extensive parametric study of steel–concrete composite moment-resisting frames with steel I-beams and CFT columns. Based on thousands of nonlinear dynamic analyses under various ground motions, they established empirical expressions for estimating maximum seismic damage using four damage indices. Their optimization process considered multiple parameters, including story count, beam strength ratio, and material properties, to develop efficient damage prediction formulas. The optimization procedure is shown in Figure 10.

STEP 1: Identification of performance level (i.e., level of damage)

STEP 2: Determining sections of the structural members

STEP 3: Evaluation of damage according to the proposed methodology

STEP 4: Has the desired level of performance been attained?

IF (YES) go to STEP 5 - IF (NO) return to STEP 2

STEP 5: FINISH

Figure 10. Optimization strategies for composite frames using advanced damage control procedures, illustrating performance enhancement methodologies (adapted from Kamaris et al. (2016) [172]).

Kaveh et al. (2022) [173] introduced a performance-based optimization design for steel–concrete composite MRFs (moment-resisting frames) using a Chebyshev chaotic map-based optimization algorithm. The study focused on 8-story and 20-story frames to minimize total weight and enhance seismic performance. The process included obtaining top designs, nonlinear pushover analysis, fragility curve plotting, and damage-margin ratio calculation to identify the optimal design for each frame, highlighting improved efficiency in weight reduction and seismic behavior assessment.

11.2. Optimal Design Strategies for Earthquake-Resistant Composite Buildings: Key Findings

Recent investigations have indicated that optimized composite systems frequently outperform traditional single-material structures in both cost-effectiveness and seismic resistance. The strategic arrangement of steel and concrete, in conjunction with innovative connection designs, can significantly enhance a building's ability to dissipate energy during earthquakes. Additionally, research has underscored the importance of considering the entire life cycle of the composite structure, including construction, maintenance, and potential retrofitting costs, in the optimization process. These findings highlight the potential for composite structures to offer superior seismic performance while achieving cost savings, making them an attractive option for earthquake-prone areas.

12. Seismic Performance of Steel–Concrete Composite Bridges

12.1. Seismic Performance of Steel–Concrete Composite Bridges: State-of-the-Art

Composite bridges, which integrate steel and concrete elements, have gained popularity in modern infrastructure due to their effectiveness and adaptability. These bridges capitalize on the advantages of both materials, with steel providing tensile strength and concrete offering compressive resistance. When exposed to seismic loads, composite bridges display intricate behavior that necessitates thorough analysis and design considerations. The interaction between steel and concrete components under dynamic loading can result in distinct stress distributions and energy dissipation mechanisms. Factors, such as shear connectors, deck–girder interactions, and material nonlinearities, significantly influence the overall seismic performance of these bridges. Understanding the behavior of composite structures during earthquake conditions is crucial for developing resilient design approaches and ensuring the safety of bridge infrastructure in seismically active areas. Ongoing research in this field aims to enhance analytical models, refine experimental techniques, and improve design guidelines to optimize the seismic performance of composite bridges.

Lin et al. (2020) [174] introduced a composite rigid-frame bridge that combines a steel–concrete composite box girder and concrete-filled double-skin steel tube piers with rigid connecting joints, showing improved static and dynamic performance over traditional bridges. Shake table tests of a scale model were conducted to assess seismic behavior and damage patterns. The results identified damage mainly at the upper and lower ends of the composite piers, with enhanced responses under near-fault ground motions. Numerical modeling confirmed the bridge's superior seismic performance, particularly in near-fault earthquakes. Furthermore, Paolacci et al. (2018) [175] examined short-to-medium-span composite I-girder bridges, which are gaining popularity due to their reduced construction time and costs, as well as their suitability for seismic areas. The cyclic behavior of pier-to-deck joints using concrete crossbeams was analyzed and experimental tests showed good seismic performance of the joints, with different typologies tested.

Zhou et al. (2020) [176] examined the behavior of dumbbell steel tube-confined reinforced concrete piers. These composite piers, which are shown in Figure 11, enhance strength and ductility compared to traditional reinforced concrete piers. Five composite piers and one reference dumbbell-reinforced concrete (RC) pier under pseudo-static loading revealed improved performance due to the confinement and enhancement effect of the steel tube. Seismic performance slightly improved with higher axial load ratios and wider webs. A simplified model accurately predicted lateral resistance, showing promising results for improving the performance of RC piers.

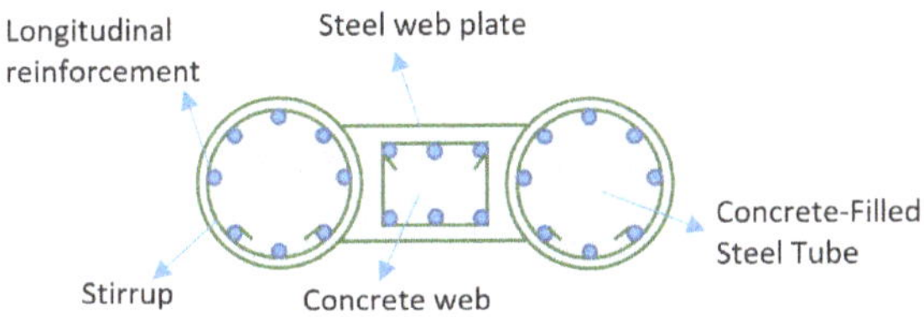

Figure 11. Dumbbell steel tube confined reinforced concrete pier design, highlighting innovative structural reinforcement techniques (adapted from Zhou et al. (2020) [176]).

Zhou et al. (2022) [177] introduced an enhanced type of concrete-encased column connections for circular CFT piers. They found that the new design eliminated the need for baseplates and anchor bolts. Experimental testing on four concrete-encased column connection specimens showed good seismic performance, with flexural failure and high-energy dissipation capacity. Calculations for strength degradation and load transfer mechanisms matched the test results well. Furthermore, Du et al. (2023) [178] examined the behavior of posttensioned precast segmental bridge piers under three-directional movements. They used shaking table tests to study an innovative posttensioned precast segmental pier under tri-directional ground motions and their results showed that adding more prestressing tendons or energy-dissipating bars reduced displacement responses. Piers with energy-dissipating bars resisted twisting better than those without. They also compared the effects of vertical ground motion on seismic behavior, finding a significant impact on compressive force and shear resistance.

Gu et al. (2024) [179] delved into the seismic performance and potential applications of continuous beam bridges with prefabricated concrete-filled steel tubular piers. Initially, two prototype bridges were compared, one with 10 m double-column reinforced concrete piers and another with 35 m double-column reinforced concrete piers. Then, seven bridges were designed with prefabricated concrete-filled steel tube piers to match the lateral stiffness of the reinforced concrete piers. The hysteretic characteristics of composite piers were verified through numerical models and seismic response analysis. Fragility curves were generated through time-history analyses.

Xiang et al. (2023) [180] proposed a fiber-based nonlinear modeling approach for the seismic analysis of partially concrete-filled steel tubular bridge piers, addressing computation time and convergence issues by considering probable local buckling of the outer steel tube and efficiently captured composite pier behavior. Comparing numerical modeling with experimental data showed the effectiveness of the fiber models in seismic assessments. Thorough cyclic pushover and time-history analyses highlighted the importance of coordinating steel tube and concrete design parameters to improve seismic performance. A new design approach involving layered concrete filling was recommended for enhancing structural seismic performance and was especially useful for retrofitting existing hollow steel piers without increasing foundation forces. Furthermore, a prefabricated steel tube-confined concrete circular pier with a grouted sleeve connection was developed by Fu et al. (2022) [181]. This composite pier could address limitations in typical precast concrete piers, enhancing their use in seismic hazard zones. Scale models of circular piers were tested under cyclic loading, showing tube-confined concrete circular piers had improved flexural strength and energy dissipation due to steel tube confinement. Ductility and stiffness also increased, with a more even distribution of pier body stiffness. Parametric analysis revealed thicker and stronger steel tubes increased flexural strength and stiffness.

Li et al. (2023) [182] examining hollow bridge piers with concrete-filled double-skin tubular sections evaluated their seismic design and cost-effectiveness. The results showed that these composite piers had higher displacement ductility, lateral force-bearing capacity, cost-performance ratio, and flexural stiffness with reduced residual drift ratio. Certain factors, such as the diameter-to-thickness ratio of outer steel tubes and hollow ratio, significantly influenced the seismic performance of composite piers and recommendations include a diameter-to-thickness ratio between 60 and 150 for effective and ductile seismic design of composite bridge piers. Similarly, Li et al. (2023) [183] executed experimental tests comparing posttensioned precast segmental concrete-filled double-skin steel tubular piers with cast-in-place concrete-filled double-skin steel tubular piers, showing superior ductility, energy dissipation, and minimal damage to precast segments. The numerical analysis explored design parameters' impact on pier behavior, revealing the crucial roles of energy-dissipating bars, gravity loads, and prestressing loads in energy dissipation and self-centering capacities. Initial stress levels in prestressing strands chiefly influenced failure mode and deformation capacity. Lin et al. (2023) [184] investigated the seismic behavior of steel–concrete composite rigid-frame bridges with posttensioned precast segmental concrete-filled double-skin steel tube piers under across-fault ground motions. Figure 12 depicts the positional relationship between the bridge and the seismic faults.

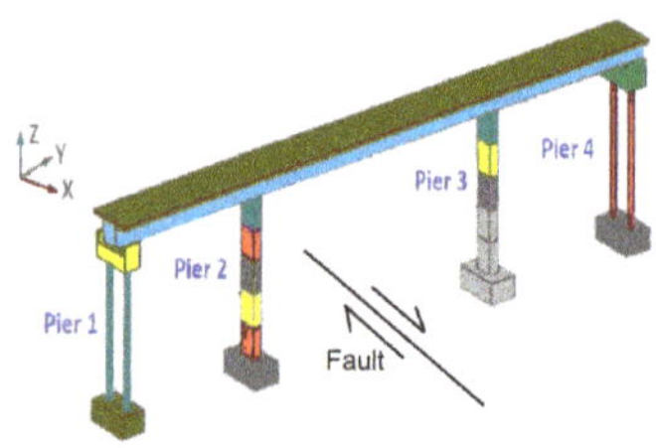

Figure 12. Spatial relationship analysis between composite bridge infrastructure and seismic fault lines, demonstrating critical geological risk assessment approach (adapted from Lin et al. (2023) [184]).

A numerical model was developed using LS-Dyna software (https://lsdyna.ansys.com/), validated by previous tests. Two types of ground motions were considered, with several fling steps. The study compared the damages, joint deformations, energy absorption, and seismic responses of the bridges and found that during strike-slip and dip-slip across-fault ground motions, posttensioned precast segmental piers sustained less damage compared to monolithic piers. Monolithic piers exhibited damage along the entire pier, while posttensioned precast segmental piers experienced localized damage at segment-to-segment joints. Under transverse strike-slip ground motions, posttensioned precast segmental piers showed excellent self-centering performance with minimal residual lateral deformations. Posttensioned precast segmental piers could accommodate displacements through torsion slip at joints and showed better seismic resilience and post-earthquake recovery performance compared to monolithic piers.

A new type of pier has been developed by Yan et al. (2024) [185] using a composite material made of glass fiber-reinforced polymer, steel, and rubber concrete, featuring a double-skin design. Two 1:4-scale specimens were created for an investigation, where the first specimen underwent a mixed loading test involving pseudo-dynamic loading and quasi-static loading. This loading simulated the phenomenon where, after an earthquake, it is common to experience multiple aftershocks, emphasizing the need to assess the seismic performance of newly constructed composite piers. On the other hand, the second specimen was exclusively subjected to a quasi-static test. It is also found that composite piers exhibited elastic behavior in frequent seismic events, with glass fiber fracture and steel tube yielding in infrequent events. Despite seismic damage, specimens still showed energy dissipation, capacity, and ductility. A correction coefficient was proposed for calculating the plastic hinge length due to the underestimation of the plastic deformation capacity in existing specifications.

Zeng et al. (2024) [186] proposed a design method for normal strength and ultra-high performance concrete composite bridge piers, optimizing ultra-high-performance concrete jacket thickness and pier diameter. They suggested that the jacket thickness should be 0.2–0.3 times the pier radius, while the composite pier diameter should be 0.8 times that of an equivalent reinforced concrete pier. The fragility analysis showed that composite piers reduced damage probabilities under small-to-moderate earthquakes across all damage states. For strong earthquakes, they effectively controlled extensive damage but were less effective for slight and moderate damage levels. Similarly, Zeng et al. (2023) [187] investigated ultra-high-performance concrete and normal strength concrete composite bridge piers under cyclic loading, finding that the ultra-high-performance concrete cover effectively prevented compressive crushing and enhanced load-carrying capacity compared to conventional reinforced concrete piers. The composite piers generally failed due to reinforcement fracturing at the bottom. Cracks developed quickly in the ultra-high-performance concrete cover, indicating a need for reinforcement. The ultra-high-performance concrete cover significantly increased the balanced axial load of the composite section, influencing failure modes and load-carrying mechanisms. An optimal axial load ratio of 0.1–0.15 was recommended for composite piers to balance performance and cost-effectiveness.

Wu et al. (2018) [188] examined the shear strength of reinforced concrete bridge piers embedded with steel tubes and braces. By conducting tests on seven specimens, they found that including steel tubes and braces greatly improves the shear resistance and flexibility of the piers compared to conventional reinforced concrete. The combined piers exhibited a higher capacity to bear loads, better distribution of cracks, and more gradual failure patterns. A mathematical model was formulated to estimate the shear strength, and it demonstrated a close correlation with the results obtained from experiments.

12.2. Seismic Performance of Steel–Concrete Composite Bridges: Key Findings

The behavior of composite bridges under seismic loads is complex due to the interaction between steel and concrete elements. Generally, these bridges outperform traditional ones by offering increased strength, ductility, and energy dissipation capabilities. The composite act between concrete and steel enhances the overall stiffness of the structure and reduces deformations, resulting in better seismic resistance. Concrete-filled steel tube piers are known for their excellent ductility and energy absorption, which help to minimize earthquake damage. When designing composite bridges for seismic regions, the focus is on capacity design principles to ensure the formation of plastic hinges in designated areas while maintaining elastic behavior elsewhere. Proper detailing of connections between steel and concrete components is essential to prevent premature failure. Advanced analysis techniques, such as pushover analysis and time-history analysis, are necessary for accurately predicting the seismic response of composite bridges and optimizing their design for enhanced earthquake resilience.

13. Smart Materials and Sensors

Recent advancements in smart materials and sensing technologies offer promising opportunities to enhance the seismic performance of steel–concrete composite structures [189]. Smart materials, such as shape memory alloys and magnetorheological fluids, can provide adaptive responses to seismic loads, while modern sensors enable real-time monitoring of structural behavior during earthquakes [190]. These technologies can potentially revolutionize how composite structures respond to and recover from seismic events. When integrated with traditional composite construction methods, smart materials, and sensors can facilitate more resilient structural systems, offering capabilities, such as self-centering, enhanced energy dissipation, and immediate post-earthquake damage assessment (Shen et al., 2024) [191]. The implementation of these innovations represents a significant step toward developing more intelligent and responsive seismic-resistant structures.

Current research demonstrates the potential of piezoelectric materials, fiber optic sensors, and wireless sensor networks in revolutionizing structural health monitoring for composite structures. Zhang et al. (2018) [192] found that these systems can detect and quantify damage in real-time, enabling prompt post-earthquake assessment and informed decision-making regarding building occupancy and required repairs. Chen et al. (2020) [193] found that embedded sensors could monitor key parameters, such as strain, displacement, and acceleration, providing valuable data for both immediate safety evaluations and long-term performance analysis. The integration of these sensing technologies with machine learning algorithms further enhances their capability to predict structural behavior and identify potential vulnerabilities before they become critical (Hu, 2015) [194].

Zhang et al. (2018) [195] examined the internal concrete damage in L-shaped concrete-filled steel tube (L-CFST) columns using piezo-ceramic smart aggregates under low-frequency cyclic loading. Wavelet packet analysis established the damage index. Experimental results showed the feasibility of using smart aggregates to monitor concrete damage in L-CFST columns.

Zhang et al. (2021) [196] found that the practical implementation of smart materials in composite structures has shown promising results in experimental studies. Chen et al. (2015) [197] reviewed a variety of smart materials' connection details to concrete-filled rectangular tubular (CFRT) columns that different researchers have developed. Shape memory alloys incorporated into beam–column connections have demonstrated superior energy dissipation and self-centering capabilities compared to conventional solutions. Thus, Hayashi et al. (2018) [198] and Zhong (2023) [199] examined the application of self-centering principles on steel–concrete composite structures, where they found that the self-centering composite frames have reduced permanent deformations while the initial posttensioned forces can control their uplifting force.

Magnetorheological dampers have been successfully employed to provide adaptive damping in response to varying seismic intensities. However, challenges remain in scaling these technologies for widespread application, including cost considerations, long-term durability, and the need for standardized design guidelines [200]. Addressing these obstacles requires collaborative efforts between researchers, practitioners, and regulatory bodies to develop comprehensive frameworks for the effective integration of smart technologies into composite structural systems [201].

14. Towards Next-Generation Composite Structural Systems: Research Perspectives

14.1. Emerging Vibration Protection Strategies

Recent developments in friction dampers and smart controllers offer intriguing insights into advanced structural protection mechanisms. Velychkovych et al. [202] proposed a novel friction damper utilizing an open shell with a helical cut and a deformable filler. This design leverages the bending effects of the shell and frictional interactions to achieve superior damping characteristics. Velychkovych et al. [202] predicted potential applications in earthquake-resistant structures, particularly in the energy and construction industries.

Smart controllers are emerging as powerful tools for vibration assessment and mitigation. A notable example is the Smart 4 controller, which demonstrates remarkable capabilities in monitoring vibration loads [203]. In drilling applications, this technology has shown potential for real-time vibration assessment, enabling operators to adjust drilling modes and prevent potential equipment damage. While initially developed for industrial contexts, such smart monitoring approaches hold promise for structural health monitoring in seismic-sensitive environments.

The development of shell-shock absorbers presents another innovative approach to vibration management. These devices, characterized by parallel spring and friction modules, exhibit exceptional damping properties under high operational loads [204]. The unique design allows for energy dissipation through structural hysteresis, a mechanism that could be particularly relevant in seismic engineering applications.

Furthermore, advanced modeling techniques are enhancing our understanding of vibration protection mechanisms. Dutkiewicz et al. [205] have developed sophisticated mechanical–mathematical models that can predict the behavior of damping systems under various loading conditions. These models, which analyzed contact interactions and energy dissipation, provide valuable insights into the design and optimization of vibration mitigation technologies.

While the aforementioned innovations represent exciting developments in vibration protection, it should be acknowledged that further research is necessary to fully integrate such technologies into steel–concrete composite structural systems. The potential for interdisciplinary collaboration between materials science, mechanical engineering, and structural design remains significant.

14.2. 3D Printing Techniques

Three-dimensional printing techniques represent a revolutionary approach in composite structural engineering, offering unprecedented opportunities for complex geometrical configurations and optimized material distributions. Advanced additive manufacturing methods enable the creation of intricate structural components with precisely controlled material properties, potentially transforming traditional fabrication limitations in steel–concrete composite systems. Current research is exploring multi-material printing strategies that can integrate metallic and cementitious materials with enhanced microstructural characteristics, allowing for localized property modulation and improved load transfer mechanisms [206–208]. Computational design optimization coupled with 3D printing technologies permits the development of novel connection geometries that could significantly enhance seismic energy dissipation capabilities [209–212]. Preliminary experimental studies suggest that 3D-printed composite connections might offer superior ductility and strength characteristics compared to conventional fabrication methods, though extensive validation is required [213–215]. The primary research challenges include achieving consistent material interfaces, ensuring long-term structural integrity, and developing standardized manufacturing protocols that meet rigorous engineering performance criteria [216–218].

14.3. Self-Healing Material Technologies

Self-healing material technologies present a transformative potential for enhancing the resilience and durability of composite structural systems, particularly in seismically active regions [219–222]. Innovative approaches involving embedded healing agents, bacterial precipitation, and advanced polymer-based autonomous repair mechanisms offer promising strategies for mitigating progressive damage accumulation during cyclic loading conditions [223–225]. Experimental investigations have demonstrated that strategically incorporated healing agents can effectively restore material integrity by autonomously sealing microcracks and preventing potential structural degradation [226,227]. Emerging research focuses on developing sophisticated self-healing concrete formulations that can respond dynamically to mechanical stress, potentially extending the service life of critical structural components [228,229]. The integration of nanoengineered healing capsules, biological agents, and responsive polymeric networks allows for targeted healing interventions at the microstructural level [230,231].

Despite significant technological promise, substantial research challenges remain, including long-term performance validation, the scalability of healing mechanisms, and a comprehensive understanding of healing efficiency under complex loading scenarios, especially for the case of steel–concrete composite structures under seismic loading conditions.

14.4. Machine Learning-Based Models

Recent advancements in machine learning have demonstrated significant potential for predicting the seismic performance of steel–concrete composite structures [1–6]. Tang et al. [79] utilized a random forest with firefly algorithm optimization to estimate the seismic performance of recycled aggregate concrete-filled steel tube columns, achieving high correlation coefficients. Zhang et al. [150] developed ensemble machine learning models for bond stress estimation, employing Bayesian optimization and SHAP interpretability techniques to enhance predictive accuracy. Li et al. [15] introduced a multi-indicator percussion method integrated with machine learning, achieving an exceptional 99.6% accuracy in interfacial debonding detection using key damage indices. Guo et al. [149] applied an extremely randomized trees algorithm to analyze damage sensitivity in steel–concrete composite beam bridges, identifying microcracks in steel beams as the most critical performance factor. Li et al. [15] developed a backpropagation neural network for predicting high-strength bolt connector strengths, achieving an over 93% goodness of fit. Yang et al. [6] proposed a novel blending fusion model with generative adversarial network augmentation for identifying failure modes of steel tube-reinforced concrete shear walls, improving prediction accuracy by an average of 13–81% across various metrics.

In conclusion, machine learning-based approaches represent a transformative paradigm in seismic performance prediction, offering unprecedented accuracy, interpretability, and insights into the complex behavior of steel–concrete composite structures under extreme loading conditions.

15. Conclusions and Future Research Directions

This comprehensive review has explored the seismic performance of steel–concrete composite structures through multiple perspectives, from component-level behavior to system-wide response strategies. The synthesis of current research reveals both significant advances and critical research gaps that warrant further investigation.

Properly detailed beam–column joints demonstrate excellent seismic performance, though complex stress distributions at interfaces remain challenging to predict. Future research should develop standardized connection details optimized for varying seismic hazard levels. For composite beams and columns, concrete significantly improves energy dissipation and delays local buckling, with the research needs focused on quantifying long-term performance under repeated seismic events. Composite shear walls offer superior lateral load resistance compared to conventional systems. System-level investigations confirm that properly designed composite structures create redundancy and alternative load paths enhancing structural resilience. Future research should address the optimization of steel–concrete proportions for different building heights and integrate performance-based methodologies with practical construction considerations.

Modeling techniques have advanced significantly, though computational models still struggle to fully represent composite action at interfaces. Future work should develop efficient models that accurately represent material interface deterioration under cyclic loading, potentially integrating machine learning with physics-based approaches.

For foundations and bridges, composite construction enhances seismic resistance with improved load transfer characteristics and ductility. Future research should investigate soil–structure interaction under extreme events and develop optimized connection details.

Progressive collapse studies confirm that composite structures possess inherent robustness against disproportionate collapse. Seismic isolation and energy dissipation devices demonstrate significant reductions in seismic demands when integrated with composite structures. Research should develop adaptive systems responsive to varying earthquake intensities.

Optimization strategies now effectively balance material efficiency with seismic objectives, while emerging smart materials offer promising directions for adaptive structural response. Future work should incorporate life-cycle considerations and explore integration of monitoring systems with traditional composite structures.

In conclusion, while substantial progress has been made, significant challenges remain in developing design methodologies that bridge theoretical advancements with practical implementation to enhance the structural resilience of steel–concrete composite structures in seismic regions.

Author Contributions: Conceptualization, P.K. and G.H.; methodology, P.K. and G.H.; investigation, P.K., G.P. and G.H.; resources, P.K. and G.H.; writing—original draft preparation, P.K.; writing—review and editing, G.H.; visualization, P.K. and G.H.; supervision, P.K., G.P. and G.H. All authors have read and agreed to the published version of the manuscript.

Funding: This research received no external funding.

Data Availability Statement: No data were created.

Conflicts of Interest: The authors declare no conflicts of interest.

References

1. Ataei, A.; Bradford, M.A.; Liu, X. Computational modelling of the moment-rotation relationship for decon-structable flush end plate beam-to-column composite joints. *J. Constr. Steel Res.* **2017**, *129*, 75–92.
2. Van-Long, H.; Jean-Pierre, J.; Jean-François, D. Extended end-plate to concrete-filled rectangular column joint using long bolts. *J. Constr. Steel Res.* **2015**, *113*, 156–168.
3. Wang, J.; Pan, X.; Peng, X. Pseudo-dynamic tests of assembly blind bolted composite frames to CFST columns. *J. Constr. Steel Res.* **2017**, *139*, 83–100.

4. Peng, Z.; Dai, S.B.; Pi, Y.L.; Li, D.; Zhang, Y.C.; Huang, J. Seismic behaviour of innovative ring-bar rein-forced connections composed of T-shaped CFST columns and RC beams with slabs. *Thin-Walled Struct.* **2018**, *127*, 1–16.
5. Peng, Z.; Dai, S.B.; Pi, Y.L.; Zhang, Y.C.; Huang, J. Seismic performance of end-plate connections between T-shaped CFST columns and RC beams. *J. Constr. Steel Res.* **2018**, *145*, 167–183.
6. Peng, S.; Xiong, Z.; Xu, C.X.; Fei, J.; Chen, X. Research on shear strength of CFSST column and H-section with beam composite joint of unequal depth. *J. Constr. Steel Res.* **2021**, *180*, 106575.
7. Mou, B.; Liu, X.; Sun, Z. Seismic behavior of a novel beam to reinforced concrete-filled steel tube column joint. *J. Constr. Steel Res.* **2021**, *187*, 106931. [CrossRef]
8. Yu, Y.; Luo, Y.; Lin, S.; Wei, B.; Jiang, L.; Liu, Q. Experimental seismic behaviors of different connections to CFST end-columns in shear walls. *J. Constr. Steel Res.* **2022**, *196*, 107385.
9. Cheng, Y.; Yang, Y.; Teng, Y.; Yu, Y.; Chen, Y.F. Seismic performance of T-shaped CFST column to U-shaped steel-concrete composite beam joint. *J. Constr. Steel Res.* **2022**, *199*, 107564.
10. Li, J.; Mou, B.; Wang, Z. Cyclic behavior of column-to-column connections in novel prefabricated H-shaped steel beam to CFST column joint. *J. Constr. Steel Res.* **2023**, *200*, 107657.
11. Chen, Q.J.; Cai, J.; Bradford, M.A.; Liu, X.; Zuo, Z.L. Seismic behaviour of a through-beam connection be-tween concrete-filled steel tubular columns and reinforced concrete beams. *Eng. Struct.* **2014**, *80*, 24–39. [CrossRef]
12. Jeddi, M.Z.; Sulong, N.R.; Khanouki, M.A. Seismic performance of a new through rib stiffener beam connection to concrete-filled steel tubular columns: An experimental study. *Eng. Struct.* **2017**, *131*, 477–491. [CrossRef]
13. Amadio, C.; Bedon, C.; Fasan, M.; Pecce, M.R. Refined numerical modelling for the structural assessment of steel-concrete composite beam-to-column joints under seismic loads. *Eng. Struct.* **2017**, *138*, 394–409. [CrossRef]
14. Liu, J.; Yang, Y.; Liu, J.; Zhou, X. Experimental investigation of special-shaped concrete-filled steel tubular column to steel beam connections under cyclic loading. *Eng. Struct.* **2017**, *151*, 68–84. [CrossRef]
15. Li, W.; Xu, L.F.; Qian, W.W. Seismic performance of 3-D steel beam to concrete-encased CFST column joints: Tests. *Eng. Struct.* **2021**, *232*, 111793. [CrossRef]
16. Wu, C.; Liu, J.; Tan, W.; Wang, P. Seismic behavior of composite interior joints of prefabricated H-shaped steel reinforced concrete column-steel beam. In *Structures*; Elsevier: Amsterdam, The Netherlands, 2020; Volume 23, pp. 558–572.
17. Wang, J.; Zhang, H.; Jiang, Z. Seismic behavior of blind bolted end plate composite joints to CFTST columns. *Thin-Walled Struct.* **2016**, *108*, 256–269. [CrossRef]
18. Zhang, Y.; Li, D. Development and testing of precast concrete-filled square steel tube column-to-RC beam connections under cyclic loading. *Constr. Build. Mater.* **2021**, *280*, 122540. [CrossRef]
19. Wang, J.X.; Yang, Y.; Xian, W.; Li, Q.Y. Progressive collapse mechanism analysis of concrete-filled square steel tubular column to steel beam joint with bolted–welded hybrid connection. *Int. J. Steel Struct.* **2020**, *20*, 1618–1635. [CrossRef]
20. Yang, Y.; Yang, W.; Tang, X.; Cheng, Y. Seismic behavior of plane frame with special-shaped CFST columns, H-shaped steel beams and vertical rib connections. In *Structures*; Elsevier: Amsterdam, The Netherlands, 2021; Volume 31, pp. 721–739.
21. Xu, X.; Cheng, R.; Yang, P.; Chen, K.; Li, J. Behavior of T-shaped CFST column and U-shaped steel-concrete composite beam joints. *J. Build. Eng.* **2021**, *43*, 103157. [CrossRef]
22. Duan, L.; Nie, X.; Su, H.; Tan, J. Seismic Response Analysis of Steel–Concrete Composite Frame Structures with URSP Connectors. *Materials* **2022**, *15*, 8655. [CrossRef]
23. Wang, J.; Wang, W.; Guo, L. Seismic tests and nonlinear model of beam-CFDST column joints with blind fasteners. *J. Build. Eng.* **2022**, *45*, 103415.
24. Zhou, J.; Chen, Z.; Xu, W.; Liao, H.; Wang, N. Seismic behavior of 3-D beam to L-shaped concrete encased steel composite column joint: An experimental study. *J. Build. Eng.* **2022**, *52*, 104437.
25. Chen, L.; Feng, J.; Xue, Y.; Liang, C. Seismic behavior of an innovative prefabricated steel-concrete composite beam-column joint. *J. Build. Eng.* **2023**, *76*, 107211.
26. Gan, D.; Zhou, Z.; Zhou, X.; Hai Tan, K. Seismic behavior tests of square reinforced concrete–filled steel tube columns connected to RC beam joints. *J. Struct. Eng.* **2019**, *145*, 04018267.
27. Lai, Z.; Fischer, E.C.; Varma, A.H. Database and review of beam-to-column connections for seismic design of composite special moment frames. *J. Struct. Eng.* **2019**, *145*, 04019023.
28. Fan, J.; Li, Q.; Nie, J.; Zhou, H. Experimental study on the seismic performance of 3D joints between concrete-filled square steel tubular columns and composite beams. *J. Struct. Eng.* **2014**, *140*, 04014094.
29. Wu, C.; Liu, J.; Shi, W. Seismic performance of composite joints between prefabricated steel-reinforced concrete columns and steel beams: Experimental study. *Bull. Earthq. Eng.* **2020**, *18*, 3817–3841.
30. Liu, J.; He, W.; Cheng, G.; Chen, Y.F. Seismic behavior of large-diameter CFTST column to steel beam connections. *J. Constr. Steel Res.* **2020**, *175*, 106338.

31. Amadio, C.; Bedon, C.; Fasan, M. Numerical assessment of slab-interaction effects on the behaviour of steel-concrete composite joints. *J. Constr. Steel Res.* **2017**, *139*, 397–410.
32. Demonceau, J.F.; Ciutina, A. Characterisation of beam-to-column steel-concrete composite joints beyond current Eurocode provisions. In *Structures*; Elsevier: Amsterdam, The Netherlands, 2019; Volume 21, pp. 167–175.
33. Mo, J.; Uy, B.; Li, D.; Thai, H.T.; Tran, H. A review of the behaviour and design of steel–concrete composite shear walls. In *Structures*; Elsevier: Amsterdam, The Netherlands, 2021; Volume 31, pp. 1230–1253.
34. Lan, T.; Wang, X.; Cui, Y.; Liu, X.; You, Y. Analysis and application of double steel plate concrete composite shear wall in the R&D Building of Zhanjiang Bay Laboratory. *Buildings* **2023**, *13*, 3055. [CrossRef]
35. Todea, V.; Dan, D.; Florut, S.C.; Stoian, V. Experimental investigations on the seismic behavior of composite steel concrete coupled shear walls with central openings. In *Structures*; Elsevier: Amsterdam, The Netherlands, 2021; Volume 33, pp. 878–896.
36. Qiao, Q.Y.; Cao, W.L.; Li, X.Y.; Dong, H.Y.; Zhang, W.W.; Yin, F. Seismic behavior of shear walls with boundary CFST columns and embedded multiple steel plates: Experimental investigation. *Eng. Struct.* **2018**, *160*, 243–256.
37. Najm, H.M.; Ibrahim, A.M.; Sabri, M.M.; Hassan, A.; Morkhade, S.; Mashaan, N.S.; Eldirderi, M.M.A.; Khedher, K.M. Modelling of cyclic load behaviour of smart composite steel-concrete shear wall using finite element analysis. *Buildings* **2022**, *12*, 850. [CrossRef]
38. Rahnavard, R.; Hassanipour, A.; Mounesi, A. Numerical study on important parameters of composite steel-concrete shear walls. *J. Constr. Steel Res.* **2016**, *121*, 441–456.
39. Zhang, X.; Qin, Y.; Chen, Z. Experimental seismic behavior of innovative composite shear walls. *J. Constr. Steel Res.* **2016**, *116*, 218–232.
40. Hossain, K.M.; Rafiei, S.; Lachemi, M.; Behdinan, K. Structural performance of profiled composite wall under in-plane cyclic loading. *Eng. Struct.* **2016**, *110*, 88–104.
41. Polat, E.; Bruneau, M. Modeling cyclic inelastic in-plane flexural behavior of concrete filled sandwich steel panel walls. *Eng. Struct.* **2017**, *148*, 63–80.
42. Zhao, W.; Guo, Q.; Huang, Z.; Tan, L.; Chen, J.; Ye, Y. Hysteretic model for steel–concrete composite shear walls subjected to in-plane cyclic loading. *Eng. Struct.* **2016**, *106*, 461–470.
43. Wang, M.Z.; Guo, Y.L.; Zhu, J.S.; Yang, X.; Tong, J.Z. Sectional strength design of concrete-infilled double steel corrugated-plate walls with T-section. *J. Constr. Steel Res.* **2019**, *160*, 23–44.
44. Luo, Q.; Wang, W.; Sun, Z.; Xu, S.; Wang, B. Seismic performance analysis of corrugated-steel-plate composite shear wall based on corner failure. *J. Constr. Steel Res.* **2021**, *180*, 106606.
45. Haghi, N.; Epackachi, S.; Kazemi, M.T. Macro modeling of steel-concrete composite shear walls. In *Structures*; Elsevier: Amsterdam, The Netherlands, 2020; Volume 23, pp. 383–406.
46. Shafaei, S.; Varma, A.H.; Broberg, M.; Klemencic, R. Modeling the cyclic behavior of composite plate shear walls/concrete filled (C-PSW/CF). *J. Constr. Steel Res.* **2021**, *184*, 106810.
47. Chen, L.; Mahmoud, H.; Tong, S.M.; Zhou, Y. Seismic behavior of double steel plate–HSC composite walls. *Eng. Struct.* **2015**, *102*, 1–12. [CrossRef]
48. Epackachi, S.; Whittaker, A.S.; Varma, A.H.; Kurt, E.G. Finite element modeling of steel-plate concrete composite wall piers. *Eng. Struct.* **2015**, *100*, 369–384. [CrossRef]
49. Huang, Z.; Liew, J.R. Compressive resistance of steel-concrete-steel sandwich composite walls with J-hook connectors. *J. Constr. Steel Res.* **2016**, *124*, 142–162. [CrossRef]
50. Hu, H.S.; Nie, J.G.; Eatherton, M.R. Deformation capacity of concrete-filled steel plate composite shear walls. *J. Constr. Steel Res.* **2014**, *103*, 148–158. [CrossRef]
51. Huang, Z.; Zhang, S.; Guo, L. Cyclic behavior of low shear-span ratio dovetailed profiled steel–concrete composite shear walls. *J. Build. Eng.* **2022**, *59*, 105094. [CrossRef]
52. Wang, S.; Wang, W.; Chen, Y.; Hou, H.; Xie, S. Seismic performance of double skin composite wall under maximum considered far-field earthquake loading protocol. *J. Build. Eng.* **2022**, *55*, 104705. [CrossRef]
53. Ji, J.; Zhang, H.; Jiang, L.; Zhang, Y.; Yuan, Z.; Zhang, Z.; Chu, X.; Li, B. Seismic Behaviors of Novel Steel-Reinforced Concrete Composite Frames Prestressed with Bonding Tendons. *Buildings* **2023**, *13*, 2124. [CrossRef]
54. Kenarangi, H.; Kizilarslan, E.; Bruneau, M. Cyclic behavior of c-shaped composite plate shear walls–concrete filled. *Eng. Struct.* **2021**, *226*, 111306.
55. Serras, D.N.; Skalomenos, K.A.; Hatzigeorgiou, G.D.; Beskos, D.E. Modeling of circular concrete-filled steel tubes subjected to cyclic lateral loading. In *Structures*; Elsevier: Amsterdam, The Netherlands, 2016; Volume 8, pp. 75–93.
56. Sahin, B.; Bravo-Haro, M.A.; Elghazouli, A.Y. Assessment of cyclic degradation effects in composite steel-concrete members. *J. Constr. Steel Res.* **2022**, *192*, 107231.
57. Liu, X.; Bradford, M.A.; Chen, Q.J.; Ban, H. Finite element modelling of steel–concrete composite beams with high-strength friction-grip bolt shear connectors. *Finite Elem. Anal. Des.* **2016**, *108*, 54–65.

58. Katwal, U.; Tao, Z.; Hassan, M.K. Finite element modelling of steel-concrete composite beams with profiled steel sheeting. *J. Constr. Steel Res.* **2018**, *146*, 1–15. [CrossRef]
59. Peng, S.; Wang, J.; Yao, Y.; Min, T.; Li, C.; Luo, L.; Zhu, Y.; Wen, X.; Zhou, M. A modified Park-Ang model for seismic damage assessment of recycled aggregate concrete-filled square steel tube columns. *Case Stud. Constr. Mater.* **2022**, *17*, e01692. [CrossRef]
60. Sahin, B.; Bravo-Haro, M.A.; Elghazouli, A.Y. Modelling and response of composite steel-concrete members under cyclic loading. *Ce/Papers* **2023**, *6*, 160–166. [CrossRef]
61. El Jisr, H.; Kohrangi, M.; Lignos, D.G. Proposed nonlinear macro-model for seismic risk assessment of compo-site-steel moment resisting frames. *Earthq. Eng. Struct. Dyn.* **2022**, *51*, 1180–1200. [CrossRef]
62. Papavasileiou, G. Analytical framework for the substitution of steel-concrete composite columns with equivalent steel columns in structural design. *ESR J.* **2017**, *2*.
63. Skalomenos, K.A.; Hatzigeorgiou, G.D.; Beskos, D.E. Modeling level selection for seismic analysis of concrete-filled steel tube/moment-resisting frames by using fragility curves. *Earthq. Eng. Struct. Dyn.* **2015**, *44*, 199–220. [CrossRef]
64. Lin, Y.; Zhang, X. The slab effect of composite frames: From modified fiber modeling to seismic fragility assessment. *J. Constr. Steel Res.* **2021**, *186*, 106877. [CrossRef]
65. Fa-xing, D.; Jing, L.; Xue-mei, L.; Zhi-wu, Y.; Yong-suo, L. Experimental investigation on hysteretic behavior of simply supported steel-concrete composite beam. *J. Constr. Steel Res.* **2018**, *144*, 153–165. [CrossRef]
66. Han, L.H.; An, Y.F.; Roeder, C.; Ren, Q.X. Performance of concrete-encased CFST box members under bending. *J. Constr. Steel Res.* **2015**, *106*, 138–153. [CrossRef]
67. Suzuki, A.; Kimura, Y. Cyclic behavior of component model of composite beam subjected to fully reversed cyclic loading. *J. Struct. Eng.* **2019**, *145*, 04019015. [CrossRef]
68. Zhu, L.; Ning, Q.; Han, W.; Zhao, C. Seismic performance of self-centering concrete-filled square steel tubular column-to-steel beam connection equipped with SMA bolts. *Thin-Walled Struct.* **2023**, *183*, 110351.
69. Huang, Y.; Yi, W.; Zhang, R.; Xu, M. Behavior and design modification of RBS moment connections with composite beams. *Eng. Struct.* **2014**, *59*, 39–48.
70. Di Cesare, A.; Belviso, P.; Ponzo, F.C.; Vitone, G. Seismic Behavior and Modeling of Ductile Composite Steel-Trussed Concrete Beam to Column Joints. *Appl. Sci.* **2023**, *13*, 11139. [CrossRef]
71. Chen, Y.; Li, W.; Fang, C. Performance of partially encased composite beams under static and cyclic bending. In *Structures*; Elsevier: Amsterdam, The Netherlands, 2017; Volume 9, pp. 29–40.
72. El Jisr, H.; Elkady, A.; Lignos, D.G. Composite steel beam database for seismic design and performance assessment of composite-steel moment-resisting frame systems. *Bull. Earthq. Eng.* **2019**, *17*, 3015–3039.
73. Suzuki, A.; Kimura, Y. Mechanical performance of stud connection in steel-concrete composite beam under reversed stress. *Eng. Struct.* **2021**, *249*, 113338. [CrossRef]
74. Suzuki, A.; Hiraga, K.; Kimura, Y. Cyclic behavior of steel-concrete composite dowel by clothoid-shaped shear connectors under fully reversed cyclic stress. *J. Adv. Concr. Technol.* **2023**, *21*, 76–91.
75. Montuori, R.; Nastri, E.; Piluso, V.; Todisco, P. Experimental and analytical study on the behaviour of circular concrete filled steel tubes in cyclic bending. *Eng. Struct.* **2024**, *304*, 117610.
76. Kim, S.B.; Lee, E.T.; Kim, J.R.; Kim, S.S. Experimental study on bending behavior and seismic performance of hybrid composite beam with new shape. *Int. J. Steel Struct.* **2016**, *16*, 477–488.
77. Wang, J.T.; Wu, X.H.; Yang, B.; Sun, Q. Bearing capacity and damage behavior of hcftst columns under cyclic loading. In *Structures*; Elsevier: Amsterdam, The Netherlands, 2021; Volume 32, pp. 1492–1506.
78. Qian, W.W.; Li, W.; Han, L.H.; Zhao, X.L. Analytical behavior of concrete-encased CFST columns under cyclic lateral loading. *J. Constr. Steel Res.* **2016**, *120*, 206–220.
79. Tang, Y.C.; Li, L.J.; Feng, W.X.; Liu, F.; Liao, B. Seismic performance of recycled aggregate concrete–filled steel tube columns. *J. Constr. Steel Res.* **2017**, *133*, 112–124.
80. Ma, H.; Qiang, J.; Xi, J.; Zhao, Y. Cyclic loading tests and horizontal bearing capacity of recycled concrete filled circular steel tube and profile steel composite columns. *J. Constr. Steel Res.* **2022**, *199*, 107572. [CrossRef]
81. Ma, H.; Liu, F.; Fang, L.; Zhao, Y.; Gan, X. Cyclic loading tests and seismic behaviors of steel-reinforced recycled aggregate concrete filled square steel tube composite columns. *J. Constr. Steel Res.* **2024**, *212*, 108270. [CrossRef]
82. Luo, C.; Wang, F.; Chen, H.; Qi, A.; Chen, Y. Study on the hysteretic behavior of recycled aggregate concrete-filled steel tube columns containing ferronickel slag. *J. Build. Eng.* **2022**, *46*, 103695. [CrossRef]
83. Zeng, L.; Li, L.; Xiao, P.; Zeng, J.; Liu, F. Experimental study of seismic performance of full-scale basalt FRP-recycled aggregate concrete-steel tubular columns. *Thin-Walled Struct.* **2020**, *151*, 106185. [CrossRef]
84. Skalomenos, K.A.; Hatzigeorgiou, G.D.; Beskos, D.E. Parameter identification of three hysteretic models for the simulation of the response of CFT columns to cyclic loading. *Eng. Struct.* **2014**, *61*, 44–60. [CrossRef]

85. Zhang, J.; Li, X.; Cao, W.; Yu, C. Cyclic behavior of steel tube-reinforced high-strength concrete composite columns with high-strength steel bars. *Eng. Struct.* **2019**, *189*, 565–579. [CrossRef]
86. Zheng, Y.; Yang, S.; Lai, P. Hysteretic behavior of multi-cell L-shaped concrete-filled steel tubular columns at different loading angles. *Eng. Struct.* **2020**, *202*, 109887. [CrossRef]
87. Kim, H.J.; Park, H.G.; Hwang, H.J. Cyclic lateral loading test for composite columns with high-strength steel angle cage. *Eng. Struct.* **2022**, *250*, 113463. [CrossRef]
88. Xiong, J.; Ren, F.; Li, S.; Tian, S.; Li, Y.; Mo, J. A study on low-frequency vibration mitigation by using the metamaterial-tailored composite concrete-filled steel tube column. *Eng. Struct.* **2024**, *305*, 117673. [CrossRef]
89. Hassan, W.M.; Farag, M. Seismic performance of steel-reinforced concrete composite columns in existing and modern construction. *Soil Dyn. Earthq. Eng.* **2021**, *151*, 106945. [CrossRef]
90. Mostafa, M.M.; Wu, T.; Liu, X.; Fu, B. The composite steel reinforced concrete column under axial and seismic loads: A review. *Int. J. Steel Struct.* **2019**, *19*, 1969–1987. [CrossRef]
91. Hassan, W.M.; Reyes, J.C.; González, C.; Pallarés, F.J.; Spinel, J.S. Seismic vulnerability and resilience of steel-reinforced concrete (SRC) composite column buildings with non-seismic details. *Eng. Struct.* **2021**, *244*, 112810. [CrossRef]
92. Campian, C.; Nagy, Z.; Pop, M. Behavior of fully encased steel-concrete composite columns subjected to monotonic and cyclic loading. *Procedia Eng.* **2015**, *117*, 439–451. [CrossRef]
93. Chen, C.; Wang, C.; Sun, H. Experimental study on seismic behavior of full encased steel-concrete composite columns. *J. Struct. Eng.* **2014**, *140*, 04014024. [CrossRef]
94. Chen, Z.; Dong, S.; Du, Y. Experimental study and numerical analysis on seismic performance of FRP confined high-strength rectangular concrete-filled steel tube columns. *Thin-Walled Struct.* **2021**, *162*, 107560. [CrossRef]
95. Gautham, A.S.D.R.; Sahoo, D.R. Behavior of steel-reinforced composite concrete columns under combined axial and lateral cyclic loading. *J. Build. Eng.* **2021**, *39*, 102305. [CrossRef]
96. Serras, D.N.; Skalomenos, K.A.; Hatzigeorgiou, G.D. A displacement/damage controlled seismic design method for MRFs with concrete-filled steel tubular columns and composite beams. *Soil Dyn. Earthq. Eng.* **2021**, *143*, 106608. [CrossRef]
97. Serras, D.N.; Skalomenos, K.A.; Hatzigeorgiou, G.D.; Beskos, D.E. Inelastic behavior of circular concrete-filled steel tubes: Monotonic versus cyclic response. *Bull. Earthq. Eng.* **2017**, *15*, 5413–5434. [CrossRef]
98. Bai, Y.; Ma, X.; Wang, B.; Cao, G.; Beer, M. Cumulative Component Damages on Collapse Capacity of Ductile Steel and CFT Moment Resisting Frames under Over-design Ground Motions. *J. Earthq. Eng.* **2022**, *26*, 3012–3033. [CrossRef]
99. Wu, Z.; Nie, X.; Zhao, J.; Wang, W.; Duan, L. Numerical Study on the Seismic Behavior of Steel–Concrete Composite Frame with Uplift-Restricted and Slip-Permitted (URSP) Connectors. *Buildings* **2023**, *13*, 2598. [CrossRef]
100. Zhang, L.; Wang, C.; Chen, C.; Cui, M. Seismic Performance Assessment of Composite Frame–High-Strength Steel Plate Wall Core Tube Resilient Structural System. *Buildings* **2024**, *14*, 301. [CrossRef]
101. Zhao, H.; Tao, M.X.; Ding, R. Experimental study on seismic behaviour of composite frames with wide floor slabs considering the effect of floor loads. *Eng. Struct.* **2020**, *220*, 111024.
102. Tondini, N.; Zanon, G.; Pucinotti, R.; Di Filippo, R.; Bursi, O.S. Seismic performance and fragility functions of a 3D steel-concrete composite structure made of high-strength steel. *Eng. Struct.* **2018**, *174*, 373–383.
103. Braconi, A.; Caprili, S.I.L.V.I.A.; Degee, H.; Guendel, M.; Hjiaj, M.; Hoffmeister, B.; Karamanos, S.A.; Rinaldi, V.; Salvatore, W.A.L.T.E.R.; Somja, H. Efficiency of Eurocode 8 design rules for steel and steel-concrete composite structures. *J. Constr. Steel Res.* **2015**, *112*, 108–129.
104. Badalassi, M.; Braconi, A.; Cajot, L.G.; Caprili, S.; Degee, H.; Gündel, M.; Hjiaj, M.; Hoffmeister, B.; Karamanos, S.A.; Salvatore, W.; et al. Influence of variability of material mechanical properties on seismic performance of steel and steel–concrete composite structures. *Bull. Earthq. Eng.* **2017**, *15*, 1559–1607.
105. Etli, S. Parametric Analysis of the Performance of Steel-Concrete Composite Structures Designed with TBDY 2018. *Int. J. Innov. Eng. Appl.* **2022**, *6*, 7–16.
106. Etli, S. Performance evaluation of steel-concrete composite structures designed in poorly graded soils. *Rev. Constr.* **2023**, *22*, 259–276.
107. Denavit, M.D.; Hajjar, J.F.; Perea, T.; Leon, R.T. Seismic performance factors for moment frames with steel-concrete composite columns and steel beams. *Earthq. Eng. Struct. Dyn.* **2016**, *45*, 1685–1703.
108. Zhao, H.; Qian, X.Y.; Zhu, Y.J. Seismic behaviour of post-earthquake composite frame structures with different damage levels. In *Structures*; Elsevier: Amsterdam, The Netherlands, 2023; Volume 58, p. 105482.
109. Ding, F.X.; Yin, G.A.; Jiang, L.Z.; Bai, Y. Composite frame of circular CFST column to steel-concrete composite beam under lateral cyclic loading. *Thin-Walled Struct.* **2018**, *122*, 137–146.
110. Etli, S.; Güneyisi, E.M. Assessment of seismic behavior factor of code-designed steel–concrete composite buildings. *Arab. J. Sci. Eng.* **2021**, *46*, 4271–4292.
111. Zhao, H. Analysis of seismic behavior of composite frame structures. *Steel Compos. Struct.* **2016**, *20*, 719–729. [CrossRef]

112. Wang, K.; Lu, X.F.; Yuan, S.F.; Cao, D.F.; Chen, Z.X. Analysis on hysteretic behavior of composite frames with concrete-encased CFST columns. *J. Constr. Steel Res.* **2017**, *135*, 176–186. [CrossRef]
113. Skalomenos, K.A.; Hatzigeorgiou, G.D.; Beskos, D.E. Seismic behavior of composite steel/concrete MRFs: Deformation assessment and behavior factors. *Bull. Earthq. Eng.* **2015**, *13*, 3871–3896. [CrossRef]
114. Karthiga, S.N.; Santhosh, R.; Kannan, V. Comparison of seismic performance of reinforced concrete frame structure and composite frame structure using response spectrum analysis. In *IOP Conference Series: Materials Science and Engineering*; IOP Publishing: Bristol, UK, 2020; Volume 764, p. 012057.
115. Andreotti, R.; Giuliani, G.; Tondini, N.; Bursi, O.S. Hybrid simulation of a partial-strength steel–concrete composite moment-resisting frame endowed with hysteretic replaceable beam splices. *Earthq. Eng. Struct. Dyn.* **2023**, *52*, 51–70. [CrossRef]
116. Li, J.; Xu, L. Seismic performance improvement of continuous rigid-frame bridges with hybrid control system under near-fault ground motions. *Soil Dyn. Earthq. Eng.* **2023**, *168*, 107858. [CrossRef]
117. Hüsem, M.; Nasery, M.M.; Okur, F.Y.; Altunişik, A.C. Experimental evaluation of damage effect on dynamic characteristics of concrete encased composite column-beam connections. *Eng. Fail. Anal.* **2018**, *91*, 129–150. [CrossRef]
118. Kanyilmaz, A.; Muhaxheri, M.; Castiglioni, C.A. Influence of repairable bolted dissipative beam splices (structural fuses) on reducing the seismic vulnerability of steel-concrete composite frames. *Soil Dyn. Earthq. Eng.* **2019**, *119*, 281–298. [CrossRef]
119. Nasery, M.M.; Hüsem, M.; Okur, F.Y.; Altunişik, A.C. Damage effect on experimental modal parameters of haunch strengthened concrete-encased composite column–beam connections. *Int. J. Damage Mech.* **2020**, *29*, 297–334. [CrossRef]
120. Qi, L.; Xue, J.; Zhai, L.; Zhao, X.; Leon, R.T. Experimental and numerical studies on seismic performance of traditional style steel-concrete composite frame. *Compos. Struct.* **2018**, *201*, 514–527. [CrossRef]
121. Xue, J.; Ma, L.; Zhao, X.; Zhang, X.; Zhai, L.; Zhao, X. Investigation on steel-concrete composite frame in Chinese traditional style buildings by pseudo-dynamic test. *J. Constr. Steel Res.* **2019**, *157*, 282–296. [CrossRef]
122. Rashidi, M.; Sharafi, P.; Alembagheri, M.; Bigdeli, A.; Samali, B. Operational modal analysis, testing and modelling of prefabricated steel modules with different LSF composite walls. *Materials* **2020**, *13*, 5816. [CrossRef]
123. Zeng, L.; Xiao, Y.; Chen, Y.; Jin, S.; Xie, W.; Li, X. Seismic damage evaluation of concrete-encased steel frame-reinforced concrete core tube buildings based on dynamic characteristics. *Appl. Sci.* **2017**, *7*, 314. [CrossRef]
124. Reena, G.C.; Gurupatham BG, A.; Tsavdaridis, K.D. Column link behavior in eccentrically braced composite 3-dimensional frames. *Buildings* **2023**, *13*, 2970. [CrossRef]
125. Liang, Q.Q. *Analysis and Design of Steel and Composite Structures*; CRC Press: Boca Raton, FL, USA, 2018.
126. Skalomenos, K.A.; Hatzigeorgiou, G.D.; Beskos, D.E. Application of the hybrid force/displacement (HFD) seismic design method to composite steel/concrete plane frames. *J. Constr. Steel Res.* **2015**, *115*, 179–190. [CrossRef]
127. Gajbhiye, P.D.; Mashaan, N.S.; Bhaiya, V.; Wankhade, R.L.; Vishnu, S.P. Inelastic Behavior of Steel and Composite Frame Structure Subjected to Earthquake Loading. *Appl. Mech.* **2023**, *4*, 899–926. [CrossRef]
128. Johnson, R.P. *Composite Structures of Steel and Concrete: Beams, Slabs, Columns and Frames for Buildings*; John Wiley & Sons: Hoboken, NJ, USA, 2018.
129. Abbas, H.; Siddiqui, N.A.; Khateeb, B.M.; Almusallam, T.H.; Al-Salloum, Y.A. Performance of new CFST square column-to-foundation connections for cyclic loads. *J. Constr. Steel Res.* **2021**, *185*, 106868. [CrossRef]
130. Khateeb, B.M.; Siddiqui, N.A.; Almusallam, T.H.; Abbas, H.; Al-Salloum, Y.A. Behavior of novel CFST circular column-to-foundation connections under cyclic loading. *Eng. Struct.* **2020**, *221*, 111051. [CrossRef]
131. Serras, D.N.; Panagaki, S.D.; Skalomenos, K.A.; Hatzigeorgiou, G.D. Inelastic lateral and seismic behaviour of concrete-filled steel tubular pile foundations. *Soil Dyn. Earthq. Eng.* **2021**, *143*, 106657. [CrossRef]
132. Zhou, M.; Yuan, W.; Zhang, Y. Seismic material properties of reinforced concrete and steel casing composite concrete in elevated pile-group foundation. *Pol. Marit. Res.* **2015**, *22*, 141–148. [CrossRef]
133. Wang, N.; Zhou, F.; Xu, H.; Xu, Z. Experimental study on steel-plate composite wall-to-foundation connections subjected to combined axial compression and cyclic lateral-force. *Eng. Struct.* **2020**, *207*, 110205. [CrossRef]
134. Vakili Sadeghi, H.; Mirghaderi, S.R.; Epackachi, S.; Asgarpoor, M.; Gharavi, A. Numerical study on split base plate connection with concentric anchors between steel-plate composite wall and concrete basemat. *Struct. Des. Tall Spec. Build.* **2022**, *31*, e1937.
135. Kurt, E.G.; Varma, A.H.; Sohn, Y.M. Direct shear strength of rebar-coupler anchor systems for steel-plate composite (SC) walls. *Int. J. Steel Struct.* **2016**, *16*, 1397–1409.
136. Siddiqui, N.A.; Abbas, H.; Khateeb, B.M.; Almusallam, T.H.; Al-Salloum, Y. Optimum embedment depths for CFST column-to-foundation connections: Analytical and reliability-based approach. *J. Build. Eng.* **2024**, *91*, 109628.
137. Feng, M.; Liu, B.; Zhang, L.; Wang, Y.; Sun, H. Seismic behavior of embedded rubberized concrete-filled corrugated steel tube column-to-foundation connections: Experimental, numerical modelling, and design. *Soil Dyn. Earthq. Eng.* **2024**, *176*, 108285.
138. Li, X.; Xiao, Y.; Xu, Y.M.; Lu, J.; Ding, B.D.; Zhou, T. Structural behavior of double-CFT-pile foundations under cyclic loads. *Soil Dyn. Earthq. Eng.* **2020**, *128*, 105863.

139. Won, D.; Lee, J.; Seo, J.; Kang, Y.J.; Kim, S. Hysteretic performance of column-footing joints with steel composite hollow RC columns under cyclic load. *J. Build. Eng.* **2020**, *29*, 101165.
140. Thusoo, S.; Obara, T.; Kono, S.; Miyahara, K. Design models for steel-encased high-strength precast concrete piles under axial-flexural loads. *Eng. Struct.* **2021**, *228*, 111465.
141. Stephens, M.T.; Lehman, D.E.; Roeder, C.W. Design of CFST column-to-foundation/cap beam connections for moderate and high seismic regions. *Eng. Struct.* **2016**, *122*, 323–337.
142. Wang, W.; Lyu, X.; Zheng, J.; Yi, S.; Li, J.; Yu, Y. Post-Fire Seismic Performance of Concrete-Filled Steel Tube Frame Structures Considering Soil-Structure Interaction (SSI). *Buildings* **2024**, *14*, 555. [CrossRef]
143. Darwish, A.Q.; Bhandari, M. Seismic response reduction of high rise steel-concrete composite buildings equipped with base isolation system. *Mater. Today Proc.* **2022**, *59*, 516–524.
144. Darwish, A.Q.; Bhandari, M. Vibration response reduction of seismic forces using lead rubber bearing isolators in composite buildings. *J. Vib. Eng. Technol.* **2022**, *10*, 1309–1324. [CrossRef]
145. Li, K.; Fa, G. The structural design and earthquake resistance assessment of composite steel–concrete frame structures with welded dissipative fuses. In *Structures*; Elsevier: Amsterdam, The Netherlands, 2023; Volume 53, pp. 742–748.
146. Li, C.; Liu, Y.; Li, H.N. Fragility assessment and optimum design of a steel–concrete frame structure with hybrid energy-dissipated devices under multi-hazards of earthquake and wind. *Eng. Struct.* **2021**, *245*, 112878. [CrossRef]
147. Zhuang, L.; Wang, J.; Nie, X.; Wu, Z. Experimental study on seismic behaviour of eccentrically braced compo-site frame with vertical LYP steel shear link. *Eng. Struct.* **2022**, *255*, 113957. [CrossRef]
148. Javaid, K.; Verma, N. Seismic performance of irregular composite buildings: A comparative study of the effectiveness of buckling restrained braces and viscous dampers. *Mater. Today Proc.* 2023; *in press*. [CrossRef]
149. Guo, L.; Wang, J. Hysteretic energy distribution model of CFDST composite frame with metallic dampers using coupled shear-flexural beam model. In *Structures*; Elsevier: Amsterdam, The Netherlands, 2023; Volume 57, p. 104971.
150. Zhuang, L.D.; Zhao, J.Z. Numerical Study on the Seismic Behavior of Eccentrically Braced Composite Frames with a Vertical Low-Yield-Point Steel Shear Link. *Buildings* **2022**, *12*, 1359. [CrossRef]
151. Thakur, R.; Tiwary, A.K. Comparative study on the effectiveness of fluid viscous dampers and base isolation: An approach toward enhancing seismic performance of composite structures. *Innov. Infrastruct. Solut.* **2023**, *8*, 267. [CrossRef]
152. Zhao, J.Z.; Tao, M.X.; Zhuang, L.D. Development of bent shear panel dampers for eccentrically braced composite frames. *J. Constr. Steel Res.* **2022**, *193*, 107292.
153. Li, B.; Wang, J.; Yang, J.; Pan, X.; Baniotopoulos, C.C. Pseudo-dynamic response and analytical evaluation of blind bolted CFT frames with BRBs. *J. Constr. Steel Res.* **2020**, *166*, 105744. [CrossRef]
154. Katsimpini, P.; Papagiannopoulos, G.; Hatzigeorgiou, G. Seismic Response of a Cable-Stayed Bridge with Concrete-Filled Steel Tube (CFST) Pylons Equipped with the Seesaw System. *GeoHazards* **2024**, *5*, 1074–1092. [CrossRef]
155. Katsimpini, P. On the Seismic Response of Composite Structures Equipped with Wall Dampers Under Multiple Earthquakes. *Modelling* **2024**, *6*, 1. [CrossRef]
156. Katsimpini, P.; Papagiannopoulos, G.; Hatzigeorgiou, G. A Simple Method to Evaluate the Bearing Capacity of Concrete-Filled Steel Tubes with Rectangular and Circular Sections: Beams, Columns, and Beam–Columns. *Appl. Sci.* **2024**, *14*, 8995. [CrossRef]
157. Chorafa, E.; Skrapalliou, E.; Katsimpini, P. On the Nonlinear Behavior of Composite Structures under Multiple Earthquakes Considering Soil–Structure Interaction. *CivilEng* **2024**, *5*, 673–693. [CrossRef]
158. Javaid, K.; Verma, N. Vibration control of seismic forces using viscous dampers and buckling restrained braces in irregular composite buildings. In *IOP Conference Series: Earth and Environmental Science*; IOP Publishing: Bristol, UK, 2023; Volume 1110, p. 012052.
159. Zandonini, R.; Baldassino, N.; Freddi, F.; Roverso, G. Steel-concrete frames under the column loss scenario: An experimental study. *J. Constr. Steel Res.* **2019**, *162*, 105527. [CrossRef]
160. Zheng, L.; Wang, W.D.; Li, H.W. Progressive collapse resistance of composite frame with concrete-filled steel tubular column under a penultimate column removal scenario. *J. Constr. Steel Res.* **2022**, *189*, 107085. [CrossRef]
161. Zheng, Y.H.; Zhong, W.H.; Tan, Z.; Meng, B.; Duan, S.C. A novel modeling method for fiber-based progressive collapse analysis of composite frames. *Eng. Fail. Anal.* **2022**, *140*, 106519. [CrossRef]
162. Wang, W.; Li, H.; Wang, J. Progressive collapse analysis of concrete-filled steel tubular column to steel beam connections using multi-scale model. In *Structures*; Elsevier: Amsterdam, The Netherlands, 2017; Volume 9, pp. 123–133.
163. Papavasileiou, G.; Pnevmatikos, N. The cost of retrofitting steel-concrete composite buildings against progressive collapse with steel cables. *Int. J. Progress. Sci. Technol. (IJPSAT)* **2017**, *6*, 103–115.
164. Wang, J.; Li, Y. Collapse Resistance of Composite Structures with Various Optimized Beam–Column Connection Forms. *J. Compos. Sci.* **2023**, *7*, 477. [CrossRef]
165. Bai, Y.; Hu, H.; Wang, J.; Sun, Q. Modeling on collapse behaviour of high-rise concrete-filled steel composite frames under over-design seismic excitations. *Struct. Infrastruct. Eng.* **2017**, *13*, 1563–1575. [CrossRef]

166. Zheng, L.; Wang, W.D. Multi-scale numerical simulation analysis of CFST column-composite beam frame un-der a column-loss scenario. *J. Constr. Steel Res.* **2022**, *190*, 107151. [CrossRef]
167. Wang, K.; Luo, H.; Guo, K.; Zhou, J.; Ma, H. Seismic collapse resistance performance on out-jacketing frames with concrete-encased CFST columns for adding storeys. In *Structures*; Elsevier: Amsterdam, The Netherlands, 2021; Volume 33, pp. 41–53.
168. Papavasileiou, G.S.; Charmpis, D.C.; Lagaros, N.D. Optimized seismic retrofit of steel-concrete composite buildings. *Eng. Struct.* **2020**, *213*, 110573. [CrossRef]
169. Papavasileiou, G.S.; Charmpis, D.C. Earthquake-resistant buildings with steel or composite columns: Comparative assessment using structural optimization. *J. Build. Eng.* **2020**, *27*, 100988.
170. Papavasileiou, G.S.; Charmpis, D.C. Seismic design optimization of multi–storey steel–concrete composite buildings. *Comput. Struct.* **2016**, *170*, 49–61.
171. Lin, Y.; Zhang, X. Life-cycle cost-oriented multiobjective optimization of composite frames considering the slab effect. *Struct. Des. Tall Spec. Build.* **2023**, *32*, e2008.
172. Kamaris, G.S.; Skalomenos, K.A.; Hatzigeorgiou, G.D.; Beskos, D.E. Seismic damage estimation of in-plane regular steel/concrete composite moment resisting frames. *Eng. Struct.* **2016**, *115*, 67–77. [CrossRef]
173. Kaveh, A.; Mahdipour Moghanni, R.; Javadi, S.M. Chaotic optimization algorithm for performance-based optimization design of composite moment frames. *Eng. Comput.* **2022**, *38*, 2729–2741.
174. Lin, Y.; Bi, K.; Zong, Z.; Hao, H.; Lin, J.; Chen, Y. Seismic performance of steel-concrete composite rigid-frame bridge: Shake table test and numerical simulation. *J. Bridge Eng.* **2020**, *25*, 04020032.
175. Paolacci, F.; Giannini, R.; Alessandri, S.; Corritore, D. On the cyclic behaviour of new pier-to-deck connections for short-medium span composite I-girder bridges. *J. Traffic Transp. Eng.* **2018**, *5*, 439–453. [CrossRef]
176. Zhou, X.; Wang, H.; Wang, X.; Liu, J. Seismic behavior of dumb-bell steel tube confined reinforced concrete piers. *Eng. Struct.* **2020**, *206*, 110126.
177. Zhou, X.; Xu, T.; Liu, J.; Wang, X.; Chen, Y.F. Seismic performance of concrete-encased column connections for concrete filled thin-walled steel tube piers. *Eng. Struct.* **2022**, *269*, 114803. [CrossRef]
178. Du, C.; Li, Z.X.; Liang, X.; Zhang, H.; Zhao, B. Shaking table test on seismic behavior of posttensioned precast segmental Concrete-Filled double skin steel tubular piers. *Eng. Struct.* **2023**, *294*, 116752. [CrossRef]
179. Gu, C.; Wang, X.; Zhou, X.; Huang, C. Performance-based assessment of the novel prefabricated CFST bridge pier by numerical approach. *Eng. Struct.* **2024**, *314*, 118362. [CrossRef]
180. Xiang, N.; Feng, Y.; Chen, X. Novel fiber-based seismic response modelling and design method of partially CFST bridge piers considering local buckling effect. *Soil Dyn. Earthq. Eng.* **2023**, *170*, 107911. [CrossRef]
181. Fu, T.; Ren, X.; Wang, K.; Zhu, Z. Seismic performance of prefabricated steel tube-confined concrete circular pier. In *Structures*; Elsevier: Amsterdam, The Netherlands, 2022; Volume 43, pp. 910–925.
182. Li, Z.X.; Du, C.; Liu, D.; Liang, X.; Zhao, B. Comparative study on seismic performance of concrete-filled double skin tubular piers and hollow concrete piers: Experimental and analytical. In *Structures*; Elsevier: Amsterdam, The Netherlands, 2023; Volume 49, pp. 1078–1092.
183. Li, Z.X.; Du, C.; Liang, X.; Zhang, H.; Zhao, B. Experimental and numerical investigation on hysteretic behavior of posttensioned precast segmental Concrete-filled double skin steel tubular piers. In *Structures*; Elsevier: Amsterdam, The Netherlands, 2023; Volume 54, pp. 1772–1787.
184. Lin, Y.; Li, Y.; Zong, Z.; Bi, K.; Xing, K.; Li, Y. Numerical study of seismic performance of steel-concrete composite rigid-frame bridge with precast segmental CFDST piers crossing fault-rupture zones. In *Structures*; Elsevier: Amsterdam, The Netherlands, 2023; Volume 56, p. 105039.
185. Yan, J.; Yang, J.; Zheng, J.; Jiang, K.; Jiang, A.; Ning, X.; Han, X.; Li, H. Experimental study on GFRP-steel double-skin rubber concrete composite piers under mixed loading. In *Structures*; Elsevier: Amsterdam, The Netherlands, 2024; Volume 64, p. 106493.
186. Zeng, X.; Zhu, S.; Deng, K.; Pang, L.; Zhao, C. Design method and seismic vulnerability of bridges using UHPC-NSC composite pier. *Eng. Struct.* **2024**, *312*, 118265. [CrossRef]
187. Zeng, X.; Zhu, S.; Deng, K.; Zhao, C.; Zhou, Y. Experimental and numerical study on cyclic behavior of a UHPC-RC composite pier. *Earthq. Eng. Eng. Vib.* **2023**, *22*, 731–745.
188. Wu, Y.; Zhang, N.; Li, J.; Gao, M.; Chen, J.; Sha, X. A research on the shear capacity of column pier embedded steel tube. In *IOP Conference Series: Materials Science and Engineering*; IOP Publishing: Bristol, UK, 2018; Volume 397, p. 012038.
189. Younesi, A.; Rezaeifar, O.; Gholhaki, M.; Esfandiari, A. Damage detection in concrete filled tube columns based on experimental modal data and wavelet technique. *Mech. Adv. Compos. Struct.* **2020**, *7*, 245–254.
190. Yue, J.; Fang, H.; Xia, Y.; Wang, H.; Wu, K. Damage Evolutions and AE Characteristics for Square Concrete-Filled Steel Tubular Columns Under Axial Load. *Int. J. Steel Struct.* **2020**, *20*, 1904–1915. [CrossRef]
191. Shen, Q.; Li, K.; Wang, J.; Wang, F.; Hu, Z.; Li, G. Cyclic behaviour of circular CFT-SG columns under axial tension-compression: Novel FE modelling and design methods. In *Structures*; Elsevier: Amsterdam, The Netherlands, 2024; Volume 62, p. 106148.

192. Zhang, J.; Li, Y.; Du, G.; Song, G. Damage detection of L-shaped concrete filled steel tube (L-CFST) columns under cyclic loading using embedded piezoceramic transducers. *Sensors* **2018**, *18*, 2171. [CrossRef]
193. Chen, D.; Montano, V.; Huo, L.; Fan, S.; Song, G. Detection of subsurface voids in concrete-filled steel tubular (CFST) structure using percussion approach. *Constr. Build. Mater.* **2020**, *262*, 119761.
194. Hu, J.W. *Smart Connection Systems: Design and Seismic Analysis*; Taylor and Francis: London, UK, 2015.
195. Zhang, J.; Xu, J.; Guan, W.; Du, G. Damage detection of concrete-filled square steel tube (CFSST) column joints under cyclic loading using piezoceramic transducers. *Sensors* **2018**, *18*, 3266. [CrossRef]
196. Zhang, D.; Li, N.; Zhang, S. Energy dissipation and resilience of precast segmented concrete-filled steel tube self-centering column: Parameter study and design methodology. *Eng. Struct.* **2021**, *244*, 112747.
197. Chen, Z.; Qin, Y.; Wang, X. Development of connections to concrete-filled rectangular tubular columns. *Adv. Steel Constr.* **2015**, *11*, 408–426.
198. Hayashi, K.; Skalomenos, K.A.; Inamasu, H.; Luo, Y.B. Self-centering rocking composite frame using double-skin concrete-filled steel tube columns and energy-dissipating fuses in multiple locations. *J. Struct. Eng.* **2018**, *144*, 04018146.
199. Zhong, C. Self-Centering, Shear-Controlling Rocking-Isolation Podium System for Enhanced Resilience of Tall Buildings. Doctoral Dissertation, University of Toronto, ON, Canada, 2023.
200. Zhang, Y.; Guo, J.; Yang, J.; Li, X. Recent structural developments and applications of magnetorheological dampers (mrd): A review. *Magnetochemistry* **2023**, *9*, 90. [CrossRef]
201. Zhai, Q.; Zhang, J.; Xiao, J.; Du, G.; Huang, Y. Feasibility of piezoceramic transducer-enabled active sensing for the monitoring cross-shaped concrete filled steel tubular (CCFST) columns under cyclic loading. *Measurement* **2021**, *182*, 109646.
202. Velychkovych, A.; Mykhailiuk, V.; Andrusyak, A. Numerical Model for Studying the Properties of a New Friction Damper Developed Based on the Shell with a Helical Cut. *Appl. Mech.* **2025**, *6*, 1. [CrossRef]
203. Landar, S.; Velychkovych, A.; Ropyak, L.; Andrusyak, A. A method for applying the use of a smart 4 controller for the assessment of drill string bottom-part vibrations and shock loads. *Vibration* **2024**, *7*, 802–828. [CrossRef]
204. Shatskyi, I.; Velychkovych, A. Analytical model of structural damping in friction module of shell shock absorber connected to spring. *Shock Vib.* **2023**, *2023*, 4140583.
205. Dutkiewicz, M.; Velychkovych, A.; Shatskyi, I.; Shopa, V. Efficient model of the interaction of elastomeric filler with an open shell and a chrome-plated shaft in a dry friction damper. *Materials* **2022**, *15*, 4671. [CrossRef]
206. Ng, W.L.; Goh, G.L.; Goh, G.D.; Ten, J.S.J.; Yeong, W.Y. Progress and opportunities for machine learning in materials and processes of additive manufacturing. *Adv. Mater.* **2024**, *36*, 2310006.
207. Mohanty, S.; Gokuldoss Prashanth, K. Metallic coatings through additive manufacturing: A review. *Materials* **2023**, *16*, 2325. [CrossRef]
208. Farid, M.I.; Wu, W.; Guiwei, L.; Yu, Z. Research on imminent enlargements of smart materials and structures towards novel 4D printing (4DP: SMs-SSs). *Int. J. Adv. Manuf. Technol.* **2023**, *126*, 2803–2823.
209. Hassan, A.; Alomayri, T.; Noaman, M.F.; Zhang, C. 3D Printed Concrete for Sustainable Construction: A Review of Mechanical Properties and Environmental Impact. *Arch. Comput. Methods Eng.* **2025**, 1–31. [CrossRef]
210. Baktheer, A.; Classen, M. A review of recent trends and challenges in numerical modeling of the anisotropic behavior of hardened 3D printed concrete. *Addit. Manuf.* **2024**, *89*, 104309.
211. Tahmasebinia, F.; Jabbari, A.A.; Skrzypkowski, K. The Application of Finite Element Simulation and 3D Printing in Structural Design within Construction Industry 4.0. *Appl. Sci.* **2023**, *13*, 3929. [CrossRef]
212. Raphael, B.; Senthilnathan, S.; Patel, A.; Bhat, S. A review of concrete 3D printed structural members. *Front. Built Environ.* **2023**, *8*, 1034020.
213. Shufrin, I.; Pasternak, E.; Dyskin, A. Environmentally friendly smart construction—Review of recent developments and opportunities. *Appl. Sci.* **2023**, *13*, 12891. [CrossRef]
214. Wu, Z.; Memari, A.M.; Duarte, J.P. State of the art review of reinforcement strategies and technologies for 3D printing of concrete. *Energies* **2022**, *15*, 360. [CrossRef]
215. Cao, X.; Wu, S.; Cui, H. Experimental study on in-situ mesh fabrication for reinforcing 3D-printed concrete. *Autom. Constr.* **2025**, *170*, 105923.
216. Menna, C.; Mata-Falcón, J.; Bos, F.P.; Vantyghem, G.; Ferrara, L.; Asprone, D.; Salet, T.; Kaufmann, W. Opportunities and challenges for structural engineering of digitally fabricated concrete. *Cem. Concr. Res.* **2020**, *133*, 106079.
217. Tian, Q.; Zhou, J.; Hou, J.; Zhou, Z.; Liang, Z.; Sun, M.; Hu, J.; Huang, J. Building the future: Smart concrete as a key element in next-generation construction. *Constr. Build. Mater.* **2024**, *429*, 136364.
218. De Schutter, G.; Lesage, K.; Mechtcherine, V.; Nerella, V.N.; Habert, G.; Agusti-Juan, I. Vision of 3D printing with concrete—Technical, economic and environmental potentials. *Cem. Concr. Res.* **2018**, *112*, 25–36. [CrossRef]
219. Amran, M.; Onaizi, A.M.; Fediuk, R.; Vatin, N.I.; Muhammad Rashid, R.S.; Abdelgader, H.; Ozbakkaloglu, T. Self-healing concrete as a prospective construction material: A review. *Materials* **2022**, *15*, 3214. [CrossRef]

220. Qian, H.; Umar, M.; Khan, M.N.A.; Shi, Y.; Manan, A.; Raza, A.; Li, F.; Li, Z.; Chen, G. A state-of-the-art review on shape memory alloys (SMA) in concrete: Mechanical properties, self-healing capabilities, and hybrid composite fabrication. *Mater. Today Commun.* **2024**, *40*, 109738.
221. Albuhairi, D.; Di Sarno, L. Low-carbon self-healing concrete: State-of-the-art, challenges and opportunities. *Buildings* **2022**, *12*, 1196. [CrossRef]
222. Fernandez, C.A.; Correa, M.; Nguyen, M.T.; Rod, K.A.; Dai, G.L.; Cosimbescu, L.; Rousseau, R.; Glezakou, V.A. Progress and challenges in self-healing cementitious materials. *J. Mater. Sci.* **2021**, *56*, 201–230.
223. Luhar, S.; Luhar, I.; Shaikh, F.U.A. A review on the performance evaluation of autonomous self-healing bacterial concrete: Mechanisms, strength, durability, and microstructural properties. *J. Compos. Sci.* **2022**, *6*, 23. [CrossRef]
224. Huseien, G.F.; Nehdi, M.L.; Faridmehr, I.; Ghoshal, S.K.; Hamzah, H.K.; Benjeddou, O.; Alrshoudi, F. Smart bio-agents-activated sustainable self-healing cementitious materials: An all-inclusive overview on progress, benefits and challenges. *Sustainability* **2022**, *14*, 1980. [CrossRef]
225. Nodehi, M.; Ozbakkaloglu, T.; Gholampour, A. A systematic review of bacteria-based self-healing concrete: Biomineralization, mechanical, and durability properties. *J. Build. Eng.* **2022**, *49*, 104038. [CrossRef]
226. Anjali, R.; Kumar, S.A.; Gangolu, J.; Abiraami, R. Experimental study on self-healing of micro-cracks in concrete with combination of environmentally friendly bacteria. *Sustain. Struct. Build.* **2024**, *95*, 109.
227. Hermawan, H.; Wiktor, V.; Gruyaert, E.; Serna, P. Experimental investigation on the bond behaviour of steel reinforcement in self-healing concrete. *Constr. Build. Mater.* **2023**, *383*, 131378.
228. Elmenshawy, Y.; Elmahdy, M.A.; Moawad, M.; Elshami, A.A.; Ahmad, S.S.; Nagai, K. Investigating the bacterial sustainable self-healing capabilities of cracks in structural concrete at different temperatures. *Case Stud. Constr. Mater.* **2024**, *20*, e03188.
229. Al-Obaidi, S.; He, S.; Schlangen, E.; Ferrara, L. Effect of matrix self-healing on the bond-slip behavior of micro steel fibers in ultra-high-performance concrete. *Mater. Struct.* **2023**, *56*, 161. [CrossRef]
230. Şimşek, B. Investigation of self-healing ability of hydroxyapatite blended cement paste modified with graphene oxide and silver nanoparticles. *Constr. Build. Mater.* **2022**, *320*, 126250.
231. VS, S.; Xavier, J.R. Effects of nanomaterials on mechanical properties in cementitious construction materials for high-strength concrete applications: A review. *J. Adhes. Sci. Technol.* **2024**, *38*, 3737–3768.

MDPI AG
Grosspeteranlage 5
4052 Basel
Switzerland
Tel.: +41 61 683 77 34

Applied Sciences Editorial Office
E-mail: applsci@mdpi.com
www.mdpi.com/journal/applsci

www.ingramcontent.com/pod-product-compliance
Lightning Source LLC
Chambersburg PA
CBHW040744250926
51850CB00007BB/1407
* 9 7 8 3 7 2 5 8 7 1 1 2 4 *